Uncertainty Quantification in Laminated Composites
A Meta-model Based Approach

Sudip Dey
Mechanical Engineering Department
National Institute of Technology, Silchar, India

Tanmoy Mukhopadhyay
Department of Engineering Science
University of Oxford
Oxford, UK

Sondipon Adhikari
Chair of Aerospace Engineering
College of Engineering
Swansea University, Wales, UK

CRC Press
Taylor & Francis Group
Boca Raton London New York

CRC Press is an imprint of the
Taylor & Francis Group, an **informa** business

A SCIENCE PUBLISHERS BOOK

CRC Press
Taylor & Francis Group
6000 Broken Sound Parkway NW, Suite 300
Boca Raton, FL 33487-2742

First issued in paperback 2021

© 2019 by Taylor & Francis Group, LLC
CRC Press is an imprint of Taylor & Francis Group, an Informa business

No claim to original U.S. Government works

Version Date: 20180626

ISBN-13: 978-0-367-78079-1 (pbk)
ISBN-13: 978-1-4987-8445-0 (hbk)

Library of Congress Cataloging-in-Publication Data

Names: Dey, Sudip, author. | Mukhopadhyay, Tanmoy, author. | Adhikari, Sondipon, author.
Title: Uncertainty quantification in laminated composites : a meta-model approach / Sudip Dey, Asst. Professor, Mechanical Engineering Department, National Institute of Technology, Silchar, India, Tanmoy Mukhopadhyay, Department of Engineering Science, University of Oxford, Oxford, UK, Sondipon Adhikari, Chair of Aerospace Engineering, College of Engineering, Swansea University, Wales, UK.
Description: Boca Raton, FL : CRC Press, Taylor & Francis Group, [2018] | "A science publishers book." | Includes bibliographical references and index.
Identifiers: LCCN 2018022527 | ISBN 9781498784450 (hardback)
Subjects: LCSH: Laminated materials--Mathematical models.
Classification: LCC TA418.9.L3 D49 2018 | DDC 620.1/18--dc23
LC record available at https://lccn.loc.gov/2018022527

Visit the Taylor & Francis Web site at
http://www.taylorandfrancis.com

and the CRC Press Web site at
http://www.crcpress.com

Preface

Who should read it and why? Natural and engineered materials have acquired an unprecedented role in the history of human civilisation. Entire eras have been named after predominant materials, such, as the stone age, bronze age, and iron age to mention a few. Composite materials are the defining advanced materials in the current era with the exceptional promise of applicability in a number of high-end areas such as aerospace, marine, automotive, construction and defence sectors. Composite materials convincingly demonstrate that their mechanical properties can be tailored to specific engineering demands and therefore perform superiorly compared to conventional metallic materials. From a design perspective, this advantage arises through a fundamental mathematical fact that the mechanical properties of composite materials are functions of significantly more parameters compared to their metallic counterparts. However, the advantages of composite materials come at a cost that the designers and analysts have to deal with many parameters. This gives rise to two major problems. Firstly, considering many parameters simultaneously makes the design process computationally more expensive. Secondly, perhaps more importantly, the increase in the number of parameters leads to an unavoidable escalation of uncertainty associated with these parameters. One way to address both of these problems simultaneously is to use metamodels which effectively 'replace' the original physics-based computationally expensive model with a data-based computationally inexpensive model. The aim of this book is to introduce predominant techniques in this direction.

This book is the first comprehensive text on the treatment of uncertainties in the modelling and analysis of composite materials. The authors have drawn on their considerable research experience to produce this book. The text is written from an engineering standpoint, comprising fundamental and complex theories that are relevant across a wide range of metamodeling techniques. The book introduces faculties, researchers and students about the confluence of composite structure theory, uncertainty modelling and propagation and metamodeling approaches. The pedagogical objective of this book is to systematically present the latest developments in the metamodeling techniques and explain how they can be used in conjunction with composite structures. The focus has been on the mathematical and computational aspects. This book will be relevant to aerospace, mechanical and civil engineering disciplines and various sub-disciplines within them. The intended readers of this book include senior undergraduate students and graduate students doing projects or doctoral research in the field of composite structures. Researchers, Professors and practicing engineers working in the field of composite structures will also find this book useful.

Existing works and the need for this book: There are some excellent books which already exist in the field of composite materials. For example, the book by Reddy (2003) covers essential details on analytical techniques and physics-based modelling of laminated composites. Uncertainty quantification has gained immense attention from the research community in the recent years. Extensive works from different disciplines such as mathematics, statistics, engineering and applied sciences have led to some excellent books on uncertainty quantification. As an example, the book by Smith (2014) gives a comprehensive account of uncertainty quantification approaches from a general multidisciplinary point of view. Meta-modelling is a classical topic which has seen a significant explosion over the past two decades due to the increasing demand for inexpensive computational tools. As a result, there are excellent books available to the readers on this topic. We refer to the books by Myers et al. (2016) and Forrester and Keane (2008) for a detailed exposure on statistical and mathematical aspects underpinning the metamodeling techniques for computational models. Although there are outstanding books available *separately* on the topics of composite structures, uncertainty quantification and metamodeling, to date there is no book which comprehensively discusses the role of metamodeling techniques for efficient uncertainty quantification and sensitivity analysis of composite structures in a unified manner. This book was conceived by us to *fill this essential gap* in the literature. We hope that this text will be an invaluable reference for next-generation engineers and researchers working in the area of design, analysis and manufacturing of composite materials for a wide range of practical applications. As significant research works have gone into uncertainty quantification in composite structures recently and many seminal papers have been published, the book also covers some of these latest developments with the introduction of fundamentals in a concise way. The attention in the book is mainly focussed on theoretical and computational aspects, although some reference to experimental works is given. Using this book, engineering and applied science graduate students and researchers will be able to implement and develop metamodels for applications in composite structural mechanics.

What will you find in this book? This book covers the essential fundamentals, applications and important references related to different metamodeling approaches specifically applicable to the aspect of uncertainty quantification in composite structures. Chapter 1 gives a general introduction to the need for considering uncertainty in engineering. The Chapter 2 of this book gives an overview of uncertainty quantification and a general review of the literature related to uncertainty quantification in composite structures. Chapter 3 presents a bottom-up approach to analyse the effect of stochasticity in material and structural parameters of a composite plate on the dynamic responses based on high dimensional model representation technique. Chapter 4 and 5 deals with the stochastic dynamic analysis of singly curved and doubly curved composite shells respectively. Chapter 6 deals with an environmental effect (thermal uncertainty) on the stochastic dynamic analysis of composite laminates, while chapter 7 addresses one of the crucial aspects of mechanical structures arising during the operational conditions (rotational uncertainty). Often application-specific requirements are needed to be met in engineering structures such as cutouts in plate and shells. Chapter 8 deals with the stochastic dynamics of composite laminates with cutouts. Chapter 9 presents a stochastic dynamic stability analysis of composite shells with uncertain material and geometric properties. Chapter 10 presents the stochastic dynamics and stability analysis of sandwich panels. Probabilistic approaches of uncertainty quantification are followed in Chapter 3 to 10. A metamodel based non-probabilistic uncertainty propagation scheme for composites is presented in Chapter 11. Different metamodel based uncertainty propagation schemes are discussed in Chapter 3 to 11, while the comparative performance of the metamodels for analysing composite structures is presented in Chapter 12 and 13. The scope of this book includes the aspect of uncertainty modelling as well as critical evaluation of the efficient metamodel-based uncertainty propagation approaches for composite structures.

Brief history and acknowledgements: This book on uncertainty quantification in laminated composites is a result of last ten years of research by the authors in the area of probabilistic engineering mechanics. The book's initial chapters began taking shape when Dr Dey, an expert in composite mechanics, joined Swansea University as a postdoctoral research fellow to work on uncertainty propagation in large composite structures. During this time, Dr Dey collaborated with Dr Mukhopadhyay, who had experience in metamodeling methods and computational mechanics. The inception of this book emerged from the fusion of three key technical areas which were historically disconnected in nature, namely, uncertainty modelling and propagation (Prof Adhikari), composite structures (Dr Dey) and metamodeling for multi-parameter systems and computational mechanics (Dr Mukhopadhyay). Without the timely merger of such complementary expertise, this book would have never taken the shape.

In this context, it could be noted that Dr Dey and Dr Mukhopadhyay have contributed equally in this book.

The authors are deeply indebted to numerous colleagues, students, collaborators and mentors. We are genuinely thankful to all of them for numerous stimulating scientific discussions, exchanges of ideas and on many occasions direct contributions towards the intellectual content of the book. Support and encouragement from colleagues within Swansea University's Zienkiewicz Centre for Computational Engineering, such as Professor M. I. Friswell, Professor P. Nithiarasu, Dr H. H. Khodaparast is greatly acknowledged. Particular thanks to Dr P. Higino, Dr G. Caprio and Dr A. Prado from Embraer (Brazil) for their intellectual contributions and discussions at different times. The authors are grateful to Professor J. E. Cooper (University of Bristol, UK), Dr R. Chowdhury (IIT Roorkee, India), Dr A. Chakrabari (IIT Roorkee, India), Dr S. Chakraborty (University of Notre Dame, USA), Professor S. K. Sahu (NIT Rourkela, India), Dr. G. Li (Princeton University, USA), Professor H. Rabitz (Princeton University, USA), Professor E. Carrera and Dr A. Pagani (Politecnico di Torino, Italy), Dr C. Scarth (University of Bath, UK) and Professor R. Banerjee (City University London, UK) for many stimulating discussions contributing towards the intellectual content of this book. The authors would like to gratefully acknowledge the contribution of Dr S. Naskar (University of Aberdeen, UK) in preparing the initial two chapters of this book. Beside the names taken here, we are thankful to many colleagues, fellow researchers and students working in this field of research around the world, whose name cannot be listed here for page limitations. The lack of explicit mentions by no means implies that their contributions are any less. The opinions presented in the book are entirely ours, and none of our colleagues, students, collaborators and mentors has any responsibility for any shortcomings.

We have been fortunate to receive grants from various companies, charities and government organisations including an Advanced Research Fellowship from UK Engineering and Physical Sciences Research Council (EPSRC), the Wolfson research merit award from The Royal Society, the Philip Leverhulme Prize from The Leverhulme Trust, research grants from Embraer, NRN Wales and a Zienkiewicz Scholarship from Swansea University. Without these funding, it would have been impossible to conduct the works leading to this book.

Sudip Dey
Tanmoy Mukhopadhyay
Sondipon Adhikari
February 2018

Contents

CHAPTER 1

Introduction

"Uncertainty is the only certainty there is, and knowing how to live with insecurity is the only security".

— John Allen Paulos

If we think deeply, sun-rise or sun-set never ever occurs. In reality, there is no new year or end of year and thus there is no century. Actually, all of us have framed everything as per our convenience. We always try to formulate our surroundings in the form of a theory-law-hypothesis or an equation or algorithm. The perpetuity of these deterministic frameworks based on the model of certainty does not exist in the universal-continuum of time scale. The Sun and earth rotate to their own tune of rotation. In many of the cases, we cannot even identify the variability/change in pattern due to the unimaginable vastness of such systems both in terms of space and time. Since the dawn of civilization all models introduced are meant for some specific purposes, such as the solar model initially proposed by Pythagoras and Aristotle (500 BC) was meant for the purpose of assessment of weather needed for crop production. Later on, Ptolemy (300 BC) proposed a new solar model which claimed to have corrected the previous solar model. Even though the earlier models served the intended purpose to some extent, both the models are found to be wrong as compared to the present solar model (refer to Fig. 1.1). Thus there exists a deficiency in the accuracy of our understanding of the physical systems in many cases. We human beings have formulated all the uncertain facts and transformed them into the cage of certainty for our convenience. Hence, many of the theories, laws, hypothesis, and formulae (from Aristotle to Hawking) become questionable if we change its domain, assumptions and boundary conditions.

The convenient simplified models have their limitations in accurate predictions and often differ from the true values. For centuries, researchers have been abiding

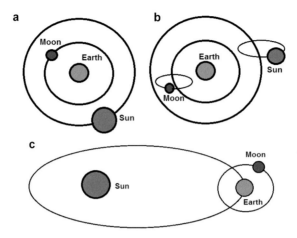

Fig. 1.1 Solar models.

by and spending their valuable time either to demonstrate or to validate those certain (/deterministic) models, which are actually uncertain in the domain of space and time. However, uncertainty being an inevitable characteristics of this universe, it is almost impossible to model a system accurately in a deterministic manner. Rather, it is a rational idea to try to quantify the effect of such uncertainty by considering it as an integral part of the model and deal with the consequences. This approach of modelling physical systems equips us to be prepared for the possible combination of outcomes by providing a detailed account of the variation from deterministic values of responses (i.e., the output quantity of interest). In case of engineering systems, if the design is carried out by considering the effect of source-uncertainties and prospective service-life conditions (such as environmental effects and damages) instead of blatantly avoiding them, the possibility of failure can be minimized (/ controlled) based on a strong scientific foundation.

Let us take a simple example to understand the influence of source-uncertainty. When we flip a fair coin, the probability of getting head or tail is 50%; but what happens if there is a soft muddy floor. It may lead to some possibilities wherein the coin will fall neither head nor tail, i.e., the coin may fall vertical. In such cases, it will not be 50% probability of getting either head or tail. Moreover, due to manufacturing uncertainties the coin may not be perfectly unbiased. This may also deviate the outcome from the general expectation. Likewise, the probability of getting one in dice having six faces is 1/6. But if the similar conditions (such as soft muddy surface, manufacturing uncertainty of the dice) are imposed here, there can be some possibilities wherein the dice will not show any specific face, or the probability of different outcomes would vary. Hence, the exact probability of getting one in those cases can't be predicted. In a similar fashion, all the physical or mathematical models are formulated based on certain boundary conditions and assumptions, which are often not strictly valid.

1.1 Source-uncertainty in engineering systems

If we think in a scale of time, the observations (/realizations) at a present time are precise and unique; while the sparsity of impreciseness (/uncertainty in prediction) increases as we go away from the present moment towards the direction of past or future (refer to Fig. 1.2). Various practical issues of life should be able to cope with the uncertainties that may originate from different aspects of design, implementation and operational conditions. The past, present and future are the unidirectional horizon of time, wherein plenty of areas can be cited to illustrate the scope of quantifying the hidden uncertainty, such as:

- Forecasting of weather in a particular place over a common span of time.
- Gain or loss on investment in financial markets, e.g., stock markets.
- Measurement error implicitly influencing the accuracy of machines or instruments.
- Response of structural systems related to various fields of engineering, such as aerospace, mechanical and civil.
- Validation and verification of material modelling in engineering.
- Success of new product, services, firms or person in the time-scale of future.
- Winning in games or gambling.
- Occurrences of events, incidents, births, deaths, accidents.
- Day to day activities like walking, sleeping, sitting, reading, writing, eating.
- Movement of particles such as atoms, molecules from Dalton's atomic theory to modern quantum mechanics.
- Actions, reactions and reflex in any biological systems.
- Human behaviour and activities.

In particular, the areas of engineering, which are most susceptible to different forms of uncertainty, are listed below:

❑ *Design of Machine elements or components (linear and non-linear model)*
- Design of joints (Any fasteners subjected to random load).
- Power transmission unit design (Shaft coupling for torque transmission).

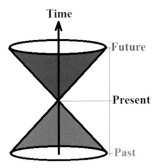

Fig. 1.2 Past-present-future domain with respect to time scale.

- Bearing design (Journal/Ball/Roller bearing).
- Vehicles Dynamics (Tire Technology).
- Pressure vessel design subjected to uncertain load.

❏ *Biomedical or Bioengineering*
- Human glucose metabolism/Diabetic Model.
  - Model for inflow of insulin by injection.
  - Model for inflow of glucose from ingested food.
- Molecular/Cellular/Organisms/Communities and Ecosystems.
- HIV model, Ebola virus model, Polio model, etc.
- Human Blood Pressure model.

❏ *Control Engineering for robust design*
- Temperature–Moisture Controller for air-conditioner.
- Speed: Fuel consumption Controller for automobiles.
- Speed: Cleanliness (dirt-removal) Controller for washing machine.
- Illumination: Power Controller for Light emitting diode (LED) light.

❏ *Aerospace and Structural Engineering*
- Bending Characteristics.
- Vibration analysis (Free/Forced).
- Inverse Problem/Design Optimization.
- Structural Health Monitoring.
- Reliability Analysis.

❏ *Environmental Engineering*
- Weather Model (rain/snowfall/humidity/temperature).
- Earthquake Model.
- Atmospheric climate (Layerwise) Model.

❏ *Coastal/Marine Engineering*
- Tide/Flood control.
- Marine traffic control.

❏ *Geo-technical Engineering*
- Contamination migration problem (by uncertain diffusion process).
- Sedimentation of clay/silt due to uncertain hydraulic transportation.
- Inaccurate measurement of the properties of soil and their spatially varying characteristics.

This book is focused on the aspect of source-uncertainties in advanced lightweight structures such as composites and efficient approaches to quantify their effects on global responses for a safe, yet economic design. A brief overview of the uncertainty quantification of composites following an efficient metamodeling approach is provided in the following sections of this chapter.

1.2 Importance of uncertainty quantification in composite structures

Composite structures are extensively used in modern aerospace, marine, construction and automobile applications due to their high strength, stiffness, lightweight and tailorable properties. Even though laminated composite structures have the advantage of modulating large number of design parameters to achieve various application-specific requirements, this concurrently brings the challenge of manufacturing the structure according to exact design specifications. Large-scale production of such structures according to the requirements of industry is always subjected to significant variability due to unavoidable manufacturing imperfections (such as intra-laminate voids, incomplete curing of resin and excess resin between plies, porosity, excess matrix voids, variations in ply thickness and fibre parameters), lack of experiences and complexity of the structural configuration. The issue aggravates further due to uncertain operational and environmental factors and the possibility of incurring different forms of damages and defects during the service life.

In general, an additional factor of safety (FOS) is incorporated by the designer to account for such unpredictable global responses, which may lead to either an ultraconservative or an unsafe design. A river and bridge model (refer to Fig. 1.3) is introduced to explain this further. Due to the presence of different forms of uncertainties (referred as the river of uncertainty in a collective form), the designed outcome and the real outcome differ for an engineering system. For a particular value of FOS, if the design is more conservative the real outcome would be less prone to failure (refer to Fig. 1.3a) and vice versa (refer to Fig. 1.3b). However a high value of FOS would be required if the design is less conservative, yet the system is required to be less prone to failure (refer to Fig. 1.3c). The above three cases normally lead to either an unsafe or uneconomic design because of the fact that the river of uncertainty is not adequately explored in this approach of analysis. An economic, yet safe design requires the in-depth analysis of the uncertainty associated with the system, as shown in Fig. 1.3d. If the level of uncertainty for a system is appropriately quantified, then the value of FOS can be adopted based on the importance of the system in a more robust manner. Moreover, probabilistic description for the response of the system could be obtained for the adopted value of FOS. As laminated composites are often used in various functionally important structures (such as aircrafts), it is important to quantify the uncertainties associated with the responses of the structure. If the actual outcome of an engineering system is considered, there could be four distinct situations in terms of attaining the design specification and variability from the target (refer to Fig. 1.4). The objective is to achieve a design outcome which is on target and has low variability. As the variability cannot be nil in case of real-life engineering systems, the objective becomes to minimize it and subsequently quantify the effect of such variability (/uncertainty) following a strong mathematical stochastic paradigm.

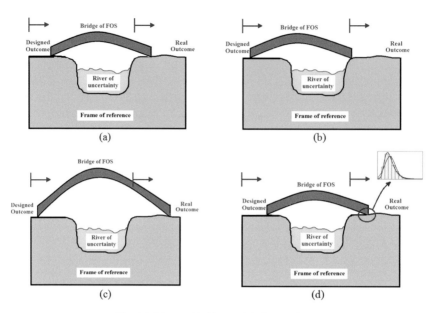

Fig. 1.3 River and bridge model of uncertainty.

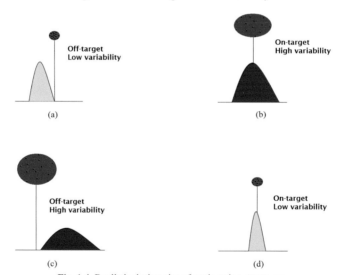

Fig. 1.4 Realistic design aim of engineering structures.

In general, uncertainties are classified into three categories, namely aleatoric (because of variability in the structural system parameters), epistemic (because of lack of information of the structural system) and prejudicial (because of the absence of variability characterization of the structural system). A detail description of different categories of uncertainty is provided in Chapter 2. Composite structures being susceptible to multiple forms of uncertainties, damages and environmental variations, the structural performances are often subjected to a significant element of risk. Thus it is of prime importance in case of composite structures to quantify the effect of source-uncertainties so that an inclusive design paradigm could be adopted to avoid any compromise in the aspects of safety and serviceability. The whole context of uncertainty quantification in various global responses of composite structures, which are increasingly being used in different industries, is summarized in Fig. 1.5. A concise discussion of the most prominent approaches for uncertainty quantification in composite structures is furnished in the next paragraph.

Following several decades of deterministic studies related to the static and dynamic responses of laminated composite structures (Reddy 2003, Chakrabarti et al. 2013, Mandal et al. 2017), the aspect of considering the effect of uncertainty in material and structural attributes have recently started receiving due attention from the scientific community. Both probabilistic (Sakata et al. 2008, Goyal and Kapania 2008, Manan and Cooper 2009) as well as non-probabilistic (Pawar et al. 2012) approaches have been investigated to analyse the effect of variability in the material and structural properties of composite structures. Plenty of researches have been reported based on intrusive methods to quantify the uncertainty of composite structures (Lal and Singh 2010, Scarth and Adhikari 2017); wherein the major drawback can be identified as the requirement of intensive analytical derivation and lack of the ability to obtain complete probabilistic description of the response quantities for systems with spatially varying attributes. Moreover, many of these approaches are valid only for a low degree of stochasticity in the input parameters. A non-intrusive method based on Monte Carlo simulation,

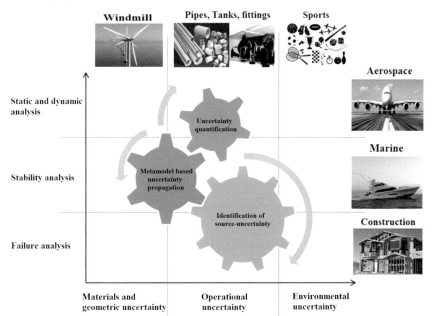

Fig. 1.5 Uncertainty quantification in various global responses of composite structures.

as adopted by many researchers (Dey et al. 2015a), can obtain comprehensive probabilistic descriptions for the response quantities of composite structures and these methods can account for much higher degree of stochasticity in the input parameters. A brief description of the Monte Carlo simulation method is provided in the next section.

1.3 Monte Carlo simulation

Uncertainty quantification is part of modern structural analysis problems. Practical structural systems are faced with uncertainty, ambiguity, and variability constantly, as discussed in the preceding sections. Even though one might have unprecedented access to information due to the recent improvement in various technologies, it is impossible to accurately predict future structural behaviour during its service life. Monte Carlo simulation, a computerized mathematical technique, lets us realize all the possible outcomes of a structural system leading to better and robust designs for the intended performances. The technique was first used by scientists working on the atom bomb; it was named after Monte Carlo, the Monaco resort town renowned for its casinos. Since its introduction in World War II, this technique has been used to model a variety of physical and conceptual systems across different fields such as engineering, finance, project management, energy, manufacturing, research and development, insurance, oil and gas, transportation and environment.

Monte Carlo simulation furnishes a range of prospective outcomes along with their respective probability of occurrence (refer to Fig. 1.6). This technique performs uncertainty quantification by forming probabilistic models of all possible results accounting for a range of values from the probability distributions of any factor that has inherent uncertainty. It simulates the outputs over and over, each time using a different set of random values from the probability distribution of stochastic input parameters. Depending upon the nature of stochasticity, a Monte Carlo simulation could involve thousands or tens of thousands of recalculations before it can provide a converged result depicting the distributions of possible outcome values of the response quantities of interest. Each set of samples is called an iteration or realization, and the resulting outcome from that sample is recorded. In this way, Monte Carlo simulation provides not only a comprehensive view of what could happen, but how likely it is to happen, i.e., the probability of occurrence.

Fig. 1.6 Schematic representation for Monte Carlo simulation based analysis of a stochastic system with three input parameters (x_i, i = 1, 2, 3) and one output parameter (y). Here $\bar{\omega}$ represents the stochastic character of a parameter.

The mean or expected value of a function $f(x)$ of an n dimensional random variable vector can be expressed as

$$\mu_f = E[f(x)] = \int_\Omega f(x)\phi(x)dx \qquad (1.1)$$

Similarly the variance of the random function $f(x)$ is given by the integral below,

$$\sigma_f^2 = Var[f(x)] = \int_\Omega (f(x) - \mu_f)^2 \phi(x)dx \qquad (1.2)$$

The above multidimensional integrals, as shown in equations (1.1) and (1.2) are difficult to evaluate analytically for many types of joint density functions and the integrand function $f(x)$ may not be available in analytical form for the problem under consideration. Thus the only alternative way is to calculate it numerically. The above integral can be evaluated using MCS approach, wherein N sample points are generated using a suitable sampling scheme in the n-dimensional random variable space. The N samples drawn from a dataset must follow the distribution specified by $\varphi(x)$. Having the N samples for x, the function in the integrand $f(x)$ is evaluated at each of the N-sampling points x_i of the sample set $\chi = \{x_1,\ldots\ldots\ldots,x_N\}$. Thus, the integral for the expected value takes the form of averaging operator as shown below

$$\mu_f = E[f(x)] = \frac{1}{N}\sum_{i=1}^{N} f(x_i) \qquad (1.3)$$

Similarly, using sampled values of MCS, the equation (1.2) leads to

$$\sigma_f^2 = Var[f(x)] = \frac{1}{N-1}\sum_{i-1}^{N}(f(x_i) - \mu_f)^2 \qquad (1.4)$$

Thus the statistical moments can be obtained using a brute force Monte Carlo simulation based approach, which is often computationally very intensive due the evaluation of function $f(x_i)$ corresponding to the N-sampling points x_i, where $N \sim 10^4$. The noteworthy fact in this context is the adoption of a metamodel based Monte Carlo simulation approach that reduces the computational burden of traditional (i.e., brute force) Monte Carlo simulation to a significant extent, as discussed in the next section.

1.4 Meta-modelling approach for uncertainty quantification

Uncertainty quantification based on Monte Carlo simulation is a popular approach because of its ability to obtain a comprehensive probabilistic description of the response quantities. However, the major lacuna of this approach is that a Monte Carlo simulation requires thousands of expensive model evaluations to be carried out corresponding to the random realizations. Thus, in case of the systems where the model evaluations are expensive (such as finite element simulation), direct Monte Carlo simulation has limited practical use because of its computational intensiveness.

In general for complex composite structures, the performance function is not available as an explicit function of the random design variables unlike various other engineering problems with closed-form solutions (Mukhopadhyay and Adhikari 2017a, Mukhopadhyay et al. 2017b, 2017e, 2018c). The performance functions or responses (such as natural frequencies, buckling loads, etc.) of the composite structure can only be evaluated numerically at the end of a structural analysis procedure such as the finite element method, which is often time-consuming and computationally expensive. To mitigate this lacuna, a meta-modelling approach can be adopted, wherein the uncertainty propagation is realized following an efficient mathematical medium. The metamodel based uncertainty propagation strategy can develop a predictive and representative mathematical model relating each response quantity of interest to a number of input variables. These metamodel equations are used to determine the global response characteristics corresponding to given values of input variables, instead of having to repeatedly run the time-consuming original simulation model (such as finite element analysis). The metamodel thus represents the result (or output) of the structural analyses encompassing (in theory) every reasonable combination of all input variables. From this, thousands of combinations of all design variables can be created (via simulation) and a pseudo analysis for each variable set can be performed by simply adopting the corresponding metamodel.

In general, the metamodels are employed to reduce the number of function evaluations based on actual simulation/experimental models in a Monte Carlo simulation, which needs large number of realizations corresponding to random set of input parameters (Metya et al. 2017, Mukhopadhyay et al. 2016f, Mahata et al. 2016). Metamodels are also quite popular in the area of optimization and inverse problems that involves multiple function evaluations (Mukhopadhyay et al. 2015b, 2016d). The development of metamodels is performed in three typical steps: selection of the representative sample points (which are able to collect all information of the whole design space in an optimal way) to construct a surrogate of the original simulation model, evaluation of responses (i.e., output) corresponding to each sample point and formulation of the mathematical or statistical model to obtain input-output relationship based on the sample set (containing a set of input parameters and corresponding output parameters). The performance of a metamodel (i.e., accuracy in prediction and computational efficiency) depends on various factors such as: dimension of the input parameter space (i.e., number of input parameters), number and quality of sample points for metamodel formation, degree of nonlinearity involved with the system and different forms of errors involved in metamodel formation (refer to Fig. 1.7). This book follows a metamodel based approach to quantify the effect of uncertainty in various global responses of composite structures.

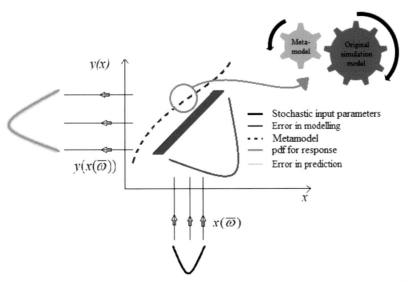

Fig. 1.7 Metamodel based analysis of stochastic systems. (Here $x(\overline{\omega})$ and $y(x(\overline{\omega}))$ are the symbolic representation of stochastic input parameters and output responses respectively. $\overline{\omega}$ denotes the stochasticity of parameters.)

1.5 Motivation and scope of the book

Uncertainty quantification in composite materials and structures, initiated from industrial needs due to inevitable variation in global responses of such structures from the deterministic predictions, has gained immense attention from the research community over the last few decades. This book aims to present an efficient uncertainty quantification scheme for laminated composite structures following meta-model based approaches for stochasticity in material and geometric parameters. Several meta-models are studied for this purpose and comparative results are presented for different global responses of composite structures including the effect of various environmental (such as temperature) and operational (such as rotation) conditions. Stochastic response of composite structures with application-specific design requirements such as cutouts is presented following the metamodel based approach. Both probabilistic and non-probabilistic approaches are discussed. Results for sensitivity analyses are presented to provide a complete understanding of the relative importance of different material and geometric parameters in global responses of the structure. To account for the effect of different forms of errors and uncertainty in metamodel formation, a collective effect of noise is presented in the metamodel based uncertainty quantification algorithms for composite structures.

Motivated by the influence of inevitable source-uncertainties in composite structures and the computational challenges involved therein, as outlined in the preceding sections, this book is written to address both the aspects of modeling different forms of uncertainties in composites as well as efficient computational approaches for uncertainty propagation. After providing a general overview of uncertainty quantification and stochastic analysis of composite structures in Chapters 1–2, rest of the chapters in this book are concentrated on various specific aspects related to uncertainty analysis of laminated composites and metamodel based algorithms for uncertainty quantification. The Chapters 3–10 of this book are focused on probabilistic approach of uncertainty quantification in composite laminates and sandwich structures, while the Chapter 11 deals with a non-probabilistic approach of uncertainty quantification in composites. The aspects of uncertainty and sensitivity of the material, geometric, environmental and operational factors for different responses of composite structures are analyzed in the Chapters 3–11. Various application-specific requirements (such as cutouts, twist and skewed geometry) in modern high-performance structural systems are analyzed. The effect of noise is analyzed in surrogate based uncertainty quantification algorithms for composites. Chapter 12 provides a critical assessment of different kriging model variants for the uncertainty quantification of composite structures. Different meta-models are used for formulating the uncertainty quantification schemes in various chapters of this book. Finally, comparative assessment of these meta-models is presented in Chapter 13. A concise summary explaining the contributions of each of the chapters in this book is given below.

The Chapter 2 of this book gives an overview of uncertainty quantification and a general review of the literature related to uncertainty quantification in composite structures. Different probabilistic and non-probabilistic methods of stochastic structural analysis are briefly presented in this chapter. After providing a concise review of the deterministic models for analyzing composite structures, the recent works on the aspect of uncertainty quantification in composite structures are discussed. The present standing of research in this area is assessed critically and the contribution of this book is justified in that context. Even though a comprehensive literature review is presented in Chapter 2, Chapters 3–13 also cite relevant studies specific to the topic of a particular chapter.

The effect of material and geometric uncertainty on the dynamic responses of composite plates is investigated in Chapter 3. A bottom up surrogate based approach is employed to quantify the variability in free vibration responses of composite cantilever plates due to uncertainty in ply orientation angle, elastic modulus and mass density. The finite element method is employed incorporating effects of transverse shear deformation based on Mindlin's theory in conjunction with a random variable approach. Parametric studies are carried out to determine the stochastic frequency response functions (SFRF) along with stochastic natural frequencies and modeshapes. In this study, a surrogate based approach using General High Dimensional Model Representations (GHDMR) is employed for achieving computational efficiency in quantifying uncertainty. This chapter also presents an uncertainty quantification scheme using commercial finite element software

(ANSYS) and thereby comparative results of stochastic natural frequencies are furnished for UQ using GHDMR approach and ANSYS.

In Chapter 4, we have concentrated on the stochastic dynamic responses of singly curved composite shells including the effect of twist angle in the geometry. The effect of transverse shear deformation is incorporated in the probabilistic finite element analysis considering an eight noded isoparametric quadratic element with five degrees of freedom at each node. The finite element model is coupled with the response surface method based on D-optimal design to achieve computational efficiency. A sensitivity analysis is carried out to address the influence of different input parameters on the output natural frequencies. The fibre orientation angle, twist angle and material properties are randomly varied to obtain the stochastic natural frequencies.

Chapter 5 presents an efficient stochastic free vibration analysis of composite doubly curved shells. The stochastic finite element formulation is carried out considering rotary inertia and transverse shear deformation based on Mindlin's theory. The sampling size and computational cost in the probabilistic analysis is reduced by employing a Kriging model based approach compared to direct Monte Carlo simulation. Besides detail investigation on the stochastic natural frequencies corresponding to low frequency vibration modes, the stochastic mode shapes and frequency response functions are also presented for a typical laminate configuration. The effect of noise on the kriging based uncertainty propagation algorithm is addressed. Results are presented for different levels of noise in a probabilistic framework to provide a comprehensive idea about stochastic structural responses under the influence of simulated noise.

Chapter 6 investigates the effect of rotational uncertainty under operating condition in the dynamic responses of composite shells. A response surface method based on central composite design algorithm is used for the quantification of rotational and ply-level uncertainties. The stochastic eigenvalue problem is solved by using QR iteration algorithm. An eight noded isoparametric quadratic element with five degrees of freedom at each node is considered in the finite element formulation. Sensitivity analysis is carried out to address the influence of different input parameters on the output natural frequencies. The sampling size and computational cost is reduced by employing the present surrogate based approach compared to direct Monte Carlo simulation. The stochastic mode shapes are also depicted for a typical laminate configuration.

Chapter 7 deals with the uncertainty caused by inevitable random variation of environmental factors such as temperature. The propagation of thermal uncertainty in composite structures has significant computational challenges. This chapter presents the thermal, ply-level and material uncertainty propagation in frequency responses of laminated composite plates by employing a surrogate model which is capable of dealing with both correlated and uncorrelated input parameters. In the present generalized high dimensional model representation (GHDMR) based approach, diffeomorphic modulation under observable response preserving homotopy (D-MORPH) regression is utilized to ensure the hierarchical orthogonality of high dimensional model representation component functions. The stochastic range of thermal field includes elevated temperatures up to 375 K and sub-zero temperatures up to cryogenic range of 125 K.

The aspect of an application-specific design requirement in engineering structures (cutouts) is illustrated in Chapter 8. This chapter deals with the effect of cutout on stochastic dynamic responses of composite laminates. Support vector regression (SVR) model in conjunction with Latin hypercube sampling is used in this investigation as a surrogate of the actual finite element model to achieve computational efficiency. The convergence of the present algorithm for laminated composite curved panels with cutout is validated with original finite element (FE) analysis along with traditional Monte Carlo simulation (MCS). Variations of input parameters (both individual and combined cases) are studied to portray their relative effect on the output quantity of interest. The layer-wise variability of structural and material properties is included considering the effect of twist angle, cutout sizes and different geometries (such as cylindrical, spherical, hyperbolic paraboloid and plate). The sensitivities of input parameters in terms of coefficient of variation are enumerated to project the relative importance of different random inputs on natural frequencies. Subsequently, the noise induced effects on SVR based computational algorithm are presented to map the inevitable variability in practical field of applications.

In Chapter 9, a stochastic dynamic stability analysis of composite panels is presented considering the effect of non-uniform partial edge loading. The system input parameters are randomized to ascertain the stochastic first buckling load and zone of resonance. Considering the effects of transverse shear deformation and rotary inertia, first order shear deformation theory is used to model the composite curved panels. Moving least square method is employed as a surrogate of the actual finite element model to reduce the computational cost. Statistical results are presented to show the effects of radius of curvatures, material properties, fibre parameters, and non-uniform load parameters on the stability boundaries.

Chapter 10 focuses on the stochastic analysis of laminated soft-core sandwich panels including the effect of skewness in the geometry. An efficient multivariate adaptive regression splines based approach for dynamics and stability analysis of sandwich plates is presented considering the random system parameters. The propagation of uncertainty in such structures has significant computational challenges due to inherent structural complexity and high dimensional space of input parameters. The theoretical formulation is developed based on a refined C0 stochastic finite element model and higher-order zigzag theory in conjunction with multivariate adaptive regression splines. A cubical function is considered for the in-plane parameters as a combination of a linear zigzag function with different slopes at each layer over the entire thickness while a quadratic function is assumed for the out-of-plane parameters of the core and constant in the face sheets. Both individual and combined stochastic effect of skew angle, layer-wise thickness, and material properties (both core and laminate) of sandwich plates are considered. Statistical analyses are carried out to illustrate the results of the first three stochastic natural frequencies and buckling load.

A non-probabilistic uncertainty propagation approach (fuzzy) for composites is presented in Chapter 11. Probabilistic descriptions of uncertain model parameters are not always available due to lack of data. This chapter investigates on the uncertainty propagation in dynamic characteristics (such as natural frequencies, frequency response function and mode shapes) of laminated composite plates by using fuzzy approach. A non-intrusive Gram–Schmidt polynomial chaos expansion (GPCE) method is adopted in the uncertainty propagation, wherein the parameter uncertainties are represented by fuzzy membership functions. A domain in the space of input data at zero-level of membership functions is mapped to a zone of output data with the parameters determined by D-optimal design. The obtained meta-model (GPCE) can also be used for higher α-levels of fuzzy membership function. The most significant input parameters such as ply orientation angle, elastic modulus, mass density and shear modulus are identified and then fuzzified. Fuzzy analysis of the first three natural frequencies is presented to illustrate the results and its performance. The proposed fuzzy approach is applied to the problem of fuzzy modal analysis for frequency response function of a simplified composite of cantilever plates. The fuzzy mode shapes are also depicted for a typical laminate configuration. The GPCE based approach is found more efficient compared to the conventional global optimization approach in terms of computational time and cost.

Chapter 12 presents a critical comparative assessment of Kriging model variants for surrogate based uncertainty propagation considering stochastic natural frequencies of laminated composite shells. The five Kriging model variants studied here are: Ordinary Kriging, Universal Kriging based on pseudo-likelihood estimator, Blind Kriging, Co-Kriging and Universal Kriging based on marginal likelihood estimator. First three stochastic natural frequencies of the composite shell are analysed by using a finite element model that includes the effects of transverse shear deformation based on Mindlin's theory in conjunction with a layer-wise random variable approach. The comparative assessment is carried out to address the accuracy and computational efficiency of the five Kriging model variants. Comparative performance of different covariance functions is also studied. Subsequently the effect of noise in uncertainty propagation is addressed by using the Stochastic Kriging. Representative results are presented for both individual and combined stochasticity in layer-wise input parameters to address performance of various Kriging variants for low dimensional and relatively higher dimensional input parameter spaces. The error estimation and convergence studies are conducted with respect to original Monte Carlo Simulation to justify merit of the present investigation. The study reveals that Universal Kriging coupled with marginal likelihood estimate yields the most accurate results, followed by Co-Kriging and Blind Kriging. As far as computational efficiency of the Kriging models is concerned, it is observed that for high-dimensional problems, CPU time required for building the Co-Kriging model is significantly less as compared to other Kriging variants.

Chapter 13 presents an exhaustive comparative investigation on different metamodels for critical comparative assessment of uncertainty in natural frequencies of composite plates on the basis of computational efficiency and accuracy. Both individual and combined variations of input parameters have been considered to account for the effect of low and high dimensional input parameter spaces in the surrogate based uncertainty quantification algorithms including the rate of convergence. Probabilistic characterization of the first three stochastic natural frequencies is carried out by using a finite element model that includes the effects of transverse shear deformation based on Mindlin's theory in conjunction with a layer-wise random variable approach. The results obtained by different metamodels have been compared with the results of direct Monte Carlo simulation (MCS) method for high fidelity uncertainty quantification. The crucial issue regarding influence of sampling techniques on the performance of metamodel based uncertainty quantification has been addressed as an integral part of this chapter.

Uncertainty Quantification in Composite Structures

An Overview

2.1 Introduction

This chapter gives an overview of uncertainty quantification and a general review of the literature related to uncertainty quantification in composite structures. Different probabilistic and non-probabilistic methods of stochastic structural analysis are briefly discussed in this chapter. After providing a concise review of the deterministic models for analyzing composite structures, the recent works on the aspect of uncertainty quantification in composite structures are discussed. The present standing of research in this area is assessed critically and the contribution of this book is justified in that context.

In general, composites are heterogeneous materials, which can be tailored according to their intended applications. Composites are widely used for various applications (aerospace, automotive, construction, marine, mechanical and many other industries) due to the fact that composites have high specific modulus, specific strength, low thermal conductivity, and high temperature resistance which are stochastic in nature. For example, about 25% to 50% of their total weight is made of composites in modern aircrafts such as Airbus A380 and Boeing 787 respectively (refer to Fig. 2.1) (Mallick 2007). A comprehensive description of the applications of composite materials can be found in Mallick (2007). In the past few decades, stochastic analyses in engineering fields have gained immense attentions from many researchers. The main focus of stochastic analysis is projected to determine the variability in responses due to uncertain randomness in material and geometrical properties and operational cum environment conditions. The

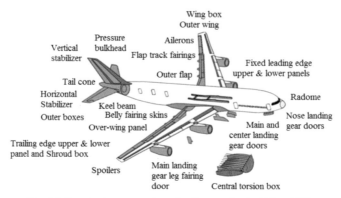

Fig. 2.1 Structural components made of composites in Airbus A380.

stochastic analysis is not a new concept, rather it started before 1990s to develop the methods of digitally generating sample functions of the stochastic field by using the Monto Carlo Simulation (MCS) for dynamic analysis of non-linear structures (Shinozuka and Deodatis 1988, Oberkampf 2000, 2001). Stochastic simulations typically include four steps: quantifying the input variable uncertainties (probabilistic characteristics of the input stochastic input parameters); development of a stochastic model which represents the variability of uncertain input parameters; implementation of a model to propagate uncertainty from the input level to the output level (global responses of the system); and finally quantifying the output parameters uncertainty (Padmanabhan and Pitchumani 1999). The propagation of uncertainty can be addresses in two ways: through simulation techniques and non-simulation techniques. Simulation techniques imply solving the deterministic system for a given number of parameter combinations and can therefore can be computationally very expensive. These samples can be used to obtain a surrogate model of the system, i.e., an equation relating the uncertain parameters to the response, whose expression is simpler and efficient to evaluate than the original equation. Thereby the surrogate model can be used for uncertainty quantification following a non-intrusive approach. Non-simulation techniques can be based on perturbation methods, which imply that the results are only valid for small variations of the uncertain parameters. Otherwise, they can be based on spectral methods, which imply solving a system of equations of size several times the size of the deterministic system.

2.2 Why and how an engineering structure becomes uncertain?

Uncertainties are unavoidable in the description of real-life engineering systems. There are several sources of uncertainties, both in the mathematical models and in the experimental results. Uncertainties can be broadly divided into following three categories. The first type of uncertainty is due to the inherent variability in the system parameters, for example, different cars manufactured from a single production line are not exactly the same. This type of uncertainty is often referred to as aleatoric uncertainty. If enough samples are present, it is possible to characterize the variability using well established statistical methods and the probable density functions of the parameters can be obtained. The second type is uncertainty due to lack of knowledge regarding a system. This type of uncertainty is often referred to as epistemic uncertainty and generally arises in the modeling of complex systems, for example the problem of predicting cabin noise in helicopters. Due to its very nature, it is difficult to quantify or model this type of uncertainties. Unlike aleatoric uncertainties, it is recognized that probabilistic models are not quite suitable for epistemic uncertainties. Several possibilistic approaches based on interval algebra, convex sets, Fuzzy sets and generalized Dempster-Schafer theory have been proposed to characterize this type of uncertainties. The third type of uncertainty is similar to the first type except that the corresponding variability characterization is not available, in which case work can be directed to gain better knowledge. This type of uncertainty often termed as prejudicial uncertainty, may consist of systematic and/or random errors, bias or other prejudices. An example of this type of uncertainty is the use of the viscous damping model in spite of knowing that the true damping model is not viscous. The total uncertainty of a system is the combination of these three types of uncertainties. Different sources of uncertainties in the modeling and simulation of dynamic systems may be attributed, but not limited, to the following factors:

a) *Mathematical models*: equations (linear, non-linear), geometry, damping model (viscous, non-viscous, fractional derivative), boundary conditions/ initial conditions, input forces.

b) *Model parameters*: Young's modulus, mass density, Poisson's ratio, damping model parameters (damping coefficient, relaxation modulus, fractional derivative order).

c) *Numerical algorithms*: weak formulations, discretization of displacement fields (in finite element method), discretization of stochastic fields (in stochastic finite element method), approximate solution algorithms, truncation and roundoff errors, tolerances in the optimization and iterative methods, artificial intelligent (AI) method (choice of neural networks).

d) *Measurements*: noise, resolution (number of sensors and actuators), experimental hardware, excitation method (nature of shakers and hammers), excitation and measurement point, data processing (amplification, number of data points, FFT), calibration.

Uncertainty related issues and reliability of computational predictions has always been a key issue ever since the computational methods were developed during 1960s. The early works in this area were mainly concentrated on a-priori error estimation arising due to the use of (coarse) discretization of the (continuum) solution field in the finite element method. Based on rigorous mathematical analysis, strong error bounds were proposed to quantify reliability of numerical predictions— see the classic book by Zienkiewicz and Taylor (1991) for more discussions.

However, with the increase in computing power, extremely high resolution modeling has become possible and consequently the 'discretization errors' have reduced over the years. Over the past two decades several studies indicated that the high-resolution finite element model alone is not sufficient for a credible prediction because of many uncertainties which exist in the process of scientific modeling. The role of uncertainty in computational mechanics is summarized in Fig. 2.2. It may be observed that there can be uncertainties in (a) input forces, (b) system description, (c) model calibration and (d) computation. In the context of Fig. 2.2, an overview of different problems and respective techniques arising in computational mechanics are tabulated in Table 2.1. This is by no means an exhaustive list, but gives an idea of the range of key research areas in the field. However, it can be realized that the effect of uncertainty can play an important role in almost all the problems related to computational structural mechanics. The process of accessing if a given set of equations are solved correctly using computational methods (e.g., finite element method) is called verification. Computational predictions of a numerical code are often compared with results from physical experiments. As noted in Fig. 2.2, there can be uncertainties in experimental results also. The process of accessing

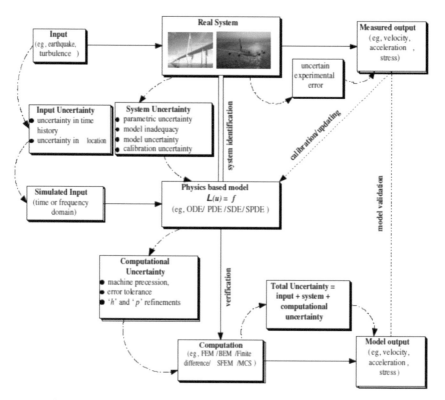

Fig. 2.2 A general overview of the role of uncertainty in computational mechanics.

Table 2.1 Problems related to computational mechanics and the main solution techniques (FEM: Finite Element Method; FDM: Finite Difference Method; SFEM: Stochastic Finite Element Method; RBDO: Reliability Based Design Optimization; SEA: Statistical Energy Analysis; RMT: Random Matrix Theory).

Input	System	Output	Problem name	Main techniques
Known (deterministic)	Known (deterministic)	Unknown	Analysis (forward problem)	FEM/FDM
Known (deterministic)	Incorrect (deterministic)	Known (deterministic)	Updating/ calibration	Modal updating
Known (deterministic)	Unknown	Known (deterministic)	System identification	Kalman filter
Assumed (deterministic)	Unknown (deterministic)	Prescribed	Design	Design optimisation
Unknown	Partially known	Known	Structural Health Monitoring (SHM)	SHM methods
Known (deterministic)	Known (deterministic)	Prescribed	Control	Modal control
Known (random)	Known (deterministic)	Unknown	Random vibration	Random vibration methods
Known (deterministic)	Known (random)	Unknown	Stochastic analysis (forward problem)	SFEM/SEA/RMT
Known (random)	Incorrect (random)	Known (random)	Probabilistic updating/ calibration	Bayesian calibration
Assumed (random/ deterministic)	Unknown (random)	Prescribed (random)	Probabilistic design	RBDO
Known (random/ deterministic)	Partially known (random)	Partially known (random)	Joint state and parameter estimation	Particle kalman filter/ensemble Kalman filter
Known (random/ deterministic)	Known (random)	Known from experiment and model (random)	Model validation	Validation methods
Known (random/ deterministic)	Known (random)	Known from different computations (random)	Model verification	Verification methods

random model outputs against random experimental results is known as model validation. Verification and validation (V&V) and Uncertainty Quantification (UQ) are fundamental for credible science based predictions.

Nowadays, the development of both numerical methods (e.g., the Finite Element method) and computational hardware makes it possible to solve deterministic high-resolution models of physical problems (e.g., as fluid mechanics or structural mechanics), represented by algebraic nonlinear systems of equations with thousands of degrees of freedom. However, spatial resolution is not enough to determine the credibility of a numerical model. A correct representation of the physical model as well as its parameters is also crucial, and both are affected by uncertainty. This uncertainty is due to several reasons, the first one is that any measurement has a limited precision. The second one is that measurement of apparently identical systems will lead to different measurements of the parameters, as, for example, the modulus of elasticity of two different samples of the same alloy, due to unaccountable effects in the production of the samples. The third one is related to errors in the mathematical model of the system considered. As a result of having uncertain parameters, the degree of confidence in a particular equation's predictions has to be assessed. This is especially true in fields of study such as reliability, where the probability of failure of a structure is calculated.

Another important field affected by uncertainty is sensitivity analysis, where knowing the effect of the variation of a parameter on the response can aid in decision making. Also, knowing the effect of uncertainty in the parameters of a model can help to quantify the degree of confidence of the model, as in cases like model updating or model validation. Finally, quantifying the effects of different sources of uncertainty on a given response can help to determine the sources of uncertainty that are not important. Examples of practical cases where these considerations are important can be elliptic partial differential equations, random vibrations, seismic activity, oil reservoir management, or composite materials.

2.3 Uncertainty analysis of composite structures

There are many complex processes involved in the manufacturing of the composites that are difficult to control leading to inevitable uncertainties in the mechanical as well as other physical properties of composites. The material and geometrical uncertainties are mainly influenced by the manufacturing process of the composites, while there are always possibilities of the involvement of different epistemic and prejudicial uncertainties in complex structural forms such as composites. Moreover various environmental factors (such as temperature, moisture), service-life conditions (such as occurrence of progressive damage in composites) and operational uncertainties (such as fluctuation in rotational speed in structural systems like turbo-machinery blades) are often involved with the structural response of composites. Moreover, application-specific requirements such as cut-outs and twist in the structural geometry make the entire system more complex and add more prospective sources of uncertainty. Analysing the individual and compound effect of various such components of uncertainty in composite structures is essential to understand their effect on the global structural responses.

Modelling of uncertainty in composites is an important issue; there are two different approaches: random variable and random field based modelling (for detail discussion see Naskar et al. 2018). In the present book a layer-wise random variable based approach is adopted for uncertainty modelling of the stochastic input parameters. The aspect of uncertainty quantification in composites can be presented by a stochastic composite tree model as shown in Fig. 2.3. Here the propagation of

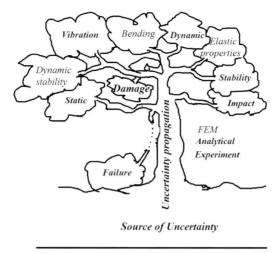

Source of Uncertainty

Fig. 2.3 Stochastic composite tree model for uncertainty analysis.

the effect of various uncertainties from the input level (source-uncertainty) to the output level (global responses such as vibration, bending, etc.) is compared to the analogy of ascent of sap in tress. Besides the probabilistic method of uncertainty quantification, non-probabilistic methods (such as fuzzy and interval analysis method) can be adopted for composite structures in cases when the statistical descriptions of the stochastic input parameters are not available. This book adopts a surrogate based approach coupled with finite element analysis for the propagation of uncertainty. In this section, a concise literature review is presented for various types of analyses carried out in composite structures covering both deterministic and stochastic approaches.

2.3.1 Deterministic analysis of composite structures

2.3.1.1 Static, dynamic and stability analysis of composite structures

In this section, a literature review concerning the static, dynamic and stability analysis of composite structures is presented. The prominent laminate theories in composites are summarized in Table 2.2 (Sayyad and Ghugal 2015). Narita and Leissa (1992) carried out a numerical and analytical approach for dynamic characteristic analysis of composite laminates along with the mode shape determination by considering the different material properties, stacking sequence and fibre angle. Jayatheertha et al. (1996) use artificial neural network for the optimum design of laminated composites considering strain and stiffness as constraints. Anlas and Goker (2001) employed the Ritz method to investigate the effect of skew angle, material properties, and lamination scheme on the free natural frequencies of the skewed composite plate. Ganapathi amd Makecha (2001) used the high order theory for dynamic analysis of thick composite plate having

Table 2.2 Various laminate theories for the modelling of composite structures.

S. N.	Theory	DOF at point	Displacement field
1.	Classical plate theory (CPT) (Love 1888)	3	$u = u_0 + z \dfrac{\partial w_0}{\partial x},$ $v = v_0 + z \dfrac{\partial w_0}{\partial y},$ $w = w_0$
2.	Reissner-Mindlin theory (First order shear deformation theory, (FSDT) (Reissner 1944, 1975) (Mindlin 1951)	5	$u = u_0 + z\theta_x,$ $v = v_0 + z\theta_y,$ $w = w_0$
3.	Second order shear deformation theory (Khdeir and Reddy 1999)	7	$u = u_0 + z\theta_x + z^2\phi_x,$ $v = v_0 + z\theta_y + z^2\phi_y,$ $w = w_0$
4.	Third order shear deformation theory (Reddy's theory) (Reddy 1984)	5	$u = u_0 + z\left[\theta_x - \dfrac{4}{3}\left(\dfrac{z}{h}\right)^2\left(\theta_x + \dfrac{\partial w_0}{\partial x}\right)\right],$ $v = v_0 + z\left[\theta_y - \dfrac{4}{3}\left(\dfrac{z}{h}\right)^2\left(\theta_y + \dfrac{\partial w_0}{\partial y}\right)\right],$ $w = w_0$

5.	Refined theories (based on Taylor series expansion) (Kant 1993)	5, 6, 7, 9, 11	$u = u_0 + z\theta_x + z^2 u_0^* + z^3 \theta_x^*$, $v = v_0 + z\theta_y + z^2 v_0^* + z^3 \theta_y^*$, $w = w_0 + z\theta_z + z^2 w_0^*$
6.	Aydogdu's shear deformation theory for laminated composite plates (Aydogdu 2009)	5	$u = u_0 - z\dfrac{\partial w_0}{\partial x} + z\alpha^{\frac{-2\left(z/h\right)^2}{\ln\alpha}}\theta_x$, $v = v_0 - z\dfrac{\partial w_0}{\partial y} + z\alpha^{\frac{-2\left(z/h\right)^2}{\ln\alpha}}\theta_y, \alpha > 0$ $w = w_0$
7.	n order shear deformation theory (Xiang et al. 2011)	5	$u = u_0 + z\theta_x - \dfrac{1}{n}\left(\dfrac{2}{h}\right)^{n-1} z^n\left[\theta_x + \dfrac{\partial w}{\partial x}\right]$, $v = v_0 + z\theta_y - \dfrac{1}{n}\left(\dfrac{2}{h}\right)^{n-1} z^n\left[\theta_y + \dfrac{\partial w}{\partial y}\right]$, $w = w_0$ $n = 3, 5, 7, 9, ...$
8.	Tangential exponential based higher order shear deformation theory (HSDT) (Mantari 2014)	5	$u = u_0 + z\left[f(z)\theta_x - \dfrac{\partial w}{\partial x}\right] + g(z)\theta_x$, $v = v_0 + z\left[f(z)\theta_y - \dfrac{\partial w}{\partial y}\right] + g(z)\theta_y$, $w = w_0$ where $g(z) = \tan\left(\dfrac{\pi z}{2h}\right) m^{\sec\left(\frac{\pi z}{2h}\right)}, f(z) = \dfrac{\pi m \sqrt{2}}{-h\sqrt{2}}(\sqrt{2}+1)$

Table 2.2 contd....

...Table 2.2 contd.

S. N.	Theory	DOF at point	Displacement field
9.	Trigonometric zigzag Theory (Sahoo and Singh 2014)	7	$u = u_0 - z\dfrac{\partial w_0}{\partial x} + \sum_{i=1}^{N_u-1}\left(z - z_i^u\right)H\left(z - z_i^u\right)\alpha_{xu}^i$ $+ \sum_{j=1}^{N_l-1}\left(z - z_j^l\right)H\left(-z + z_j^l\right)\alpha_{xl}^j + \left[g(z) + \Omega^{(k)}z\right]\beta_x$ $v = v_0 - z\dfrac{\partial w_0}{\partial y} + \sum_{i=1}^{N_u-1}\left(z - z_i^u\right)H\left(z - z_i^u\right)\alpha_{yu}^i$ $+ \sum_{j=1}^{N_l-1}\left(z - z_j^l\right)H\left(-z + z_j^l\right)\alpha_{yl}^j + \left[g(z) + \Omega^{(k)}z\right]\beta_y$ $w = w_0,\ \text{where } g(z) = z\sec\left(\dfrac{rz}{h}\right),$ $\Omega = -\sec(r/2)[1 + (r/2)\tan(r/2)]$
10.	Inverse hyperbolic shear deformation theory (Grover et al. 2013)	5	$u = u_0 - z\dfrac{\partial w_0}{\partial x} + \left[\sinh^{-1}\left(\dfrac{rz}{h}\right) - z\dfrac{2r}{h\sqrt{r^2+4}}\right]\theta_x$ $v = v_0 - z\dfrac{\partial w_0}{\partial y} + \left[\sinh^{-1}\left(\dfrac{rz}{h}\right) - z\dfrac{2r}{h\sqrt{r^2+4}}\right]\theta_y,\quad \alpha > 0$ $w = w_0$
11.	Polynomial based quasi-3D HSDT (Mantari 2016)	4	$u = u_0 - z\left[\dfrac{\partial w_b}{\partial x} + \left(\dfrac{4}{3}h^2 + \dfrac{h^2}{8}\right)\dfrac{\partial w_s}{\partial x}\right] + z^3\dfrac{\partial w_s}{\partial x},$ $v = v_0 - z\left[\dfrac{\partial w_b}{\partial y} + \left(\dfrac{4}{3}h^2 + \dfrac{h^2}{8}\right)\dfrac{\partial w_s}{\partial y}\right] + z^3\dfrac{\partial w_s}{\partial y},$ $w = w_b + \dfrac{z^2}{2}w_s$

| 12. | Non-polynomial based higher and normal deformation theory (Gupta and Talha 2016) | 4 | $u = u_0 - z\dfrac{\partial w_b}{\partial x} - \psi\left[\sinh^{-1}\left(\dfrac{kz}{h}\right) - \left(\dfrac{k}{h}\right)z\right]\dfrac{\partial w_s}{\partial x},$

 $v = v_0 - z\dfrac{\partial w_b}{\partial y} - \psi\left[\sinh^{-1}\left(\dfrac{kz}{h}\right) - \left(\dfrac{k}{h}\right)z\right]\dfrac{\partial w_s}{\partial y},$

 $w = w_b + k\cosh^2\left(\dfrac{kz}{h}\right)w_s$

 $where\ \psi = -\dfrac{h\cosh^2\left(k/2\right)}{\sqrt{\left(1+k^2/4\right)}-1}$ |

multi-layer, and found the effect of ply-angle, aspect ratio, degree of orthotropicity, and boundary condition on the response output. Banerjee (2001a, 2001b) used a computing package REDUCE to determine the mode shape and frequency response of composite beams under free vibration subjected to bending and torsion rotation.

The work was extended further for determining the exact expression for mode shape and frequency of cantilever composite Timoshenko beams by considering different boundary conditions and material properties. Ostachowicz et al. (2001) employed FEM to find the effect of delamination on the natural frequencies of a composite plate, when subjected to supersonic flow on the plate. Rikards et al. (2001) used first-order shear deformation theory for buckling and vibration analysis of composite plate and shell. Lee et al. (1995) found the effect of delamination of composite beam and column on the natural frequency and buckling load, when subjected to an axially compressive load, and found that both responses are significantly affected by delamination size, number and location. Nayak et al. (2002) carried out the parametric study to determine the effect of degree of orthotropy, aspect ratio, length-thickness ratio, number of layer on natural frequencies of the sandwich composite plate which is made of glass fibre skin, by using the Reddy's higher order theory. Lee et al. (2002) found the effect of fibre angle, height-thickness ratio, modulus ratio, and boundary condition on free vibration and mode shape of the I-section composite by using the finite element model. Hu et al. (2002) conducted the same analysis for the twisted conical composite cell, wherein effect of twist angle, taper ratio, and fibre orientation angle were determined. Zhang et al. (2002) carried out experimental and ANN model analysis for dynamic analysis of PTFE based short carbon fibre reinforced composites. Chen et al. (2003) applied the Galerkin method to determine the natural frequencies involving complex shapes of the composite structures. Free vibration analysis of laminate composite structures was conducted by Zhao et al. (2003) by using the mess-free approach. Lee et al. (2003) performed vibration analysis for multiple delamination composite beam structure, wherein single delamination was carried out by experiment, whereas multiple delaminations were conducted by finite element analysis. Kisa (2004) investigates the effect of cracks on the dynamic characteristic such as vibration and mode shape of the cantilever composite beams, wherein the effect of depth, fibre angle, and volume fraction on the dynamic behaviour of the composite beam were determined. The same analysis for two non-overlapping delaminations of the composite beam was conducted by Shu and Della (2004). Bezerra et al. (2007) used ANN to predict the shear stress of the epoxy composites considering different reinforcements such as carbon and E-glass fibre. Relatively recent studies are reported in the following paragraph.

Qin et al. (2017) developed an isogeometric approach based on the non-uniform rational B-Splines for the static and dynamic analysis curvilinearly stiffened plates in which static load and free vibration are determined. Hou and Guanghui (2018) applied weak-form quadrature element (WDE) approach for the static and dynamic analysis of Timoshenko composite beams with two layers. In this study, firstly the deflection and natural frequency are predicted, secondly stress and force obtained by WDE method are compared with the FEM, and lastly, a convergence study for higher order natural frequency is carried out. Li et al. (2016a) used layerwise/solid element method for static and free vibration analysis of the sandwich composite structures having a multilayer core. The FE co-rotational theory is used by An et al. (2017) to investigate aeroelastic analysis of laminated composite curved panels with considering the different size of curvature and ply orientation angle. Wattanasakulpong and Arisara (2015) presented an exact solutions for buckling, bending and vibration analyses of carbon nano-tube reinforced plates resting on the Pasternak elastic foundation showing the effect of carbon nanotube fraction volume, plate thickness, spring stiffness coefficient and aspect ratio. A static and fatigue analysis of multilayer composites with different shape of hole subjected to tensile load was carried out by Muc and Romanowicz (2017) to show the effect of material properties, ply orientation angle and geometry on the fatigue behaviour. Gao and Xing (2017) applied multiscale asymptotic expansion method for the static analysis of 3-D composite structures. Gunay and Taner (2017) employed an analytical approach for the static analysis of laminated thin-walled composites considering warping effect, flexural–torsional coupling and variable stiffness. Hasim (2017) applied the zigzag theory for static analysis of laminated composite beams. An experimental static and dynamic analysis of cylindrical composite shells subjected to static compressive load and the dynamic load is carried out by Bisagni (1999) to investigate the buckling of composite shell. A comparative study is carried out by Lugovy et al. (2017) for static and cyclic fatigue of ceramic composites made of different composition, while two composites with different types of fibres are compared experimentally for static and fatigue behaviours by Szebényi et al. (2017).

Deflection, free vibration and transient characteristics of the composite laminated plate and curved shallow shells are studied experimentally and numerically by Sahoo and Singh (2013, 2014) and Sahoo et al. (2016). Kishore et al. (2011) investigated static behaviour of composite laminated plates. Zappino et al. (2017) employed refined beam theory for analysis of tapered composite structure. Torabi et al. (2016) carried out theoretical analysis and experiments to investigate the effects of delamination location and size on the first three natural frequencies of composite beams. Thinh and Nguyen (2016) carried out a study for the free vibration analysis of composite cell with partially and fully filled fluid by employing the dynamic stiffness method and compared the results obtained from the experimental data. Tan et al. (2016) used the residual method to determine natural frequencies of variable stiffness composites. The effects of plate geometry, material properties, and external load on the free and forced vibration analysis were determined. Ghafari and Rezaeepazhand (2016) used the dimensional reduction method, wherein 3D elasticity problem of the composite beam was decomposed into 2D and 1D beam analysis. Rotating as well as nonrotating beams were analysed to find the effect of rotation speed and fibre angle on the free vibration. Vo et al. (2017) conducted the vibration analysis of composite beam by using normal and shear deformation theories, when subjected to axial load, as well as found the effect of normal strain, Poisson's ratio, and lay-ups on natural frequencies. Mehar et al. (2017) applied the analytical, numerical, and experimental methods for vibration analysis of reinforced polymer composites. Mori-Tanaka scheme was applied to distribute material properties in the composite. The results concluded that with the increase in the aspect ratio and volume fraction ratio, fundamental frequency increases, while it decreases with increase in thickness and curvature ratio. A vibration study of laminated composites was conducted by Shankar et al. (2017) under the hygrothermal condition, while for sandwich composites, vibration analysis was reported by Hwu et al. (2017) based on FSDT. Kumar et al. (2017a) superimposed radial and bending mode of the laminated composites to enhance the resonance bandwidth. Eftekhar et al. (2011) employed the diversity guided evolutionary algorithm to determine the natural frequency and mode shape of composite Timoshenko beams by considering the different boundary conditions. By using this approach all natural frequencies can be determined in a single run. Moreno-Garcia et al. (2014) used Ritz method based optimal sampling for determining the error associated with finite difference method while calculating the mode shapes of composite structures. The derivatives of the mode shape such as rotations and curvatures are used to localise the damage of composite structure. Qiao et al. (2007) used dynamic response based damage detection by using piezoelectric materials and scanning laser vibrometer (SLV) for composite laminated plates. Liu (2015) used an analytical method for determining the frequency and mode shape with respect to fibre volume fraction and orientation of composite structures along with considering different complex boundary conditions.

Allahyari et al. (2018) measured the dynamic behaviour of composite slabs such as natural frequencies, damping ratio and FRF experimentally by using the non-destructive technique (NDT) in which excitation was given by hammer. An FEM code was also developed for the same study and they found that for natural frequency the difference is less than 13% between experimental and numerical results. De Lima et al. (2010) carried out the sensitivity analysis of FRF for composite sandwich plates by using the finite element discretization. Kayikci and Fazil (2012) presented a design of composite laminates for optimal frequency responses by considering the ply-orientation angles as the design variable with different plate aspect ratio. Dynamic analyses of composite plate were carried out employing the SFEM by Park and Lee (2015, 2017). Nanda and Kapuria (2015) used FSDT and classical laminate theory (CLT) for formulation of SFEM to investigate wave propagation in composite curved beams and concluded that CLT gives large deviation in prediction of wavenumber as compared to FSDT. Samaratunga et al. (2015) employed wavelet spectral SFEM and FSDT for wave propagation analysis in adhesively bonded composite joints while Li and Soares (2015) used SFEM for dynamic analysis of composites with fibre reinforcement.

Song et al. (2015) carried out dynamic analysis of smart composite Timoshenko beams having piezoelectric transducer layers by using SFEM. Nanda et al. (2014) applied SFEM along with zigzag theory for wave propagation analysis of nonhomogeneous composite and sandwich beams. Optimization of composite and sandwich structures were carried out considering various static and dynamic criteria as constraints (Mukhopadhyay et al. 2015a, 2015c, Dey et al. 2015d, 2016a). Mallela and Upadhyay (2016) used ANN to predict the buckling load of composite laminated plates under the action of in-plan shear load. Mukhopadhyay (2018a) developed a multivariate adaptive regression splines based damage identification methodology for web core composite bridges including the effect of noise. Sarvestan et al. (2017) utilised spectral FEM for forced vibration analysis of Timoshenko cracked beams subjected to moving load with constant velocity and acceleration.

2.3.1.2 Impact analysis of composite structures

In this section, a brief literature review concerning the impact analysis of composites is presented, wherein it can be found that the effect of different parameters such as impact load, impact angle, impact velocity, thickness of laminates and laminate configuration are widely investigated. Different approaches and models are utilised to analyse the impact behaviour of composites. Khashaba and Othman (2017) carried out low-velocity impact testing on the woven CFRE composites considering the effect of temperature. Liao and Liu (2017) presented a FEM based low-velocity impact analysis considering the progressive failure in the laminated composites. Jiang and Hu (2017) performed experimental low-velocity impact analysis of composite structures including auxetic effect. The results show that auxetic composites have better shock resistance and energy absorption capacity as compared to non-auxetic composites. Choi (2017) studied the influence of pre-stress in composite cylinder subjected to low-velocity impact by using the shear deformation theory, Karman's large deflection theory and strain displacement relationship. Chen et al. (2017) investigated the effect low-velocity impact on the intra-laminar damage, delamination and strain rate of material in sandwich composite structures.

A comparative study was carried out to investigate the behaviour of composite and steel plate subjected to low-velocity impact loading by Park (2017). The result illustrates that laminated composite has two times more displacement compared to a steel plate. He et al. (2018) carried out an experimental and numerical study to investigate the effect of impact location and impact energy on the structural damage and flexural strength of sandwich plates with corrugated core. Jagtap et al. (2017) presented a low-velocity impact damage analysis of laminated composites by using the FEM simulation and investigated the effect of boundary conditions and impact velocity including matrix cracking and delamination. Rozylo et al. (2017) presented a model for low-velocity impact damage of composite plates subjected to Compression-After-Impact (CAI) testing. Martins et al. (2018) determined the effect of density and ply angle on the impact behaviour of the laminated composite. Arjangpay (2018) demonstrated an experimental and numerical investigation by using the FE model and nonlinear progressive damage model for low-velocity impact analysis of composites. Debski et al. (2018) employed a simplified damage model to investigate the influence of low-velocity impact on the stability and post critical state of composite columns subjected to compression.

Experimental and numerical analysis of low and high-velocity impact analysis of composites are presented by Mata-Diaz et al. (2017), Ribeiro et al. (2015) and Kursun et al. (2015). Wu et al. (2018) worked on experimental and numerical study on soft-hard-soft (SHS) cement-based composite system under multiple impact loads. Thorsson et al. (2018) presented an experimental investigation of composite laminates subjected to low-velocity edge-on impact and compression after impact. Hazimeh et al. (2015) investigated the effect of material and geometrical properties of adhesive bonding of composite double lap joint when subjected to impact loading by using the FEM. A numerical investigation on the impact behaviour of convex and concave composite cylindrical shells is carried out by Choi (2018). Neogi et al. (2017) investigated oblique impact of laminated skewed composite shells by employing modified Hertzian contact law in FE analysis considering friction due to different impact angle and velocity. A continuum damage mechanics along with FEM is employed to damage development and dynamic mechanical behaviour analysis of cross-ply composite laminates under transverse impact loading by Zhang et al. (2017). Habibi et al. (2018) assessed the influence of low-velocity impact on residual tensile properties of nonwoven flax/epoxy composite. Zhao et al. (2017) studied the localization of impact on composite plates based on integrated wavelet transform and hybrid minimization algorithm. Wang and Gunnion (2008) investigated the effect of adhesive thickness, scarf angle and the patch off-axis angle on the scarf repaired laminated composites. Wang et al. (2018) investigated the delamination behaviour of thick laminated composites subjected to low-energy impact. ANN based approach is found to be used for impact analysis in composites (Artero-Guerrero et al. 2017, Malik and Arif 2013). Yang et al. (2018) carried out failure modes analysis of C/SiC composite subjected to different impact loads.

2.3.1.3 Failure analysis of composite structures

In this section, a short literature review is presented covering the scientific articles concerning deterministic failure analysis of composites. The mechanics of cracking in the laminated composite is analysed by Dvorak and Norman (1986), wherein it was found that the strength of thin ply depends on the crack in the longitudinal direction of fibre while the strength of thick ply depends on crack in the perpendicular direction of the fibre axis. Reddy and Pandey (1987) employed FSDT and tensor polynomial failure criterion for the failure analysis of composite plates by using the FE simulation. Kam et al. (1996) utilised the principle of minimum total potential energy theory and von Karman-Mindlin plate theory for the first-ply failure and deformation analysis of laminated composite plates. Layer-wise displacement theory in FE formulation was applied for first-ply failure analysis of laminated composite plates by Kam and Jan (1995). The work was extended for theoretical and experimental analysis of first ply failure subjected to different loadings by Kam et al. (1999). Li and Fan (2018) investigated multi-failure analysis such as instability, material failure, buckling, rib crippling and delamination of composite cylinders. Chen et al. (2014) developed a model to determine the effect of progressive failure of the laminated composites subjected to tensile and impact loading using FE simulation. Błachut (2004) investigated the influence of wall thickness, boundary conditions, stacking sequence and geometric imperfection on the failure and bucking of externally pressurized composite toroidal pressure hull.

An experimental and numerical analysis was carried out by Farooq and Peter (2014) to predict the first ply failure of laminated composite panels subjected to low-velocity impact loading by employing adaptive meshing techniques. Numerical and experimental studies were carried out by Luo et al. (2017) to determine the energy absorption capacity of composite beams to predict the progressive failure subjected to dynamic loading. Maimı et al. (2011) developed a model to determine the delamination and matrix cracking in laminated composites by using plastic-damage model. First-ply failure analysis of laminated stiffened plates was carried out by Prusty et al. (2001) and Kumar et al. (2003). Dong and Wang (2015) studied the first ply failure of angle-ply composite laminates subjected to an in-plane tensile load. Chowdhury et al. (2016) investigated the matrix failure in composite laminates subjected to tensile load. Sarvestani et al. (2017) employed high order displacement method (HODM) for the progressive failure analysis of thick composite curved tubes subjected to pure bending moment. Experimental and analytical studies were carried out by Kam et al. (1997) and Chang (2000) to investigate the first-ply failure of composite pressure vessels. First ply failure of thin composite conoidal shells under constant pressure was investigated by Bakshi and Chakravorty (2014) and it was found that cross-ply laminates have higher failure load compared to angle-ply composite shells. Banat and Mania (2018) experimented and analysed the progressive failure of thin-walled fibre metal laminate columns under axial compressive loading. Longbiao et al. (2017) investigated the damage and failure of fiber-reinforced ceramic-matrix composites subjected to cyclic fatigue, dwell fatigue and thermomechanical fatigue.

A micro-mechanical model was developed by Wei et al. (2017) for failure analysis of carbon/epoxy composites subjected to compressive load. Gadade et al. (2016a) employed Puck's failure criterion as well as HSDT in FE formulation for failure analysis of laminated plates under biaxial loading and compared the results with Tsai–Hill, Tsai–Wu and Lee's failure criteria. Kumar (2013) presented a coupled ply damage and delamination failure analysis for ceramic matrix composites. Adali and Izzet (2011) investigated the effect of fibre orientation angle on the ply failure and buckling failure of curved composite panels subjected to uniaxial compressive loading. Neural network based approaches are found to be applied for fatigue analysis of composites (Al-Assaf and Kadi 2001, 2007, Kadi and Al-Assaf 2002, Kumar et al. 2017b).

2.3.2 Non-deterministic analysis of composite structures

In this section, a brief review of the non-deterministic studies concerning various static and dynamic responses concerning composite structures is presented. Venini and Mariani (1997) used the Rayleigh-Ritz method for dynamic free vibration analysis of composite plates considering the uncertainties in material properties, mass density and stiffness of the Winker foundation. Oh and Librescu (1997) used stochastic FEM for vibration and reliability analysis of the composite cantilever plates with uncertain material properties, thickness and ply orientation angle. The effects of variable material properties on the free vibration of composite panels are analysed by using perturbation based approach by Singh et al. (2002). Wang and Zhao (2001) combined the MCS and modified shear-lag model for development of an analytical framework to investigate the mechanical behaviour of composites such as tensile strength, stiffness and failure process at room and low temperatures. Onkar and Yadav (2005) analysed the forced vibration of the cross-ply laminated composite plates by using analytical methods with the consideration of random material properties. Allegri et al. (2006) employed FEM along with MCS for stochastic vibration analysis of composite truss for space applications and compared the results obtained from the perturbative approximated approaches, while Gao et al. (2009) used the Random Factor Method (RFM) for the dynamic characteristic analysis of the stochastic composite truss, wherein mode shapes and natural frequencies are determined.

An analytical model along with the shear-lag theory was applied by Chiang et al. (2005) for tensile property analysis of hybrid composites using the MCS. The effects of various parameters of composites such as strength of fibre, properties of matrix and shear strength of fiber–matrix interface on the tensile and failure strength are determined by using the MCS (Kim et al. 2003). Liu and Zheng (2006) employed MCS for the damage and failure analysis of composite, wherein fibre properties were considered as source of uncertainty in the equivalent material properties. Singh et al. (2009) determined the influence of random material properties, different thickness and aspect ratio, and other geometrical properties on the free vibration of composite plates by using the HSDT along with the MCS. Probabilistic dynamic analysis of a composite beam subjected to axial loads is conducted out by Cheng and Xiao (2007) by using the theory of probability. Qiu and Hu (2008) employed non-probabilistic interval analysis method for the transient vibrations analysis of cross-ply composite plates considering different sources of uncertainties. Chen et al. (2008) employed SFEM for probabilistic analysis of composites considering the effect of number of layers and different material properties. Relatively more recent studies are presented in the following paragraphs.

Pawar et al. (2012) presented a fuzzy approach for analysing thin-walled composite beams. Secgin et al. (2013) employed extreme value model for the stochastic dynamic analysis (free and forced vibration) of symmetrically composite laminated plates considering different orientation angle, damping ratio, and boundary conditions. Sepahvand et al. (2015) carried out analytical and experimental analysis of the free vibration and acoustic power of composites, considering the uncertainty in material and structural properties by using the generalized polynomial chaos expansions method. Jiang et al. (2015) predicted the uncertainty in elastic modulus of braided composite materials by using homogenization method coupled with the first-order perturbation approach. Sepahvand (2016) investigated the effect of random ply orientation angle on the vibration characteristic of composites by using the spectral stochastic finite element method. Scarth et al. (2014) employed the Rayleigh-Ritz method along with modified strip theory aerodynamics in the aero-elastic model for stochastic stability analysis of a wing made of composites with different ply orientation angle and a parametric study was conducted to find the effect of bend-twist coupling. Mao et al. (2013) conducted an uncertainty quantification of both phase and magnitude of the frequency response function by using the Gaussian bivariate statistical model and the probability density function (PDF) plot is derived for phase and magnitude. Manan and Cooper (2010) used Polynomial Chaos Expansion (PCE) method for probabilistic determination of uncertain frequency response functions. Li et al. (2016a) carried out stochastic thermal buckling analysis of laminated composite plates considering temperature depended parameters by employing the perturbation technique. Sepahvand and Marburg (2017) conducted the vibroacoustic analysis of composites by using the spectral finite element method (SFEM) considering the uncertainty in damping and elastic parameters.

Bahadori et al. (2018) employed Monto Carlo Simulation (MCS) for the heat conduction analysis of composite structure with temperature depended material properties. The results obtained from MCS are compared with FEM and experimental results. Bhat (2015) applied MCS for the probabilistic determination of stress variation in composite single lap joints and found the effect of various parameters for the maximum stress. Chandrashekhar and Ganguli (2010) applied MCS for stochastic dynamic analysis of sandwich composites considering random material properties. Lee et al. (2014) determined the equivalent material properties of glass/epoxy composites by applying the homogenization method and stochasticity in the equivalent material properties were assessed by MCS. Dey et al. (2015a, 2016c, 2016e) presented stochastic dynamic analyses of composite structures using various surrogate based approaches. Naskar et al. (2017) presented a stochastic natural frequency analysis of damaged thin-walled laminated composite beams with uncertainty in micromechanical properties. Sepahvand et al. (2011) employed polynomial chaos expansion (PCE) for stochastic dynamic analysis of composite plates with uncertain modal damping parameters. Balokas et al. (2017) used an ANN based approach for prediction of elastic properties of braided composites considering the manufacturing uncertainties by carrying out MCS. Chen and Qiu (2018) used the polynomial chaos expansion method for the combined uncertainty analysis of composite structure wherein normal random and interval variable uncertainties are considered. Reliability-based impact localization in composite panels using Bayesian updating and the Kalman filter is carried out by Morse et al. (2018). Patel and Soares (2017) carried out a study on delamination damage and matrix cracking in composite plates subjected to low-velocity impact in a probabilistic environment. Reliability analysis of composite laminates was carried out for the first-ply failure by Kam and Chang (1997). Stochastic failure behaviour of composites with unidirectional ply was studied by Whiteside et al. (2012). Kumar et al. (2015) presented a probabilistic failure analysis of laminated sandwich shells based on higher order zigzag theory. Karsh et al. (2018) investigated the spatial vulnerability for stochastic first ply failure strength of composite laminates including the effect of delamination.

2.4 Summary

Starting from a generalized overview for the uncertainty quantification in computational structural mechanics, this chapter goes on to address the aspect of uncertainty concerning composite structures. A concise review of literature concerning different aspects of analyses in composite structures is presented for deterministic as well as non-deterministic approaches. The literature review presented in this chapter shows two prominent approaches for uncertainty quantification in composite structures: perturbation based approach and Monte Carlo simulation based approach. The major drawback in a perturbation based approach can be identified as the requirement of intensive analytical derivation and lack of the ability to obtain complete probabilistic description of the response quantities. Moreover, this approach is valid only for a low degree of stochasticity in the input parameters. A Monte Carlo simulation approach for uncertainty quantification does not have these critical lacunas. But the MCS approach is computationally very demanding because of the requirement of carrying out large number ($\sim 10^4$) of repetitive simulations corresponding to a random set of input parameters. For the analyses of composite structures, even one such simulation is normally very computationally intensive and time consuming. In such situation the panacea is a surrogate based Monte Carlo simulation, which is adopted throughout this book. The remaining chapters in this book deal with various specific aspects of uncertainty in the global responses of composite structures.

CHAPTER 3

Stochastic Dynamic Analysis of Laminated Composite Plates

3.1 Introduction

The previous chapters highlight the importance of carrying out stochastic analysis of composite structures to quantify the effect of uncertainty in various global responses. The effect of material and geometric uncertainty on the dynamic responses of composite plates is investigated in this chapter. A bottom up surrogate based approach is employed to quantify the variability in free vibration responses of composite cantilever plates due to uncertainty in ply orientation angle, elastic modulus and mass density. The finite element method is employed incorporating effects of transverse shear deformation based on Mindlin's theory in conjunction with a random variable approach. Parametric studies are carried out to determine the stochastic frequency response functions (SFRF) along with stochastic natural frequencies and modeshapes. In this study, a surrogate based approach using General High Dimensional Model Representations (GHDMR) is employed for achieving computational efficiency in quantifying uncertainty. This chapter also presents an uncertainty quantification scheme using commercial finite element software (ANSYS) and thereby comparative results of stochastic natural frequencies are furnished for UQ using GHDMR approach and ANSYS.

Laminated composite plates are typically made from different combinations of polymer prepregs and metals. In the serial production of a composite plate, any small changes in fibre orientation angle and differences in bonding of layers may drastically affect critical responses such as modal vibration characteristics of the composite structure. New aircraft developments (e.g., reusable launch vehicle, high-speed civil transport) are departing dramatically from traditional environments. Application of historical uneconomic uncertainty factors may not be sufficient to provide adequate safety. Conversely, the trend to design for all possible unfavorable events occurring simultaneously could produce an unacceptable dynamic response. Moreover, composite materials have more intrinsic variables than metals due to their heterogeneity and are subjected to more manufacturing process sources of variation. For composite materials, the properties of constituent material vary statistically due to lack of precision and accuracy to maintain the exact properties for each layer of the laminate. Hence the uncertainties incurred during manufacturing process are due to the misalignment of ply-orientation, intralaminate voids, incomplete curing of resin, excess resin between plies, excess matrix voids and porosity resulting from machine, human and process inaccuracy. As a result, free vibration responses of such laminated composite plates show volatility from its deterministic mean value. Because of its inherent complexity, laminated composite structures can be difficult to manufacture accurately according to its exact design specification which results undesirable uncertainties in responses. The design and analysis of conventional materials is easier than that of composites because for conventional materials both materials and most geometric properties have either little or well-known variation from their nominal value. In contrast, the same does not hold good for design of structures made of laminated composites. Hence, the uncertainty calibration for structural reliability of such composite structures is essential to ensure operational safety by means of safe as well as economic design. The prime sources of random structural uncertainty considered in this study are material properties and fiber orientation of the individual constituent laminae. Because of the randomness in these input parameters, the mass matrices and the stiffness matrices of the composite structure become stochastic in nature. Thus it causes the statistical variation in the eigenvalues and eigenvectors and subsequently the dynamic response as well. Therefore, a realistic analysis of composite laminated plates is presented in this chapter to quantify the uncertainties in dynamic responses arising from the randomness in the variation of parameters like ply-orientation angle, elastic modulus and mass density, etc. A brief literature review on uncertainty quantification (UQ) of laminated composite plates is presented in the next paragraph.

The free vibration characteristics of laminated composites have been extensively considered in the literature. The pioneering work using finite element method (FEM) in conjunction with laminated composite plates is reviewed by Reddy (1985). The deterministic analyses of free vibration for laminated plate structures, incorporating a wide spectrum of approaches are reported in the open literature (Leissa and Narita 1989, Liew and Lim 1995, Liew 1996, Chow et al. 1992, Zhang et al. 2003, Lanhe et al. 2005, Liew et al. 2004). The vibration analysis of rectangular laminated composite plates is carried out by Wang et al. (2002) employing the first order shear deformation theory (FSDT) while the mesh-free method for static and free vibration analysis of shear deformable laminated composite plates is introduced by Dai et al. (2004) and successive investigation is carried out by Liu et al. (2007). The natural frequencies of composite plates with

RBF-pseudo spectral method is studied by Ferreira and Fasshauer (2007) while free vibrations of uncertain composite plates via stochastic Rayleigh-Ritz approach is investigated by Venini and Mariani (1997). Stochastic dynamic analysis of composite laminates is carried out by many researchers (Yadav and Verma 1992, Oh and Librescu 1997, Mehrez et al. 2012a, 2012b, Soize 2013, Singh et al. 2002). The stochastic vibration analysis of laminated composites with uncertain random material properties is investigated by Onkar and Yadav (2005) and Allegri et al. (2006). The natural frequencies and vibration modes of laminated composite plates with arbitrary curvilinear fiber shape paths were investigated by Honda and Narita (2012). Sepahvand et al. (2012) studied the stochastic free vibration of orthotropic plates using generalized polynomial chaos expansion. António and Hoffbauer (2007 and 2013) studied uncertainty in angle-ply composite structures followed by a global sensitivity analysis. Not surprisingly, there is continued interest in implementing stochastic concepts in material characterization, in structural response assessment, and in developing rational design and effective utilization procedures for laminated composites. Considered as a broad area within stochastic mechanics, such analyses require the identification of uncertainties and the selection of appropriate techniques for uncertainty propagation up to different modelling scales, depending on the response of interest.

Monte Carlo simulation technique in conjunction with finite element (FE) method is found to be widely used for quantifying uncertainties of laminated composite structures, wherein thousands of finite element simulations are needed to be carried out. Thus this approach is of limited practical value due to its computational intensiveness unless some form of model-based extrapolation can be used to make the method more efficient. In view of above, the present investigation attempts to quantify the uncertainty in free vibration responses of laminated composite plates using a bottom up surrogate based approach, where the computationally expensive finite element model can be effectively replaced by an efficient mathematical model. In this approach the effect of uncertainty (such as ply orientation angle, variation in material and geometric properties, etc.) is accounted in the elementary level first and then this effect is propagated towards the global responses via surrogates of the actual finite element model. A generalized high dimensional model representation (GHDMR) (Li and Rabitz 2012) is employed for surrogate model formation, wherein diffeomorphic modulation under observable response preserving homotopy (D-MORPH) regression is utilized to ensure the hierarchical orthogonality of high dimensional model representation component functions.

Other forms of high dimensional model representation approaches (cut-HDMR and RS-HDMR) are found to be recently utilized in the area of reliability analysis, optimization and sensitivity analysis of different structures (Chowdhury and Rao 2009, Mukhopadhyay et al. 2015c, Mukherjee et al. 2011). The HDMR based analyses of composite plates have been presented considering different forms of material and geometric uncertainties (Dey et al. 2016b, Dey et al. 2015a). A limitation of the studies on UQ of laminated composites as presented in the literature review section is that most of the investigations are based on finite element codes written in scientific programming languages like FORTRAN (Press et al. 1992) or MATLAB®. This restricts application of such uncertainty quantification methods to large-scale complex structures, for which commercially available finite element modelling packages are commonly used in industry. In this chapter, we present a useful industry oriented uncertainty quantification scheme using commercial finite element software (ANSYS) in conjunction with MATLAB and thereby comparative results of stochastic natural frequencies are furnished for uncertainty quantification using GHDMR approach and ANSYS. This chapter hereafter is organized as follows, Section 3.2: detail theoretical formulation for development of finite element code using FORTRAN; Section 3.3: formulation of GHDMR; Section 3.4: proposed surrogate based bottom up stochastic approach and uncertainty quantification scheme using ANSYS; Section 3.5: results and discussion; and Section 3.6: summary.

3.2 Theoretical formulation for dynamics composite plates

A rectangular composite laminated cantilever plate of uniform length L, width b, and thickness h is considered having three plies located in a three-dimensional Cartesian coordinate system (x, y, z), where the x-y plane passes through the middle of the plate thickness with its origin placed at the corner of the cantilever plate as shown in Fig. 3.1. An eight noded isoparametric plate bending element is considered for finite element formulation. The composite plate is considered with uniform thickness with the principal material axes of each layer being arbitrarily oriented with respect to mid-plane. If the mid-plane forms the x-y plane of the reference plane, then the displacements can be computed as

$$u(x, y, z) = u^0(x, y) - z\,\theta_x(x, y)$$
$$v(x, y, z) = v^0(x, y) - z\,\theta_y(x, y)$$
$$w(x, y, z) = w^0(x, y) = w(x, y)$$

(3.1)

Assuming u, v and w are the displacement components in x-, y- and z-directions, respectively and u^0, v^0 and w^0 are the mid-plane displacements, and θ_x and θ_y are rotations of cross-sections along the x- and y-axes. The strain-displacement relationships for small deformations can be expressed as

$$\varepsilon_{xx} = \varepsilon_x^0 + zk_x$$
$$\varepsilon_{yy} = \varepsilon_y^0 + zk_y$$
$$\gamma_{xy} = \gamma_{xy}^0 + zk_{yy}$$
$$\gamma_{xz} = w_{,x}^0 - \theta_x$$
$$\gamma_{yz} = w_{,y}^0 - \theta_y$$

(3.2)

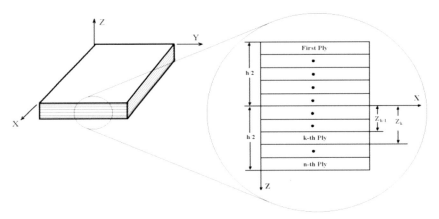

Fig. 3.1 Laminated composite plate.

where mid-plane components are given by

$$\varepsilon_x^0 = u_{,x}^0, \quad \varepsilon_y^0 = u_{,y}^0 \text{ and } \gamma_{xy}^0 = u_{,y}^0 + v_{,x}^0 \qquad (3.3)$$

and plate curvatures are expressed as

$$k_x = -\theta_{x,x} = -w_{,xx} + \gamma_{xz,x}$$
$$k_y = -\theta_{y,y} = -w_{,yy} + \gamma_{yz,y}$$
$$k_{xy} = -(\theta_{x,y} + \theta_{y,x}) = -2w_{,xy} + \gamma_{xz,y} + \gamma_{yz,x} \qquad (3.4)$$

Therefore the strains in the k-th lamina can be expressed in matrix form

$$\{\varepsilon\}^k = \begin{Bmatrix} \varepsilon_x^0 \\ \varepsilon_y^0 \\ \gamma_{xy}^0 \end{Bmatrix} + z \begin{Bmatrix} k_x^0 \\ k_y^0 \\ k_{xy}^0 \end{Bmatrix} = \{\varepsilon^0\} + z\{k\} \text{ and } \{\gamma\}^k = \begin{Bmatrix} \gamma_{yz} \\ \gamma_{xz} \end{Bmatrix} = \{\gamma\} \qquad (3.5)$$

In general, the force and moment resultants of a single lamina are obtained from stresses as

$$\{F\} = \{N_x N_y N_{xy} M_x M_y M_{xy} Q_x Q_y\}^T = \int_{-h/2}^{h/2} \{\sigma_x \ \sigma_y \ \tau_{xy} \ \sigma_x z \ \sigma_y z \ \tau_{xy} z \ \tau_{xz} \ \tau_{yz}\}^T \, dz$$

$$(3.6)$$

In matrix form, the in-plane stress resultant $\{N\}$, the moment resultant $\{M\}$, and the transverse shear resultants $\{Q\}$ can be expressed as

$$\{N\} = [A]\{\varepsilon^0\} + [B]\{k\} \text{ and } \{M\} = [B]\{\varepsilon^0\} + [D]\{k\} \qquad (3.7)$$

$$\{Q\} = [A^*]\{\gamma\} \qquad (3.8)$$

where

$$[A_{ij}^{*}] = \int_{-h/2}^{h/2} \bar{Q}_{ij} \, dz \qquad where \ i, j = 4, 5$$

$$[\bar{Q}_{ij}(\bar{\omega})] = \begin{vmatrix} m^4 & n^4 & 2m^2n^2 & 4m^2n^2 \\ n^4 & m^4 & 2m^2n^2 & 4m^2n^2 \\ m^2n^2 & m^2n^2 & (m^4 + n^4) & -4m^2n^2 \\ m^2n^2 & m^2n^2 & -2m^2n^2 & (m^2 - n^2) \\ m^3n & mn^3 & (mn^3 - m^3n) & 2(mn^3 - m^3n) \\ mn^3 & m^3n & (m^3n - mn^3) & 2(m^3n - mn^3) \end{vmatrix} [Q_{ij}]$$

Here $m = Sin\theta(\bar{\omega})$ and $n = Cos\theta(\bar{\omega})$, wherein $\theta(\bar{\omega})$ is the random fibre orientation angle where the symbol $(\bar{\omega})$ indicates the stochasticity of parameters. However, laminate consists of a number of laminae wherein $[Q_{ij}]$ and $[\bar{Q}_{ij}(\bar{\omega})]$ denotes the On-axis elastic constant matrix and the off-axis elastic constant matrix, respectively. The elasticity matrix of the laminated composite plate is given by,

$$[D'(\bar{\omega})] = \begin{vmatrix} A_{ij}(\bar{\omega}) & B_{ij}(\bar{\omega}) & 0 \\ B_{ij}(\bar{\omega}) & D_{ij}(\bar{\omega}) & 0 \\ 0 & 0 & S_{ij}(\bar{\omega}) \end{vmatrix} \qquad (3.9)$$

where

$$[A_{ij}(\bar{\omega}), \ B_{ij}(\bar{\omega}), \ D_{ij}(\bar{\omega})] = \sum_{k=1}^{n} \int_{z_{k-1}}^{z_k} [\bar{Q}_{ij}(\bar{\omega})]_k \ [1, \ z, \ z^2] \, dz \qquad i, j = 1, 2, 6 \qquad (3.10)$$

$$[S_{ij}(\bar{\omega})] = \sum_{k=1}^{n} \int_{z_{k-1}}^{z_k} \alpha_s \ [\bar{Q}_{ij}(\bar{\omega})]_k \ dz \qquad i, j = 4, 5 \qquad (3.11)$$

where α_s is the shear correction factor and is assumed as 5/6.

Now, mass per unit area is denoted by P and is given by

$$P(\bar{\omega}) = \sum_{k=1}^{n} \int_{z_{k-1}}^{z_k} \rho(\bar{\omega}) \, dz \qquad (3.12)$$

where $\rho(\bar{\omega})$ denotes the stochastic density parameter at each element. The mass matrix is expressed as

$$[M(\bar{\omega})] = \int_{Vol} [N][P(\bar{\omega})][N] d(vol) \qquad (3.13)$$

The stiffness matrix is given by

$$[K(\bar{\omega})] = \int_{-1}^{1} \int_{-1}^{1} [B(\bar{\omega})]^T \ [D(\bar{\omega})] \ [B(\bar{\omega})] \, d\xi \, d\eta \qquad (3.14)$$

where ξ and η are the local natural coordinates of the element.

3.2.1 Governing equations

The Hamilton's principle (Meirovitch 1992) is employed to study the dynamic nature of the composite structure. The principle used for the Lagrangian which is defined as

$$L_f = T - U - W \qquad (3.15)$$

where T, U and W are total kinetic energy, total strain energy and total potential of the applied load, respectively. The Hamilton's principle applicable to non-conservative system can be expressed as,

$$\delta H = \int_{t_i}^{t_f} [\delta T - \delta U - \delta W] dt = 0 \qquad (3.16)$$

The energy functional for Hamilton's principle is the Lagrangian (L_f) which includes kinetic energy (T) in addition to potential strain energy (U) of an elastic body. The expression for kinetic energy of an element is given by

$$T = \frac{1}{2}\{\dot{\delta}_e\}^T [M_e(\bar{\omega})] + [C_e(\bar{\omega})] \{\dot{\delta}_e\} \qquad (3.17)$$

The potential strain energy for an element of a plate can be expressed as,

$$U = U_1 + U_2 = \frac{1}{2}\{\delta_e\}^T [K_e(\bar{\omega})]\{\delta_e\} + \frac{1}{2}\{\delta_e\}^T [K_{\sigma e}(\bar{\omega})]\{\delta_e\} \quad (3.18)$$

The Langrange's equation of motion is given by

$$\frac{d}{dt}\left[\frac{\partial L_f}{\partial \dot{\delta}_e}\right] - \left[\frac{\partial L_f}{\partial \delta_e}\right] = \{F_e\} \qquad (3.19)$$

where $\{F_e\}$ is the applied external element force vector of an element and L_f is the Lagrangian function. Substituting $L_f = T - U$, and the corresponding expressions for T and U in Lagrange's equation, one obtains the dynamic equilibrium equation for each element in the following form

$$[M(\bar{\omega})]\{\ddot{\delta}_e\} + [C]\{\dot{\delta}_e\} + ([K_e(\bar{\omega})] + [K_{\sigma e}(\bar{\omega})])\{\delta_e\} = \{F_e\} \quad (3.20)$$

After assembling all the element matrices and the force vectors with respect to the common global coordinates, the resulting equilibrium equation is obtained. For the purpose of the present study, the finite element model is developed for different element types and finite element discretization and nodal positions of the driving point and measurement point. Considering randomness of input parameters like ply-orientation angle, elastic modulus and mass density, etc., the equation of motion of a linear damped discrete system with n degrees of freedom can expressed as

$$[M(\bar{\omega})]\ddot{\delta}(t) + [C]\dot{\delta}(t) + [K(\bar{\omega})]\delta(t) = f(t) \qquad (3.21)$$

where $[C]$ is the damping co-efficient matrix and $[K(\bar{\omega})]$ is the stiffness matrix wherein $[K(\bar{\omega})] = [K_e(\bar{\omega})] + [K_{\sigma e}(\bar{\omega})]$ in which $K_e(\bar{\omega}) \in R^{n \times n}$ is the elastic stiffness matrix, $K_{ge}(\bar{\omega}) \in R^{n \times n}$ is the geometric stiffness matrix while $M(\bar{\omega}) \in R^{n \times n}$ is the mass matrix, $\delta(t) \in R^n$ is the vector of generalized coordinates and $f(t) \in R^n$ is the forcing vector. The governing equations are derived based on Mindlin's Theory incorporating rotary inertia, transverse shear deformation. The equation represents a set of coupled second-order ordinary-differential equations. The solution of this equation also requires the knowledge of the initial conditions in terms of displacements and velocities of all the coordinates. The initial conditions can be specified as

$$\delta(0) = \delta_0 \in R^n \text{ and } \dot{\delta}(0) = \dot{\delta}_0 \in R^n \tag{3.22}$$

Equation (3.21) is considered to solve together with the initial conditions of equation (3.22) using modal analysis. The present study assumes that all initial conditions are zero and the forcing function is a harmonic excitation applied at a particular point of composite cantilever plate.

3.2.2 Modal analysis and dynamic response

Lord Rayleigh (1877) showed that undamped linear systems are capable of so-called natural motions. This essentially implies that all the system coordinates execute harmonic oscillation at a given frequency and form a certain displacement pattern. The oscillation frequency and displacement pattern are called natural frequencies and normal modes, respectively. The natural frequencies (ω_j) and the mode shapes (x_j) are intrinsic characteristic of a system and can be obtained by solving the associated matrix eigenvalue problem

$$[K(\bar{\omega})] \, x_j = \omega_j^2 \, [M(\bar{\omega})] \, x_j; \quad \text{where, } j = 1,2\ldots\ldots n \tag{3.23}$$

It can be also shown that the eigenvalues and eigenvectors satisfy the orthogonality relationship that is

$$X_i^T \, [M(\bar{\omega})] \, x_j = \lambda_{1j} \text{ and } x_i^T \, [K(\bar{\omega})] \, x_j = \omega_j^2 \, \lambda_{ij} \text{ where } i, j = 1,2\ldots\ldots n \tag{3.24}$$

Note that the Kroneker delta functions is given by $\lambda_{ij} = 1$ for $i = j$ and $\lambda_{ij} = 0$ for $i \neq j$. The property of the eigenvectors in (3.24) is also known as the mass orthonormality relationship. The solution of undamped eigenvalue problem is now standard in many finite element packages. There are various efficient algorithms available for this purpose (Bathe 1990). This orthogonality property of the undamped modes is very powerful as it allows transforming a set of coupled differential equations to a set of independent equations. For convenience, we construct the matrices

$$\Omega(\bar{\omega}) = \text{diag} \, [\omega_1, \omega_2, \omega_3 \ldots\ldots\ldots\ldots \omega_n] \in R^{n \times n}$$
$$\text{and } X(\bar{\omega}) = [x_1, x_2 \ldots\ldots\ldots\ldots x_n] \in R^{n \times n} \tag{3.25}$$

where the eigenvalues are arranged such that $\omega_1 < \omega_2$, $\omega_2 < \omega_3$, ...
$\omega_k < \omega_{k+1}$. The matrix $\mathbf{X}$ is known as the undamped modal matrix. Using these matrix notations, the orthogonality relationships (3.24) can be rewritten as

$$\mathbf{x}_i^T \,[\mathbf{M}(\bar{\omega})]\, \mathbf{x}_j = \lambda_{1j} \quad \text{and} \quad \mathbf{x}_j^T\,[\mathbf{K}(\bar{\omega})]\,\mathbf{x}_j = \omega_j^2\,\lambda_{1j} \quad \text{where } i,j = 1,2........n \quad (3.26)$$

$$\mathbf{X}^T\,[\mathbf{M}(\bar{\omega})]\,\mathbf{X} = \mathbf{I} \quad \text{and} \quad \mathbf{X}^T\,[\mathbf{K}(\bar{\omega})]\,\mathbf{X} = \Omega^2 \qquad (3.27)$$

where $\mathbf{I}$ is a (n x n) identity matrix. We use the following coordinate transformation (as the modal transformation)

$$\delta\,(\bar{\omega})\,(\mathrm{t}) = \mathbf{X}\,\mathbf{y}(\mathrm{t}) \qquad (3.28)$$

Using the modal transformation in equation (3.28), pre-multiplying equation (3.21) by $\mathbf{X}^T$ and using the orthogonality relationships in (3.27), equation of motion of a damped system in the modal coordinates may be obtained as

$$\ddot{y}(t) + X^T\,C\,X\,\dot{y}(t) + \Omega^2\,y(t) = \tilde{f}(t) \qquad (3.29)$$

Clearly, unless $\mathbf{X}^T\,[\mathbf{C}]\,\mathbf{X}$ is a diagonal matrix, no advantage can be gained by employing modal analysis because the equations of motion will still be coupled. To solve this problem, it is common to assume proportional damping. With the proportional damping assumption, the damping matrix $\mathbf{C}$ is simultaneously diagonalisable with $\mathbf{M}$ and $\mathbf{K}$. This implies that the damping matrix in the modal coordinate is

$$\mathbf{C}' = \mathbf{X}^T\,[\mathbf{C}]\,\mathbf{X} \qquad (3.30)$$

where $\mathbf{C}'$ is a diagonal matrix. This matrix is also known as the modal damping matrix. The damping factors ζ_j are defined from the diagonal elements of the modal damping matrix as

$$\mathbf{C}'_{jj} = 2\,\zeta_j\,\omega_j \qquad \text{where } j = 1,2,............n \qquad (3.31)$$

Such a damping model, introduced by Lord Rayleigh (1877) allows analyzing damped systems in very much the same manner as undamped systems since the equation of motion in the modal coordinates can be decoupled as

$$\ddot{y}_j(t) + 2\zeta_j\,\omega_j\,\dot{y}_j(t) + \omega_j^2\,y(t) = \tilde{f}_j(t) \qquad \text{where } j = 1,2,............n \qquad (3.32)$$

The generalized proportional damping model expresses the damping matrix as a linear combination of the mass and stiffness matrices, that is

$$\mathbf{C}(\bar{\omega}) = \alpha_1\,\mathbf{M}(\bar{\omega})\,\mathbf{f}\,(\mathbf{M}^{-1}(\bar{\omega})\,\mathbf{K}(\bar{\omega}))\ \text{where } \alpha_1 = 0.005 \text{ is constant damping factor}$$
$$(3.33)$$

The transfer function matrix of the system can be obtained as

$$H(i\omega)\,(\bar{\omega}) = X\,[-\omega^2\,I + 2\,i\,\omega\,\zeta\Omega + \Omega^2\,]^{-1}\,X^T = \sum_{j=1}^{n} \frac{X_j\,X_j^T}{-\omega^2 + 2\,i\,\omega\,\zeta_j\omega_j + \omega_j^2}$$
$$(3.34)$$

Using this, the dynamic response in the frequency domain with zero initial conditions can be conveniently represented as

$$\bar{\delta}(i\omega)\,(\overline{\omega}) = H(i\omega)\,\overline{f}(i\omega) \;=\; \sum_{j=1}^{n} \frac{X_j^T\,\overline{f}(i\omega)}{-\omega^2 + 2i\omega\zeta_j\omega_j + \omega_j^2}\,X_j \quad (3.35)$$

Therefore, the dynamic response of proportionally damped system can be expressed as a linear combination of the undamped mode shapes.

3.3 Formulation of GHDMR

The general high dimensional model representation (GHDMR) can construct a proper model for prediction of the output (say natural frequency) corresponding to a stochastic input domain. The present approach can treat both independent and correlated input variables, and includes independent input variables as a special case. The role of D-MORPH is to ensure the component functions' orthogonality in hierarchical manner. The present technique decomposes the function $\lambda(S)$ with component functions by input parameters, $S = (S_1, S_2, \dots, S_{kk})$. As the input parameters are independent in nature, the component functions are specifically projected by vanishing condition. Hence, it has limitation for general formulation. In contrast, a numerical analysis with component functions is portrayed in the problem of the present context wherein a unified framework for general HDMR dealing with both correlated and independent variables are established. For different input parameters, the output is calculated as (Li and Rabitz 2012)

$$\lambda(S) = \lambda_0 + \sum_{i=1}^{kk}\lambda_i(S_i) + \sum_{1 \le i < j \le kk}\lambda_{ij}(S_i, S_j) + \dots + \lambda_{12\dots kk}(S_1, S_2, \dots, S_{kk}) \tag{3.36}$$

$$\lambda(S) = \sum_{u \subseteq kk}\lambda_u(S_u) \tag{3.37}$$

where λ_0 (zeroth order component function) represents the mean value. $\lambda_i(S_i)$ and $\lambda_{ij}(S_i, S_j)$ denote the first and second order component functions, respectively while $\lambda_{12\dots kk}(S_1, S_2, \dots, S_{kk})$ indicates the residual contribution by input parameters. The subset $u \subseteq \{1, 2, \dots, kk\}$ denotes the subset where $u \subseteq kk$ for simplicity and empty set, $\Gamma \in u$. As per Hooker's definition, the correlated variables are expressed as,

$$\{\lambda_u(S_u \mid u \subseteq kk)\} = Arg\,\min_{\{g_u \in L^2(R^u),\, u \subseteq kk\}} \int \left(\sum_{u \subseteq k} g_u(S_u) - \lambda(S)\right)^2 w(S)\,dS \tag{3.38}$$

$$\forall u \subseteq kk \;, \qquad \forall i \in u \;, \qquad \int \lambda_u(S_u)\,w(S)\,dS_i\,dS_{-u} = 0 \tag{3.39}$$

and

$$\forall v \subset u \ , \qquad \forall g_v : \ \int \lambda_u(S_u) \, g_v(S_v) \, w(S) \, dS \ = \ \langle \lambda_u(S_u), g_v(S_v) \rangle \ = 0 \tag{3.40}$$

The function $\lambda(S)$ can be obtained from sample data by experiments or by modelling. To minimise the computational cost, the reduction of the squared error can be realised easily. Assuming H in Hilbert space is expanded on the basis $\{h_1, h_2, \ldots, h_{kk}\}$, the bigger subspace $\bar{H} (\supset H)$ is expanded by extended basis $\{h_1, h_2, \ldots, h_{kk}, h_{kk+1}, \ldots, h_m\}$. Then $\bar{H}$ can be decomposed as

$$\bar{H} = H \oplus H^\perp \tag{3.41}$$

where $H^\perp$ denotes the complement subspace (orthogonal) of H within $\bar{H}$. In past works, the component functions are calculated from basis functions. The component functions of Second order HDMR expansion are estimated from basis functions $\{\varphi\}$ as (Li et al. 2006)

$$\lambda_i(S_i) \approx \sum_{r=1}^{kk} \alpha_r^{(0)i} \, \varphi_r^i(S_i) \tag{3.42}$$

$$\lambda_{ij}(S_i, S_j) \approx \sum_{r=1}^{kk} [\alpha_r^{(i\,j)i} \, \varphi_r^i(S_i) + \alpha_r^{(i\,j)j} \, \varphi_r^j(S_j)] + \sum_{p=1}^{l} \sum_{q=1}^{l} \beta_{pq}^{(0)\,ij} \, \varphi_p^i(S_i) \varphi_q^j(S_j) \tag{3.43}$$

i.e., the basis functions of $\lambda_{ij}(S_i, S_j)$ contain all the basis functions used in $\lambda_i(S_i)$ and $\lambda_j(S_j)$.

The HDMR expansions at N_{samp} sample points of S can be represented as a linear algebraic equation system

$$\Gamma J = \hat{R} \tag{3.44}$$

where Γ denotes a matrix $(N_{samp} \times \tilde{t})$ whose elements are basis functions at the N_{samp} values of S; J is a vector with $\tilde{t}$ dimension of all unknown combination coefficients; $\hat{R}$ is a vector with N_{samp}-dimension wherein l-th element is $\lambda(S^{(l)}) - \lambda_0$. $S^{(l)}$ denotes the l-th sample of S, and λ_0 represents the average value of all $\lambda(S^{(l)})$. The regression equation for least squares of the above equation can be expressed as

$$\frac{1}{N_{samp}} \Gamma^T \Gamma J = \frac{1}{N_{samp}} \Gamma^T \hat{R} \tag{3.45}$$

Due to the use of extended bases, some rows of the above equation are identical and can be removed to give an underdetermined algebraic equation system

$$A J = \hat{V} \tag{3.46}$$

It has many of solutions for J composing a manifold $Y \in \Re^{\tilde{t}}$. Now the task is to find a solution J from Y to force the HDMR component functions satisfying the hierarchical orthogonal condition. D-MORPH regression provides a solution to ensure additional condition of exploration path represented by differential equation

$$\frac{dJ(l)}{dl} = \chi \, v(l) = (I_t - A^+ \, A) \, v(l) \tag{3.47}$$

wherein χ denotes orthogonal projector ensuring

$$\chi^2 = \chi \quad \text{and} \quad \chi^T = \chi \tag{3.48}$$

$$\chi = \chi^2 = \chi^T \chi \tag{3.49}$$

The free function vector may be selected to ensure the wide domain for $J(l)$ as well as to simultaneously reduce the cost $\kappa(J(l))$ which can be expressed as

$$v(l) = -\frac{\partial \kappa(J(l))}{\partial J} \tag{3.50}$$

Then we obtain

$$\frac{\partial \kappa(J(l))}{\partial l} = \left(\frac{\partial \kappa(J(l))}{\partial J}\right)^T \frac{\partial J(l)}{\partial l} = \left(\frac{\partial \kappa(J(l))}{\partial J}\right)^T P \, v(l)$$

$$= -\left(P\frac{\partial \kappa(J(l))}{\partial J}\right)^T \left(P\frac{\partial \kappa(J(l))}{\partial J}\right) \leq 0 \tag{3.51}$$

The cost function can be expressed in quadratic form as

$$\kappa = \frac{1}{2}J^T \, B \, J \tag{3.52}$$

where B denotes the positive definite symmetric matrix and J_∞ can be expressed as

$$J_\infty = V_t \, (U_{\tilde{i}-r}^T \, V_{\tilde{i}-r})^{-1} \, U_{\tilde{i}-r}^T \, A^+ \, \hat{V} \tag{3.53}$$

where the last columns $(\tilde{i}-r)$ of U and V are denoted as $U_{\tilde{i}-r}$ and $V_{\tilde{i}-r}$ which can be found by decomposition of $\chi \, B$

$$\chi B = U\begin{bmatrix} \overline{S}_r & 0 \\ 0 & 0 \end{bmatrix} V^T \tag{3.54}$$

This unique solution J_∞ in Y indicates the minimized cost function. D-MORPH regression is used to find the J which ensures the HDMR component functions' orthogonality in hierarchical manner. The construction of the corresponding cost function κ can be found in previous literature (Li and Rabitz 2012).

3.4 Bottom-up stochastic approach

In the present study, the following cases are considered wherein the random variables in each layer of laminate are investigated:

(a) Variation of ply-orientation $g\{\theta(\bar{\omega})\} = \{\theta_1 \ \theta_2 \ \theta_3........\theta_i......\theta_l \}$ angle only:

(b) Variation in elastic modulus: $g\{E_1(\bar{\omega})\} = \{E_{1(1)} \; E_{1(2)} \; E_{1(3)} \; .... E_{1(i)} ...... E_{1(l)}\}$

(c) Variation of mass density only: $\quad g\{\rho(\bar{\omega})\} = \{\rho_1 \; \rho_2 \; \rho_3 .... \rho_i ...... \rho_l\}$

(d) Combined variation of ply orientation angle, elastic modulus and mass density:

$$g\{\theta(\bar{\omega}), E_1(\bar{\omega}), \rho(\bar{\omega})\} = \{\Phi_1(\theta_1 .... \theta_l), \; \Phi_2(E_{1(1)} ...... E_{1(l)}), \Phi_3(\rho_1 .... \rho_l), \}$$

where θ_i, $E_{1(i)}$ and ρ_i are the ply orientation angle, elastic modulus along longitudinal direction and mass density of ith layer, respectively and 'l' denotes the number of layer in the laminate. The stochastic responses investigated in this chapter corresponding to layer-wise variations of the above sources of uncertainties are natural frequency, frequency response function and mode shape. Monte Carlo Simulation (MCS) based random variable approach is employed in this study for uncertainty quantification. Though this technique is computationally expensive but it is a quite versatile technique which is well established in many fields for uncertainty quantification over a long period of time.

Monte Carlo Simulation methods are based on the use of random numbers and probability statistics to investigate the problems. The results obtained by MCS are generally used as benchmarking results to compare the results obtained from other methods. In the present study, a random number generator is employed to generate the possible ply orientation angle, elastic modulus and mass density variation with a standard range of $\pm 10\%$ from respective deterministic mean values for the material properties and $\pm 5°$ for ply orientation angle. Larger the number of sample size, more the confidence in the results obtained for the MCS approach. The number of samples for MCS analysis is generally selected based on convergence study of standard deviation of the performance parameters. In the present analysis ten thousand realizations (i.e., sample size of 10,000) have been performed in all the analyses.

The flowchart for uncertainty quantification of laminated composite plates based on direct MCS using finite element FORTRAN code is presented in Fig. 3.2. As mentioned above, uncertainty quantification using MCS is generally computationally very expensive because it requires the expensive finite element code to run thousands of times. To mitigate this lacuna GHDMR is employed in the present study. In this approach few finite element simulations are generally required to be carried out corresponding to algorithmically chosen design points. The Sobol sequence (Mukhopadhyay et al. 2015c) is used in this study. On the basis of the information captured in the design domain through the design points, a fully functional mathematical model is constructed using GHDMR approach as discussed in Section 3.3. Once the computationally efficient surrogate is constructed, it is then used for subsequent analysis of the structure. A flowchart describing GHDMR based uncertainty quantification scheme of composite plates is furnished in Fig. 3.3.

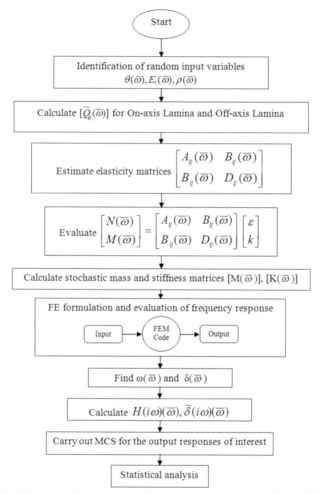

Fig. 3.2 Propagation of uncertainty in bottom up approach using direct MCS.

As mentioned in the introduction section, this study also includes an uncertainty quantification scheme using commercial finite element software ANSYS. For that purpose, the APDL script generated after modelling the composite plate in ANSYS environment is integrated with MATLAB. A fully automated MATLAB code is developed capable of rewriting the APDL script in each iteration containing the random values of stochastic input parameters, then running the APDL script to obtain desired outputs (refer to the flowchart presented in Fig. 3.4) and saving the results for each sample. Thus MCS can be carried out using ANSYS in conjunction with MATLAB for any number of samples following the proposed approach. Comparative results are furnished in the following section for stochastic analysis of free vibration responses using the FORTRAN code developed according to the formulation presented in Section 3.2 and the UQ scheme using ANSYS.

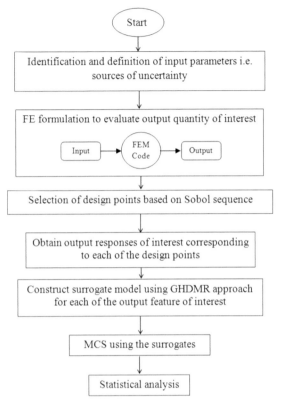

Fig. 3.3 Uncertainty quantification scheme using GHDMR approach.

3.5 Results and discussion

In the present study, the uncertainty analysis of the laminated composite plate is carried out for ply orientation angle, elastic modulus and inertia properties of laminates. The performance parameters of interest are natural frequencies, mode shapes and frequency response functions of the composite plate. The finite element code developed in FORTRAN language following the theoretical formulation described in Section 3.2 is validated with published literature as shown in Table 3.1. Table 3.2 presents comparative deterministic results for first five natural frequencies obtained using the FORTRAN code and the commercial finite element analysis software (ANSYS). The results are found to be in good agreement. The mean value of plate material and geometric properties of the graphite-epoxy composite laminated cantilever plate used in the present analysis are given below:

a) Mean deterministic values of material properties:

Material density, $\rho = 3.202 \times 10^4$ Kg/m^3

Major in-plane Poisson's ratio, $v = 0.3$

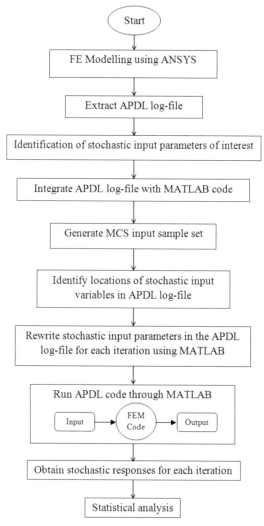

Fig. 3.4 Uncertainty quantification scheme using ANSYS.

Young's modulus, $E_1 = 138 \times 10^9$ N/m^2 and $E_2 = 8.9 \times 10^9$ N/m^2

Shear modulus, $G_{12} = G_{13} = 7.1 \times 10^9$ N/m^2 and $G_{23} = 2.84 \times 10^9$ N/m^2

b) Dimension of composite plate:

Width, $b_o = 1$ m

Thickness, $h = 0.004$ m

Length, $L = 1$ m

Monte Carlo simulation (MCS) is employed in conjunction to the finite element (FE) model for quantifying uncertainty in the aforementioned three

Table 3.1 Non-dimensional fundamental natural frequencies $[\omega = \omega_n \, L^2 \, \sqrt{(\rho/E_1 h^2)}]$ of three layered $[\theta, -\theta, \theta]$ graphite-epoxy plates, $L/b = 1$, $b/h = 20$, $\psi = 30°$.

Fibre Orientation Angle, θ	FEM code	Qatu and Leissa (1991a)
15°	0.8618	0.8759
30°	0.6790	0.6923
45°	0.4732	0.4831
60°	0.3234	0.3283

Table 3.2 Fundamental natural frequencies $[\omega = \omega_n \, L^2 \, \sqrt{(\rho/E_1 h^2)}]$ of three layered $[90°, -90°, 90°]$ graphite-epoxy plates, $L = 1$ m, $h = 0.004$ m, $L/b = 1$, $E_1 = 138$ GPa, $E_2 = 8.9$ GPa, $G_{12} = G_{13} = 7.1$ GPa, $G_{23} = 2.84$ GPa.

Frequency Number	FEM code	ANSYS
1	2.1629	2.1630
2	7.3953	7.3955
3	13.5623	13.5623
4	24.9111	24.9112
5	38.2231	38.2231

response parameters due to layer-wise individual and combined variations of the stochastic input parameters. Figure 3.5 presents comparative results of the mean values and response bounds for first thirty natural frequencies using direct MCS and the proposed GHDMR based approach, wherein a good agreement is noticed between the two plots. To illustrate the validity of the proposed GHDMR based approach further with respect to direct MCS, natural frequencies corresponding to fundamental and twentieth modes are selected arbitarily. Probability density function plots and the scatter plots corresponding to the two selected modes are presented in Fig. 3.6, wherein it is evident that the results of the two approaches are in quite good agreement corroborating the accuracy of the GHDMR based approach.

Figure 3.7 shows the probability density function plots corresponding to first three natural frequencies obtained for combined variation of the stochastic input parameters using the proposed uncertainty quantification scheme using ANSYS. The results are compared with GHDMR based approach of uncertainty quantification. Good agreement between the two approaches establishes the accuracy of the uncertainty quantification scheme using the commercial finite element software ANSYS.

The first three stochastic mode shapes are furnished in Fig. 3.8, wherein layer-wise combined variation of ply orientation angle (45°/–45°/45°), elastic modulus and mass density is considered for the graphite–epoxy laminated composite cantilever plate. From the figure it can be noticed that bending is the predominant factor in the first mode and torsion is predominant in the second mode, whereas a mixed behaviour of bending and torsion is present in the third mode of vibration.

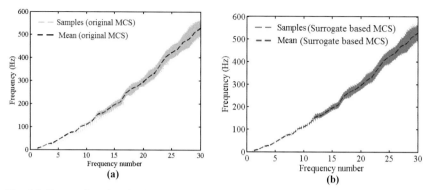

Fig. 3.5 Comparative plots for frequency responses of first thirty natural frequencies using direct MCS and GHDMR approach for combined variation of stochastic input parameters of a angle-ply (45°/−45°/45°) composite cantilever plate.

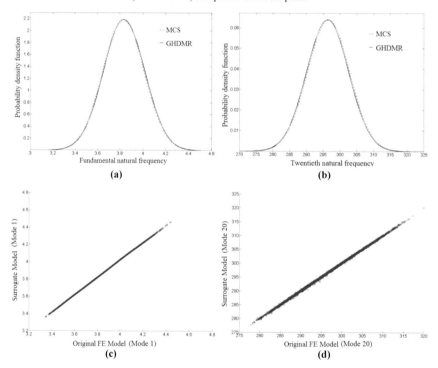

Fig. 3.6 (a, b) Probability density function and (c, d) Scatter plot for fundamental and twentieth modes with respect to direct MCS and GHDMR based approach for angle-ply (45°/−45°/45°) composite cantilever plates.

Frequency response functions (FRF) are obtained in this study for both angle-ply and cross-ply laminated composite plates. Driving point (Point 2) and the cross points (Points 1, 3 and 4) considered in the present analysis are described in

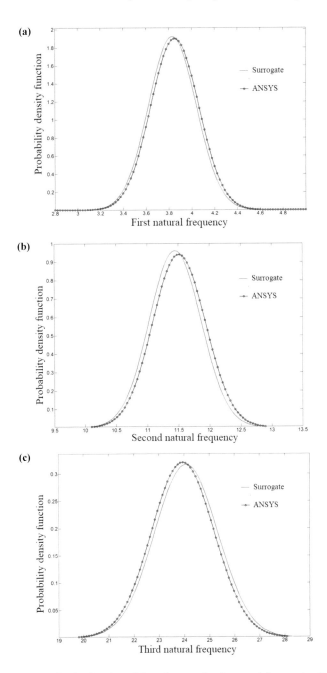

Fig. 3.7 Probability density function of first three stochastic natural frequencies for MCS using surrogate based approach and MCS using commercial finite element package ANSYS for combined variation of ply angle, longitudinal elastic modulus and mass density of an angle-ply (45°/–45°/45°) composite cantilever plate.

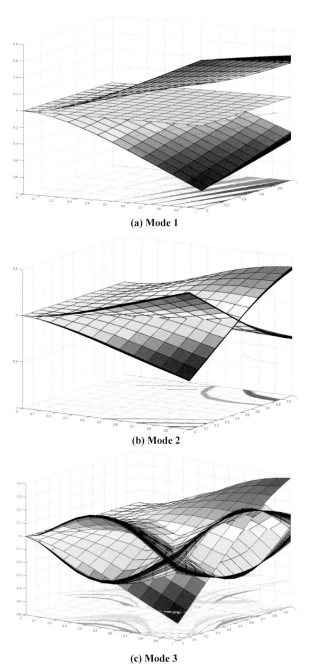

(a) Mode 1

(b) Mode 2

(c) Mode 3

Fig. 3.8 First three stochastic mode shapes considering combined variation of ply orientation angle (45°/–45°/45°), elastic modulus and mass density for graphite–epoxy laminated composite cantilever plate.

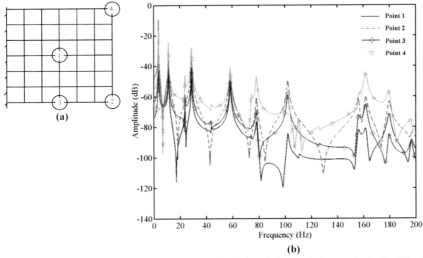

Fig. 3.9 (a) Driving point (Point 2) and cross point (Points 1, 3 and 4) for amplitude (in dB) of frequency response function (b) Amplitude with respect to frequency corresponding to driving and cross points (deterministic responses).

Fig. 3.9a. Figure 3.9b shows the deterministic FRFs in a frequency range of 200 Hz. For the purpose of numerical calculations, 0.5% damping factor is assumed for all the modes. A typical validation plot for the FRF obtained for point 3 using GHDMR approach is furnished in Fig. 3.10, wherein a comparison with direct MCS is presented. From the figure it is evident that GHDMR is quite accurate with respect to direct MCS approach while high level of computational efficiency can be achieved using GHDMR in terms of number of finite element simulations.

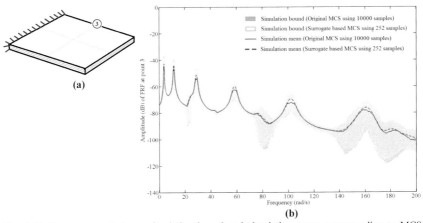

Fig. 3.10 Convergence study on simulation bound and simulation mean corresponding to MCS and surrogates three stochastic mode shapes considering combined variation of ply orientation angle (45°/–45°/45°), elastic modulus and mass density for the simulation bounds and mean and deterministic value for point 3.

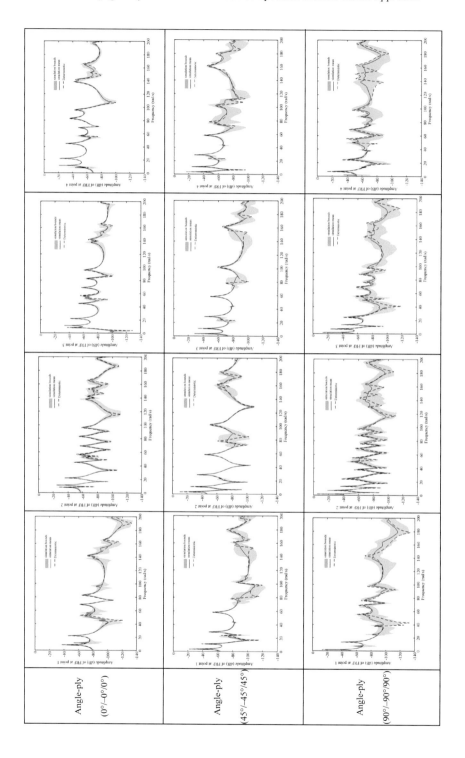

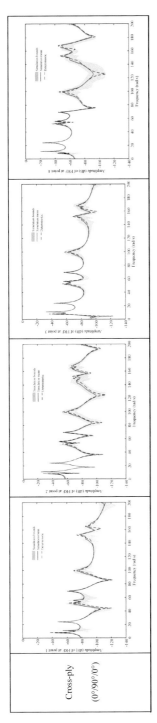

Fig. 3.11 Variation of only ply-orientation angle [θ(ω̄)] in each layer to plot the direct simulation bounds, direct simulation mean and deterministic values for amplitude (dB) with respect to frequency (rad/s) of points 1, 2, 3 and 4 considering graphite–epoxy composite laminated plate with $L = b = 1$ m, $h = 0.004$ m, $E_1 = 138$ GPa, $E_2 = 8.9$ GPa, $G_{12} = G_{13} = 7.1$ GPa, $G_{23} = 2.84$ GPa, $\rho = 3202$ Kg/m^3, $\nu = 0.3$.

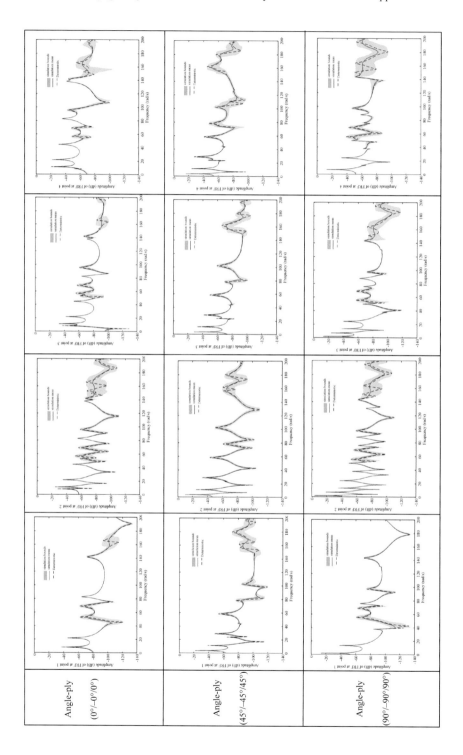

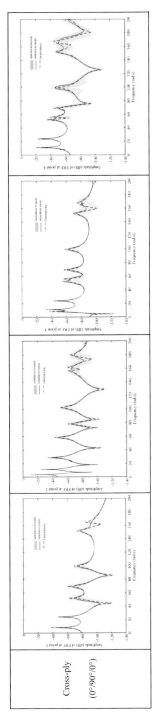

Fig. 3.12 Variation of only elastic modulus [$E_1(\bar{\omega})$] in each layer to plot the direct simulation bounds, direct simulation mean and deterministic values for amplitude (dB) with respect to frequency (rad/s) of points 1, 2, 3 and 4 considering graphite-epoxy composite laminated plate with $L = b = 1$ m, h = 0.004 m, $E_1 = 138$ GPa, $E_2 = 8.9$ GPa, $G_{12} = G_{13} = 7.1$ GPa, $G_{23} = 2.84$ GPa, $\rho = 3202$ Kg/m³, $v = 0.3$.

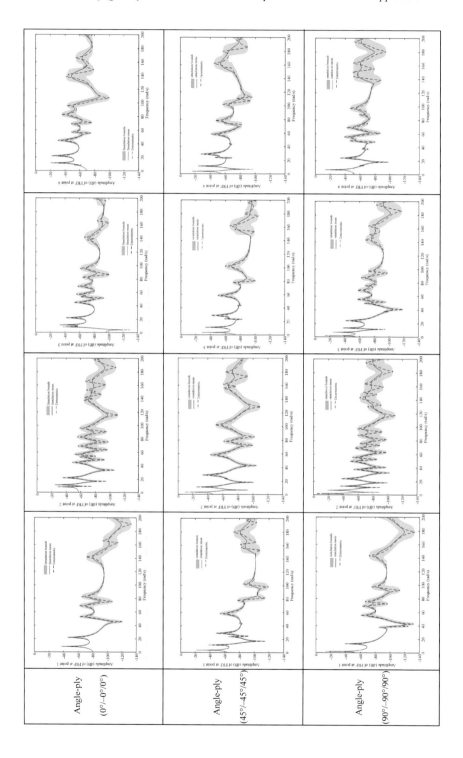

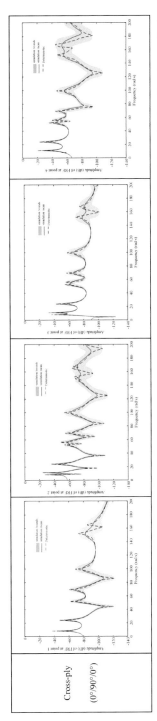

Fig. 3.13 Variation of only mass density $[\rho(\varpi)]$ in each layer to plot the direct simulation bounds, direct simulation mean and deterministic values for amplitude (dB) with respect to frequency (rad/s) of points 1, 2, 3 and 4 considering graphite-epoxy composite laminated plate with $L = b = 1$ m, $h = 0.004$ m, $E_1 = 138$ GPa, $E_2 = 8.9$ GPa, $G_{12} = G_{13} = 7.1$ GPa, $G_{23} = 2.84$ GPa, $\rho = 3202$ Kg/m^3, $v = 0.3$.

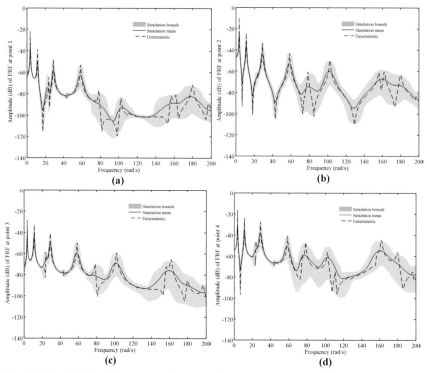

Fig. 3.14 Amplitude with respect to frequency for points 1 to 4 for combined variation of ply
orientation angle (45°/–45°/45°), elastic modulus and mass density.

The sources of uncertainty in the FRFs are systematically investigated in this
study following the proposed bottom up approach of uncertainty propagation
with different configurations of ply orientation angles. The simulation bound,
simulation mean and deterministic values of FRFs are portrayed in Figs. 3.11–3.12
for individual variation of ply orientation angle and elastic modulus (causing
stochasticity in the stiffness matrix), while Fig. 3.13 shows the same for individual
variation of density (causing stochasticity in the mass matrix). For a given amount
of variability for a particular input parameter, more volatility is observed in the
higher frequency ranges. Figure 3.14 shows the effect of combined random variation
of ply orientation angle, elastic modulus and mass density (causing stochasticity
in both the stiffness and the mass matrix) in the FRFs of an angle ply composite
plate. Relative standard deviations are plotted for combined variation of stochastic
input parameters as shown in Fig. 3.15, which gives a clear idea about the relative
volatility of FRFs in different frequency ranges for the four considered points.

It is worthy to mention here that all the results presented in this chapter are obtained using 10,000 simulations. Application of surrogate based approach using GHDMR allows to obtain these results by means of virtual simulations instead of actual finite element simulation. For layer-wise combined variation and individual variation of the stochastic input parameters, 252 and 64 samples, respectively are utilized to construct the GHDMR model. Thus for the purpose of uncertainty quantification, same number of actual finite element simulations are needed in the proposed approach, in contrast with 10,000 finite element simulations needed in the direct MCS approach. Therefore, the proposed GHDMR based approach for uncertainty quantification in composite structures is much more computationally efficient than conventional direct MCS approach in terms of finite element simulations.

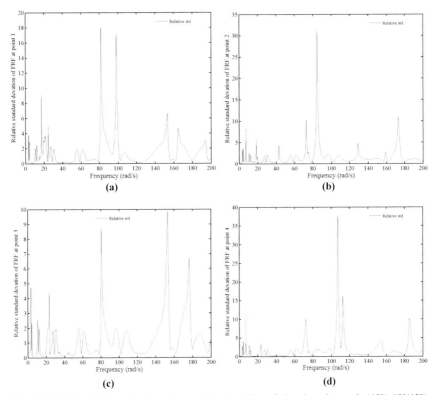

Fig. 3.15 Relative SD for points 1 to 4 for combined variation of ply orientation angle (45°/–45°/45°), elastic modulus and mass density.

3.6 Summary

This chapter presents a bottom up uncertainty propagation scheme for laminated composite plates. An efficient surrogate (GHDMR) based approach is proposed to characterize uncertainty in different free vibration responses of the structure. It is found that high level of computational efficiency can be achieved following the proposed approach compared to direct MCS without compromising the accuracy of results. The GHDMR based uncertainty propagation algorithm can be applied to more complex configurations of laminated composite structures and to quantify uncertainty for various other structural responses using the knowledge shared in this chapter. Other sources of uncertainties in stochastic analysis of different responses can also be incorporated following the GHDMR approach. Another contribution of this chapter is the proposed industry oriented uncertainty quantification scheme using commercial finite element analysis software ANSYS, which is validated with the present problem of laminated composite plates using frequency responses. This approach has immense potential to be extended towards uncertainty quantification of large-scale complex structures in future investigations.

CHAPTER 4

Stochastic Free Vibration Analysis of Pre-twisted Singly Curved Composite Shells

4.1 Introduction

In the last chapter, the effect of uncertainty on the dynamic responses of composite plates was discussed. In this chapter, we have concentrated on the stochastic dynamic responses of singly curved composite shells including the effect of twist angle in the geometry. The effect of transverse shear deformation is incorporated in the probabilistic finite element analysis considering an eight noded isoparametric quadratic element with five degrees of freedom at each node. The finite element model is coupled with the response surface method based on D-optimal design to achieve computational efficiency. A sensitivity analysis is carried out to address the influence of different input parameters on the output natural frequencies. The fibre orientation angle, twist angle and material properties are randomly varied to obtain the stochastic natural frequencies.

Composite singly curved shells are widely used in various engineering applications. Analyzing the effect of twist angle in such structures is immensely important for applications in turbomachinery blades, wind turbines and various aircraft components. Composite structures are often pretwisted due to design and operational needs such as the wing twist of an aircraft is provided to maintain optimum angle of attack preventing negative lift or thrust and maximizing the aerodynamic efficiency. The production of composite structures is always subjected to large variability due to manufacturing imperfection (both structural and material attributes) and operational factors. It is essential to estimate the probabilistic variability in the dynamic responses to ensure the safety and serviceability conditions.

The Monte Carlo simulation technique is commonly utilized to generate the variable output frequency dealing with large sample size. Although the uncertainty in material and geometric properties can be computed by direct Monte Carlo Simulation, it is inefficient and incurs high computational cost. To mitigate this lacuna, D-optimal design technique is employed in the present analysis to map the inadvertent uncertainties efficiently (Mitchell 1974, Michael and Norman 1974, Craig 1978, Montgomery 1991, Unal et al. 1996). Its purpose is to estimate response quantities in a cost-effective manner based on randomness considered in input parameters (ply orientation angle, mass density, shear modulus and elastic modulus). The proposed procedure is based on constructing response surfaces which represent an estimate for the relation between the input parameters of the finite element model and response quantities of interest. D-optimality criterion is found to provide the best rational means among all other optimality criteria for creating experimental designs in case of irregularly shaped response surfaces (Giunta et al. 1996). A multiplicative method is utilized for computing D-optimal stratified designs of experiments by Radoslav (2014). The multi-response linear models with a qualitative factor are studied by Yue et al. (2014) using D-optimal design while the optimal sampling frequency is investigated for high frequency using a finite mixture model (Choi and Kang 2014). Xu et al. (2014) studied the uncertainty propagation in statistical energy analysis for structural-acoustic coupled systems with non-deterministic parameters while a finite element based modal method is employed for determination of plate stiffness considering uncertainties (Kuttenkeuler 1999). Stochastic finite element method is employed in conjunction to laminated composites and other applications (Ghanem and Spanos 2002, Kishor et al. 2011, Shaker et al. 2008). Dey et al. (2015c) presented the effect of material and structural uncertainty on the natural frequencies of composite shells including the effect of twist.

The aspect of uncertainty quantification in composite structures is increasingly getting its due importance with the rapid adoption of composites by various engineering industries (Goyal and Kapania 2008, Shaker et al. 2008, Fang and Springer 1993, Sasikumar et al. 2014, Park et al. 1995, Ganesan and Kowda 2005). In this chapter, an efficient algorithm is presented to quantify the stochastic natural frequencies of cantilever composite conical shells using D-optimal design technique and its efficacy compared with direct Monte Carlo simulation (MCS) technique. A sensitivity analysis is also carried out to map the effect of individual input parameter on output responses. This chapter hereafter is organized as follows, Section 4.2: theoretical formulation for the dynamic analysis of twisted singly curved composite shells; Section 4.3: formulation of response surface method based on D-optimal design; Section 4.4: proposed surrogate based stochastic approach; Section 4.5: results and discussion; and Section 4.6: summary.

4.2 Governing equations for pre-twisted singly curved composite shells

The present study is carried out as an extension of previous work to investigate the effect of stochasticity in twist angle, ply orientation angle and material properties applied to the laminated composite singly curved conical shells (Dey et al. 2015f). The composite cantilever conical shell with length Lo, reference width b, thickness t, vertex angle ϕ_{ve} and base subtended angle of cone ϕ_o are furnished in Fig. 4.1. The component of radius of curvature in the chordwise direction $r_y(x,y)$ is a parameter varying both in the x- and y-directions. The variation in x-direction is considered to be linear and there is no curvature along the spanwise direction ($r_x = \infty$). The cantilever shell is clamped along x=0 with radius of twist r_{xy} which is assumed to vary depending on angle of twist. Thus a shallow conical shell of uniform thickness, made of laminated composite is considered. A shallow shell is characterized by its middle surface and is defined by the equation (Dey and Karmakar 2012a),

$$z(\overline{\omega}) = -0.5 \left[(x^2/r_x) + \{2xy/r_{xy}(\overline{\omega})\} + (y^2/r_y) \right] \qquad (4.1)$$

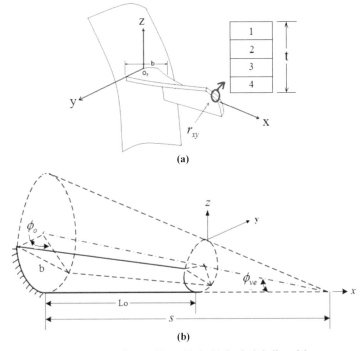

(a)

(b)

Fig. 4.1 (a) Turbo-machinery blade (b) Conical shell model.

The radius of twist (r_{xy}), length (Lo) of the shell and twist angle (Ψ) (Liew et al. 1994) can be expressed as

$$r_{xy}(\bar{\omega}) = -Lo/\tan\psi(\bar{\omega}) \tag{4.2}$$

where $\Psi(\bar{\omega})$ denotes the random angle of twist wherein the symbol ($\bar{\omega}$) indicates the stochasticity of parameters. The constitutive equation for composite shell (Jones 1975) is,

$$\{F\} = [D(\bar{\omega})]\{\varepsilon\} \tag{4.3}$$

where internal force resultant vector

$$\{F\} = \{N_x, N_y, N_{xy}, M_x, M_y, M_{xy}, Q_x, Q_y\}^T \tag{4.4}$$

$$\{F\} = \left[\int_{-h/2}^{h/2} \{\sigma_x, \quad \sigma_y, \quad \tau_{xy}, \quad \sigma_x z, \quad \sigma_y z, \quad \tau_{xy} z, \quad \tau_{xz}, \quad \tau_{yz}\} z\,dz\right]^T$$

$$\{\varepsilon\} = \{\varepsilon_x, \varepsilon_y, \varepsilon_{xy}, k_x, k_y, k_{xy}, \gamma_{xz}, \gamma_{yz}\}^T$$

$$\text{and} \quad [D(\bar{\omega})] = \begin{bmatrix} A_{ij}(\bar{\omega}) & B_{ij}(\bar{\omega}) & 0 \\ B_{ij}(\bar{\omega}) & D_{ij}(\bar{\omega}) & 0 \\ 0 & 0 & S_{ij}(\bar{\omega}) \end{bmatrix} \tag{4.5}$$

The elements of elastic stiffness matrix $[D(\bar{\omega})]$ (Jones 1975) is expressed as

$$[A_{ij}(\bar{\omega})] = \sum_{kk=1}^{nl} \left[\bar{Q}_{ij}(\bar{\omega})\right]_{kk} (z_{kk} - z_{kk-1}) \quad \text{for } i,j = 1, 2, 6$$

$$[B_{ij}(\bar{\omega})] = \frac{1}{2}\sum_{kk=1}^{nl} \left[\bar{Q}_{ij}(\bar{\omega})\right]_{kk} (z_{kk}^2 - z_{kk-1}^2) \quad \text{for } i,j = 1, 2, 6$$

$$[D_{ij}(\bar{\omega})] = \frac{1}{3}\sum_{kk=1}^{nl} \left[\bar{Q}_{ij}(\bar{\omega})\right]_{kk} (z_{kk}^3 - z_{kk-1}^3) \quad \text{for } i,j = 1, 2, 6 \tag{4.6}$$

$$[S_{ij}(\bar{\omega})] = \alpha_s \sum_{kk=1}^{nl} \left[\bar{Q}_{ij}(\bar{\omega})\right]_{kk} (z_{kk} - z_{kk-1}) \quad \text{for } i,j = 4, 5$$

where nl is the number of layers of laminate and α_s is the shear correction factor and is assumed as 5/6. $[A_{ij}(\bar{\omega})]$, $[B_{ij}(\bar{\omega})]$ and $[D_{ij}(\bar{\omega})]$ are the extension, bending-extension coupling and bending stiffness coefficients, respectively. $[\bar{Q}_{ij}]$ is the off-axis elastic constant matrix which is formulated as

$$[\bar{Q}_{ij}(\bar{\omega})]_{off} = [T_m(\bar{\omega})]^{-1} [\bar{Q}_{ij}]_{on} [T_m(\bar{\omega})]^{-T} \quad \text{for } i,j = 1, 2, 6$$

$$[\bar{Q}_{ij}(\bar{\omega})]_{off} = [\bar{T}_m(\bar{\omega})]^{-1} [\bar{Q}_{ij}]_{on} [\bar{T}_m(\bar{\omega})]^{-T} \quad \text{for } i,j = 4, 5 \tag{4.7}$$

where

$$[T_m(\bar{\omega})] = \begin{bmatrix} Sin^2\theta(\bar{\omega}) & Cos^2\theta(\bar{\omega}) & 2\,Sin\theta(\bar{\omega})\,Cos\theta(\bar{\omega}) \\ Cos^2\theta(\bar{\omega}) & Sin^2\theta(\bar{\omega}) & -2\,Sin\theta(\bar{\omega})\,Cos\theta(\bar{\omega}) \\ -Sin\theta(\bar{\omega})\,Cos\theta(\bar{\omega}) & 2\,Sin\theta(\bar{\omega})\,Cos\theta(\bar{\omega}) & Sin^2\theta(\bar{\omega}) - Cos^2\theta(\bar{\omega}) \end{bmatrix}$$

$$\text{and } [\bar{T}_m(\bar{\omega})] = \begin{bmatrix} Sin\theta(\bar{\omega}) & -Cos\theta(\bar{\omega}) \\ Cos\theta(\bar{\omega}) & Sin\theta(\bar{\omega}) \end{bmatrix} \tag{4.8}$$

in which $\theta(\bar{\omega})$ is random ply orientation angle.

$$[Q_{ij}(\bar{\omega})]_{on} = \begin{bmatrix} Q_{11}(\bar{\omega}) & Q_{12}(\bar{\omega}) & 0 \\ Q_{12}(\bar{\omega}) & Q_{22}(\bar{\omega}) & 0 \\ 0 & 0 & Q_{66}(\bar{\omega}) \end{bmatrix} \text{ for } i, j = 1, 2, 6 \tag{4.9}$$

$$[\bar{Q}_{ij}(\bar{\omega})]_{on} = \begin{bmatrix} Q_{44}(\bar{\omega}) & Q_{45}(\bar{\omega}) \\ Q_{45}(\bar{\omega}) & Q_{55}(\bar{\omega}) \end{bmatrix} \text{ for } i, j = 4, 5 \tag{4.10}$$

where

$$Q_{11}(\bar{\omega}) = \frac{E_1(\bar{\omega})}{1 - v_{12}v_{21}} \quad Q_{22}(\bar{\omega}) = \frac{E_2(\bar{\omega})}{1 - v_{12}v_{21}} \quad \text{and} \quad Q_{12}(\bar{\omega}) = \frac{v_{12}\,E_2(\bar{\omega})}{1 - v_{12}v_{21}}$$

$$Q_{66}(\bar{\omega}) = G_{12}(\bar{\omega}) \quad Q_{44}(\bar{\omega}) = G_{23}(\bar{\omega}) \quad \text{and} \quad Q_{55}(\bar{\omega}) = G_{13}(\bar{\omega})$$

The stochastic mass per unit area for each element of the singly curved composite shell (Cook et al. 1989) can be expressed as

$$P_e(\bar{\omega}) = \sum_{k=1}^{n} \int_{z_{k-1}}^{z_k} \rho_e(\bar{\omega})\, dz \tag{4.11}$$

where the suffix 'e' represents the respective parameters for an element and $\rho_e(\bar{\omega})$ denotes the stochastic density parameter at each element. The element mass matrix is expressed as

$$[M_e(\bar{\omega})] = \int_V [N]\,[P_e(\bar{\omega})]\,[N]\,dV \tag{4.12}$$

The random stiffness matrix is given by

$$[K(\bar{\omega})] = \int_{-1}^{1}\int_{-1}^{1} [B(\bar{\omega})]^T \; [D(\bar{\omega})] \; [B(\bar{\omega})] \, d\xi \, d\eta \tag{4.13}$$

where ξ and η are the local natural coordinates of the element. Applying Hamilton's principle (Meirovitch 1992), the dynamic equilibrium equation of each element for free vibration (Karmakar and Sinha 2001) can be expressed as

$$[M_e(\overline{\omega})][\ddot{\delta}_e] + [K_e(\overline{\omega})]\{\delta_e\} = 0 \qquad (4.14)$$

After assembling all the element matrices and the force vectors with respect to the common global coordinates, the equation of motion of a free vibration system with n_{dof} degrees of freedom given by Bathe (1990) and can expressed as

$$[M(\overline{\omega})][\ddot{\delta}] + [K(\overline{\omega})]\{\delta\} = 0 \qquad (4.15)$$

In the above equation, $M(\overline{\omega}) \in R^{n \times n}$ is the mass matrix, $[K(\overline{\omega})] \in R^{n \times n}$ is the stiffness matrix while $\{\delta\} \in R^n$ is the vector of generalized coordinates. The governing equations are derived based on Mindlin's theory (Cook et al. 1989) incorporating rotary inertia, transverse shear deformation. For free vibration, the stochastic natural frequencies $[\omega_n(\overline{\omega})]$ are determined from the standard eigenvalue problem given by Bathe (1990) and is solved by the QR iteration algorithm.

4.3 D-optimal design based polynomial regression method

In general, a statistical measure of goodness of a model obtained by least squares regression analysis is the minimum generalized variance of the estimates of the model coefficients (Montgomery 1991). D-optimal design method is employed to provide a mathematical and statistical approach in portraying the input-output mapping by construction of meta-model with small representative number of samples. The problem of estimating the coefficients of a linear approximation is modelled by least squares regression analysis

$$Y = X\beta + \varepsilon \qquad (4.16)$$

where 'Y' is a vector of observations of sample size, 'ε' is the vector of errors having normal distribution with zero mean, 'X' is the design matrix and 'β' is a vector of unknown model coefficients and can be estimated by using the least squares method as

$$\beta = (X^T X)^{-1} X^T Y \qquad (4.17)$$

A measure of accuracy of the column of estimators, β is the variance-covariance matrix which is defined as

$$V(\beta) = \sigma^2 (X^T X)^{-1} \qquad (4.18)$$

where σ^2 is the variance of the error. The $V(\beta)$ matrix is a statistical measure of the goodness of the fit. $V(\beta)$ is a function of $(X^T X)^{-1}$ and therefore, one would want to minimize $(X^T X)^{-1}$ to improve the quality of the fit. If X denotes the design matrix as a set of value combinations of coded parameters and X^T is the transpose of X, then D-optimality is achieved if the determinant of $(X^T X)^{-1}$ is minimal. The letter "D' stands for the determinant of the $(X^T X)$ matrix associated with the model. In the present study, the constructed meta-models provide an approximate meta-model equation which relates the input random parameters 'x_i' (say ply orientation angle, elastic modulus, etc., of each layer of laminate) and output 'Y' (say natural frequency) for a particular system. The response surface model developed is actually an approximate mathematical model representing a certain inherent property of a physical system and it maps the input parameters 'x_i' to the corresponding responses 'Y' by an explicit function

$$Y = f(x_1, x_2, x_3, x_4, \dots x_i \dots x_k) + \varepsilon \qquad (4.19)$$

where 'f' denotes the approximate response function, 'ε' is the statistical error term having a normal distribution with null mean value and 'k' is the number of input variables. The input variables are usually coded as dimensionless variables with zero as mean value and a standard deviation of 'x_i'. The first order and second order polynomials (Montgomery 1991) are expressed as

First order model (interaction), $Y = \beta_o + \sum\limits_{i=1}^{k} \beta_i\, x_i + \sum\limits_{i=1}^{k}\sum\limits_{j>i}^{k} \beta_{ij}\, x_i\, x_j + \varepsilon$

$$(4.20)$$

Second order model, $Y = \beta_o + \sum\limits_{i=1}^{k} \beta_i\, x_i + \sum\limits_{i=1}^{k}\sum\limits_{j>i}^{k} \beta_{ij}\, x_i\, x_j + \sum\limits_{i=1}^{k} \beta_{ii}\, x_i^2 + \varepsilon$

$$(4.21)$$

where β_o, β_i, β_{ij} and β_{ii} are the coefficients of the polynomial equation. The meta-model is employed to fit approximately for a set of points in the design space using a multiple regression fitting scheme. The position of design points is chosen algorithmically according to the selected number of input variables and their range of variability. Hence the design points are not considered at any specific positions; instead, they are selected in such a fashion so that it meets the optimality criteria. It provides the good accuracy of approximation compared to direct Monte Carlo Simulation method. In D-optimal design, the total sample size (n) is the summation of the minimum number of design points [$n_d = \dfrac{1}{2}(k+1)\,(k+2)$], additional model points ($n_a = k$) and lack-of-fit points (n_l) (i.e., $n = n_d + n_a + n_l$) where k is the number of stochastic input parameter. For model construction in the present study, an over-determined D-optimal design (number of additional samples n_a, along with the minimum point design and $n_l = 10$ samples to estimate the lack of fit) has been used. Output feature (natural frequency) is selected in such a way that it is found to be sensitive enough corresponding to chosen input features. The insignificant input features are screened out and not considered in the model formation. A quantitative evaluation for effect of each parameter on the total model variance is carried out using analysis of variance (ANOVA) method according to its F-test value

$$F_p = \frac{n-k-1}{k}\left(\frac{SS_R}{SS_E}\right) \qquad (4.22)$$

where F_p denotes the F-test value of any input parameter 'p' while n, SS_E and SS_R are the number of samples used in the design procedure, sum of squares due to the model and the residual error, respectively. The meta-model constructed is checked by three basic criteria such as coefficient of determination or R^2 term (measure of the amount of variation around the mean explained by the model) and R^2_{adj} term (measure of the amount of variation with respect to mean value explained by the model, adjusted for the number of terms in the model) and R^2_{pred} (measure of the prediction capability of the response surface model) which are expressed as

$$R^2 = \left(\frac{SS_R}{SS_T}\right) = 1 - \left(\frac{SS_E}{SS_T}\right) \text{ where } 0 \le R^2 \le 1 \qquad (4.23)$$

$$R^2_{adj} = 1 - \frac{\left(\dfrac{SS_E}{n-k-1}\right)}{\left(\dfrac{SS_T}{n-1}\right)} = 1 - \left(\frac{n-1}{n-k-1}\right)\left(1 - R^2\right) \text{ where } 0 \le R^2_{adj} \le 1 \qquad (4.24)$$

$$R^2_{pred} = 1 - \left(\frac{PRESS}{SS_T}\right) \text{ where } 0 \le R^2_{pred} \le 1 \qquad (4.25)$$

where $SS_T = SS_E + SS_R$ is the total sum of square and *PRESS* (predicted residual error sum of squares) measures the goodness of fit of the model corresponding to chosen samples in the design space.

There are many sampling techniques such as, Factorial designs, Central composite design, Optimal design, Taguchi's orthogonal array design, Plackett-Burman designs, Koshal designs, Box-Behnken design, Latin hypercube sampling, and sobol sequence (Giunta et al. 2003, Santner et al. 2003, Koehler and Owen 1996) and surrogate model formation methods such as, Polynomial regression method, Moving least squares method, Kriging, Artificial neural networks, Radial basis function, Multivariate adaptive regression splines, Support vector regression, and High dimensional model representation (Koziel and Yang 2011). Sampling method and surrogate modelling technique for a particular problem should be chosen depending on the complexity of the model, presence of noise in sampling data, nature and dimension (number) of input parameters, desired level of accuracy and computational efficiency. Many comparative studies have been performed over the years to guide the selection of surrogate model types (Jin et al. 2001, Kim et al. 2009, Li et al. 2010). The application of D-optimal design for surrogate model formation in the area of structural engineering and mechanics (specifically in FRP composite structures) is still very scarce, in spite of the fact that it has an immense potential to be used as surrogate of expensive finite element simulation or experimentations of repetitive nature. D-optimal design method is suitable to handle large number of input parameters efficiently for composite structures compared to others as the present model under consideration is linear in nature and all the data used are noise free.

4.4 Stochastic approach using D-optimal design

The stochasticity in material properties of laminated composite conical shells, such as elastic modulus, mass density and geometric properties such as ply-orientation angle, thickness, section inertia as input parameters is considered depending on boundary condition for natural frequency analysis of composite conical shells. In the present study, frequency domain feature (first three natural frequencies) is considered as output. It is assumed that the distribution of randomness of input parameters exists within a certain band of tolerance with their central deterministic values. The cases wherein the random variables considered in each layer of laminate are investigated for

a) Variation of ply-orientation angle only: $\theta(\bar{\omega}) = \{\theta_1 \ \theta_2 \ \theta_3 ....\theta_i......\theta_l\}$

b) Variation of mass density only: $\rho(\bar{\omega}) = \{\rho_1 \ \rho_2 \ \rho_3 ....\rho_i......\rho_l\}$

c) Variation of longitudinal shear modulus: $G_{12}(\bar{\omega}) = \{G_{12\,(1)} \ G_{12\,(2)} \ G_{12\,(3)} ....G_{12\,(i)} ...G_{12\,(l)}\}$

d) Variation of transverse shear modulus: $G_{23}(\bar{\omega}) = \{G_{23\,(1)} \ G_{23\,(2)} \ G_{23(3)} ....G_{23\,(i)} ...G_{23\,(l)}\}$

e) Variation in elastic modulus: $E_1(\bar{\omega}) = \{E_{1(1)} \ E_{1(2)} \ E_{1(3)} ....E_{1(i)}......E_{1(l)}\}$

f) Combined variation of ply orientation angle, mass density, shear modulus along longitudinal and transverse direction and elastic modulus:

$g_1 \{\theta(\bar{\omega}), \rho(\bar{\omega}), G_{12}(\bar{\omega}), G_{23}(\bar{\omega}), E_1(\bar{\omega})\}$

$$= \{ \Phi_1(\theta_1...\theta_l), \Phi_2(\rho_1...\rho_l), \Phi_3(G_{12(1)}...G_{12(l)}), \Phi_4(G_{23(1)}...G_{23(l)}), \Phi_5(E_{1(1)}...E_{1(l)})\}$$

g) Combined variation of ply orientation angle, mass density, shear modulus along longitudinal and transverse direction, elastic modulus and angle of twist:

$g_2 \{\theta(\bar{\omega}), \rho(\bar{\omega}), G_{12}(\bar{\omega}), G_{23}(\bar{\omega}), E_1(\bar{\omega})\}$

$$= \{ \Phi_1(\theta_1...\theta_l), \Phi_2(\rho_1...\rho_l), \Phi_3(G_{12(1)}...G_{12(l)}), \Phi_4(G_{23(1)}...G_{23(l)}), \Phi_5(E_{1(1)}...E_{1(l)}), \Phi_6(\psi_1...\psi_l)\}$$

where $\theta_i, \rho_i, G_{12(i)}, G_{23(i)}, E_{1(i)}$ and ψ_i are the ply orientation angle, mass density, shear modulus along longitudinal direction, shear modulus along transverse direction, elastic modulus along longitudinal direction and angle of twist, respectively and '*l*' denotes the number of layer in the laminate. The tolerances of $\pm 5°$, $\pm 10°$ and $\pm 15°$ for ply orientation angle with subsequent $\pm 10\%$, $\pm 20\%$ and $\pm 30\%$ for material properties respectively from deterministic mean value are considered in the present study. In addition to above, the uncertainty in natural frequency due to random angle of twist $[\psi(\bar{\omega})]$ ($\pm 5°$ variation considered for $\psi = 0°$, 15°, 30° and 45°) is also investigated for a typical laminate. Figure 4.2 describes the D-optimal design based Monte Carlo simulation approach for uncertainty quantification.

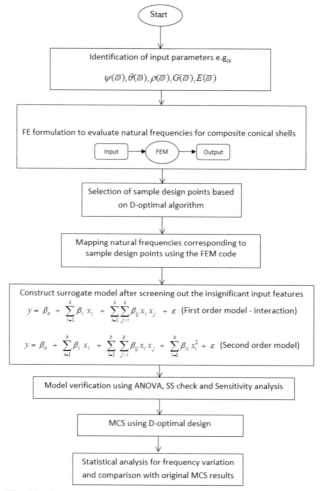

Fig. 4.2 Flowchart of stochastic natural frequency using D-optimal design.

4.5 Results and discussion

In the present study, four layered graphite–epoxy symmetric angle-ply and cross-ply laminated composite cantilever shallow conical shells are considered with aspect ratio of 0.7, thickness ratio of 280. The parameters of graphite–epoxy composite cantilever singly curved conical shells (Qatu and Lissa 1991b) are considered with deterministic value as $E_1 = 138$ GPa, $E_2 = 8.9$ GPa, $G_{12} = G_{13} = 7.1$ GPa, $G_{23} = 2.84$ GPa, $\rho = 3202$ kg/m^3, t$=0.002$ m, $v=0.3$, Lo/s$=0.7$, $\phi_o = 45°$, $\phi_{ve} = 20°$. A typical discretization of (6×6) mesh on plan area with 36 elements, 133 nodes with natural coordinates of an isoparametric quadratic element are considered for the present finite element method. For full scale MCS, the number of original FE analysis is same as the sampling size. In general for complex composite structures, the performance function is not available as an explicit function of the random design variables. The random response in terms of natural frequencies of the composite structure can only be evaluated numerically at the end of a structural analysis procedure such as the finite element method which is often time-consuming. The present D-optimal design methodology is employed to find a predictive and representative meta-model relating each natural frequency to a number of input variables. The meta-models are used to determine the first three natural frequencies corresponding to given values of input variables, instead of time-consuming deterministic FE analysis. The probability density function is plotted as the benchmark of bottom line results. Due to paucity of space, only a few important representative results are furnished.

4.5.1 Validation

The present computer code is validated with the results available in the open literature. Table 4.1 presents the non-dimensional fundamental frequencies of graphite-epoxy composite twisted plates with different ply-orientation angle (Qatu and Leissa 1991a). The present formulation is further validated for non-dimensional first three natural frequencies of composite conical shells as furnished in Table 4.2. The differences between the results by Liew et al. (1994) and the present FEM approach can be attributed to consideration of transverse shear deformation and rotary inertia in the present FEM approach. The comparative study depicts an excellent agreement with the previously published results and

Table 4.1 Non-dimensional first three natural frequencies $[\omega = \omega_n\ L^2\ \sqrt{(\rho/E_1 h^2)}]$ of three layered $[\theta, -\theta, \theta]$ graphite-epoxy twisted plates, L/b=1, b/h=20, $\psi = 30°$.

Fibre Orientation Angle, θ	f_1		f_2		f_3	
	Present FEM	Qatu and Leissa (1991a)	Present FEM	Qatu and Leissa (1991a)	Present FEM	Qatu and Leissa (1991a)
15°	0.8618	0.8759	2.6551	2.6840	3.7661	3.8113
30°	0.6790	0.6923	2.5594	2.5952	3.7464	3.7969
45°	0.4732	0.4831	2.2287	2.2560	2.7493	2.7938
60°	0.3234	0.3283	1.7797	1.8088	1.9528	2.0015

Table 4.2 First three non-dimensional natural frequencies $[\omega = \omega_n\ b^2\ \sqrt{(\rho h/D)}, D = Eh^3/12(1-v^2)]$ for the untwisted shallow conical shells with v=0.3, s/t=1000, $\phi_o = 30°$, $\phi_{ve} = 15°$.

Aspect ratio (L/s)	f_1		f_2		f_3	
	Present FEM	Liew et al. (1994)	Present FEM	Liew et al. (1994)	Present FEM	Liew et al. (1994)
0.6	0.3552	0.3599	1.1889	1.2037	1.4602	1.4840
0.7	0.3013	0.3060	0.8801	0.8963	1.4090	1.4323
0.8	0.2731	0.2783	0.6984	0.7122	1.3795	1.4065

hence it demonstrates the capability of the computer codes developed and insures the accuracy of analyses. In the present D-optimal design method, a sample size of 29 is considered for each layer of individual variation of ply-orientation angle, mass density, longitudinal and transverse shear modulus and logitudinal elastic modulus and twist angle, respectively. Due to the increment of number of input variables for combined random variation of ply-orientation angle, mass density, longitudinal and transverse shear modulus and elastic modulus, the subsequent sample size of 261 is adopted to meet the convergence criteria. Figure 4.3 depicts a representative plot describing the relationship between the original FE model and the constructed D-optimal design meta-model for fundamental natural frequency signifying the accuracy of the present meta-model. In contrast, Fig. 4.4 shows a sample comparison of the probability density functions (PDF) for both original MCS and D-optimal design meta-model using sample size of 10,000 corresponding to first three natural frequencies. The low scatterness of the points found around the diagonal line in Fig. 4.3 and the low deviation obtained between the probability density function estimations of original MCS and D-optimal design responses in Fig. 4.4 corroborates the fact that D-optimal design meta-models are accurately formed. These two plots are checked and are found in good agreement ensuring the efficiency and accuracy of the present constructed meta-model. The values of R^2, R^2_{adj} and R^2_{pred} are found to be close to one ensuring the best fit. The difference between R^2_{adj} and R^2_{pred} is found less than 0.2 which indicates that the model can be used for further prediction. In addition to above, another check is carried out namely, adequate precision which compares the range of the predicted values at the design points to the average prediction error. For all cases of the present D-optimal design meta-model, its value is consistently found greater than four which indicates that the present model is adequate to navigate the design space.

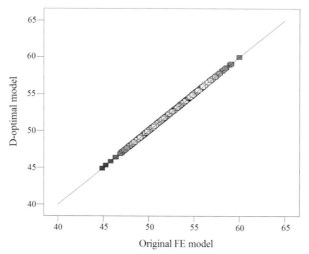

Fig. 4.3 Surrogate model validation for fundamental natural frequency for individual variation of ply-orientation angle of angle-ply (45°/–45°/–45°/45°) composite cantilever conical shells.

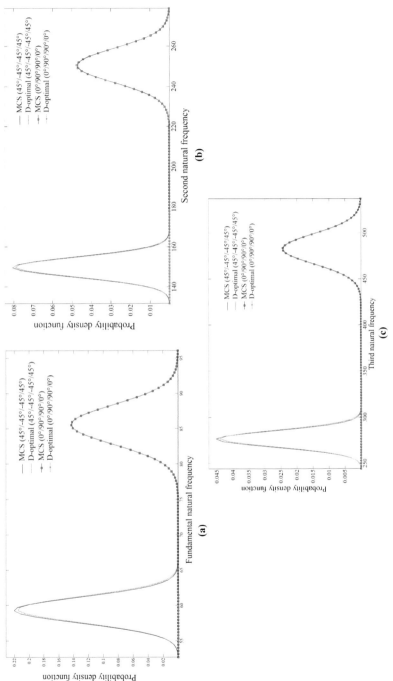

Fig. 4.4 (a,b,c) Comparative study by PDF plot of original MCS and D-optimal design results with respect to first three natural frequencies considering combined variation $[\rho(\bar{\omega}), G_{12}(\bar{\omega}), G_{23}(\bar{\omega}), E_1(\bar{\omega})]$ for angle-ply and cross-ply composite conical shells.

While evaluating the statistics of responses through full scale MCS, computational time is exorbitantly high because it involves number of repeated FE analysis. However, in the present method, MCS is conducted in conjunction with the D-optimal design model. Here, although the same sampling size as in direct MCS (with sample size of 10,000) is considered, the number of actual FE analysis is much less compared to original MCS and is equal to the number of representative samples required to construct the D-optimal design meta-model. The D-optimal design meta-model is formed on which the full sample size of direct MCS is conducted. Hence, the computational time and effort expressed in terms of FE calculation is reduced compared to full scale direct MCS. This provides an efficient affordable way for simulating the uncertainties in natural frequency. The computational time required in the present study is observed to be around (1/345) times (for individual variation of inputs) and (1/38) times (for combined variation of inputs) of direct Monte Carlo simulation. The sensitivity of a given material or geometric property to each random variable is also quantified in the present meta-model context.

4.5.2 Statistical analysis

The material properties such as mass density $[\rho(\bar{\omega})]$, shear modulus (logitudinal and transverse) $[G_{12}(\bar{\omega}), G_{23}(\bar{\omega})]$ and elastic modulus in longitudinal direction, $[E_1(\bar{\omega})]$ are assumed to be varied randomly in each layer of the laminate. The composite conical shell is scaled randomly with uniform distribution of input variables in the range of having the lower and the upper limit as $\pm 10\%$ variability with respective mean values while for ply orientation angle $[\theta(\bar{\omega})]$ the bound is considered as within $\pm 5°$ fluctuation with respective mean values of each layer of laminates. Although the uniform distribution of stochastic input variables are considered, the stochastic outputs (i.e., natural frequencies) are very close to Gaussian distribution. The D-optimal design meta-models are formed to generate first three natural frequencies for angle-ply and cross-ply composite conical shells. The natural frequencies of tested angle-ply (45°/–45°/–45°/45°) laminate are found to be lower than that of the same for cross-ply (0°/90°/90°/0°) laminate irrespective of stochasticity considered in random input parameters.

Table 4.3 presents the comparative study between MCS and D-optimal design for maximum values, minimum values and percentage of deviation for first three natural frequencies obtained due to individual stochaticity in each layer due to randomness considered in only ply-orientation angle for angle-ply and cross-ply composite cantilever conical shells. In this case, the percentage of deviation between original MCS and D-optimal design of maximum value, minimum value, mean value and standard deviation for fundamental natural frequencies of angle-ply are found to be higher than that of the same for cross-ply composite conical shells. Similar to previous case, the percentage of deviation between original MCS and D-optimal design of maximum value, mean value and standard deviation for fundamental natural frequencies of angle-ply are found to be higher than that of

Table 4.3 Comparative study between MCS (10,000 samples) and D-optimal design (29 samples) results for maximum values, minimum values and percentage of deviation for first three natural frequencies obtained due to individual stochasiticity in ply-orientation angle $[\theta(\bar{\omega})]$ for angle-ply $(45°/-45°/-45°/45°)$ and cross-ply $(0°/90°/90°/0°)$ composite conical shells.

Parameter	Type of analysis	Angle-ply (45°/−45°/−45°/45°)			Cross-ply (0°/90°/90°/0°)		
		f_1	f_2	f_3	f_1	f_2	f_3
Max value	MCS	41.0610	108.3828	185.9936	52.4799	156.4330	287.6360
	D-optimal	41.1975	108.9131	186.8028	52.4771	156.4974	287.9516
	Deviation (%)	−0.332%	−0.489%	−0.435%	0.005%	−0.041%	−0.110%
Min value	MCS	37.0048	94.8742	163.5500	47.7011	133.0286	242.0915
	D-optimal	36.9809	94.7998	163.4015	47.6972	133.1006	242.2256
	Deviation (%)	0.065%	0.078%	0.091%	0.008%	−0.054%	−0.055%
Mean value	MCS	39.0706	101.7120	174.9002	49.9388	144.9036	265.9599
	D-optimal	39.0839	101.7428	174.9548	49.9392	144.9268	265.9982
	Deviation (%)	−0.034%	−0.030%	−0.031%	−0.001%	−0.016%	−0.014%
Standard deviation (SD)	MCS	0.6866	2.2887	3.7995	0.9905	4.9971	9.6838
	D-optimal	0.6756	2.2998	3.8160	0.9920	4.9112	9.5052
	Deviation (%)	1.602%	−0.485%	−0.434%	−0.151%	1.719%	1.844%

the same for cross-ply composite conical shells expect for minimum value wherein a reverse trend is identified. All the results obtained using original MCS and D-optimal design are observed to meet good in agreement. Due to the cascading effect resulting from combined stochasticity considered in five input parameters in each layer, the bandwidth of variation of natural frequency is found to be higher than the stochasticity considered for variation of any single input parameter.

Considering individual stochasticity in Young's modulus, shear modulus and mass density, the parametric studies are carried out for maximum values, minimum values, means and standard deviations of first three natural frequencies as depicted in Table 4.4 wherein the maximum and minimum volatility in natural frequency are observed due to individual randomness of mass density and transverse shear modulus, respectively while longitudinal shear modulus trailed by elastic modulus are found to be intermediate. The sensitivity analysis using D-optimal design is performed for significant input parameter screening. The effect of stochasticity of the ply orientation angle of each and individual layer separately for angle-ply composite conical shells on respective fundamental natural frequency are shown in Fig. 4.5 wherein for two outermost layers (i.e., layer 1 and layer 4), it is found to have inverse relationship with ply orientation angle, but in contrast due to negative value of ply orientation angle for two inner layers (i.e., layer 2 and layer 3), the reverse trend is observed. The variation in elastic stiffness due to randomness of ply orientation angle predominantly influence the fundamental natural frequency. The interaction effect corresponding to randomness in ply orientation angle of respective pair of layers for four layered angle-ply laminate are furnished in Fig. 4.6(a–f) wherein two outermost layers are found to play a predominant role to influence the trend of volatility of fundamental natural frequency. The bandwidth of fundamental natural frequency of combined ply angle variation of all four layers at a time is found in lower range compared to the range found corresponding to randomness considered in separate each and individual layer of the angle-ply laminate. Figure 4.7 presents the probability density function plot with stochastic angle of twist ($\pm 5°$ variation) for first three natural frequencies of angle-ply (45°/–45°/45°/–45°) conical shells. Needless to mention that as the twist angle increases, due to reduction of stochastic elastic stiffness, the stochastic first three natural frequencies are found to decrease subsequently. For a particular value of twist angle, as the degree of variation of input parameters increases, the respective mean natural frequency values are found to reduce (Fig. 4.8), but the volatility (standard deviation) in stochastic natural frequencies are found to increase. In conformity to the same, the probability density function plots with respect to first three natural frequencies are furnished in Fig. 4.8 considering combined variation for cross-ply composite singly curved conical shells.

The sensitivity index obtained of material and geometric properties to each random variable at the constituent and ply levels is used as a guide to increase the structural reliability or to reduce the cost. Furthermore, the probabilistic structural analysis and risk assessment is performed once the uncertain laminate properties are computationally simulated. The sensitivity of different random variables for

Table 4.4 Maximum value, minimum value, mean value and standard deviation (SD) of first three natural frequencies obtained by D-optimal design method (29 samples) due to individual stochasticity in $[\rho(\bar{\omega})]$, $[G_{12}(\bar{\omega})]$, $[G_{23}(\bar{\omega})]$ and $[E_1(\bar{\omega})]$ in each layer for four layered graphite-epoxy angle-ply $(45°/-45°/-45°/45°)$ and cross-ply $(0°/90°/90°/0°)$ composite conical shells.

Laminate	Parameter	$\rho(\bar{\omega})$			$G_{12}(\bar{\omega})$			$G_{23}(\bar{\omega})$			$E_1(\bar{\omega})$		
		f_1	f_2	f_3	f_1	f_2	f_3	f_1	f_2	f_3	f_1	f_2	f_3
Angle-ply	Max	56.6011	147.7441	263.9452	55.4453	139.1648	258.7947	54.0149	135.2645	251.8812	54.6998	136.7561	254.8680
	Min	51.6930	129.4530	241.0577	52.4898	131.1424	244.5731	54.0064	135.0492	251.8499	53.2482	133.5960	248.5535
	Mean	54.0328	135.3123	251.9686	53.9973	135.2167	251.7860	54.0100	135.2569	251.8656	53.9938	135.2197	251.7920
	SD	0.7842	1.9639	3.6571	0.4765	1.3220	2.4287	0.0013	0.0025	0.0051	0.2567	0.5659	1.1415
Cross-ply	Max	90.1346	264.7853	510.6328	86.9052	253.8686	489.5655	86.0162	252.6815	487.2938	89.0553	263.1691	507.5916
	Min	82.4120	242.0987	466.8847	85.0697	251.3858	484.8193	86.0010	252.6479	487.2215	82.8033	241.6712	465.9084
	Mean	86.0389	252.7533	247.4294	85.9966	252.6479	487.2274	86.0086	252.6647	487.2580	85.9591	252.4768	486.8506
	SD	1.2491	3.6695	7.0761	0.3493	0.4525	0.8575	0.0031	0.0069	0.0148	1.2723	4.3821	8.4763

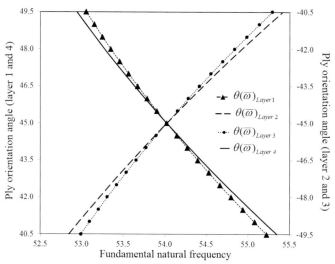

Fig. 4.5 Typical variation of fundamental natural frequency with the change in ply orientation angle for angle-ply (45°/–45°/–45°/45°) composite conical shells.

the laminate frequencies of different modes may be different. In case of combined random variation of all input parameters, the percentage contribution to the sensitivity of each and individual input parameter influencing output responses of angle-ply and cross-ply composite conical shells are shown in Fig. 4.9 and Fig. 4.10, respectively wherein first three natural frequencies are considered. It is observed for angle-ply laminate that the predominant effect on sensitivity of tested natural frequencies is influenced by ply orientation angle followed by mass density, trailed by subsequently longitudinal shear modulus, elastic modulus and least contribution is observed for transverse shear modulus. In contrast for cross-ply composite conical shells, the sensitivity effect of tested natural frequencies is found to be dominated by the ply orientation angle followed by the elastic modulus of two outmost layers (null sensitivity effect is identified for the inner two layers due to coupling effect) while subsequently sensitivity contribution is trailed by mass density followed by longitudinal shear modulus and the null sensitivity effect is identified for transverse shear modulus.

4.6 Summary

This chapter presents the uncertainty quantification in natural frequencies of composite singly curved shells considering the stochastic effect of twist angle, ply orientation angle and material properties. The metamodel formed from a small set of samples based on D-optimal design is found to establish an adequate level of accuracy and computational efficacy compared to the direct Monte Carlo Simulation method. Ply-orientation angle is found to be most sensitive among

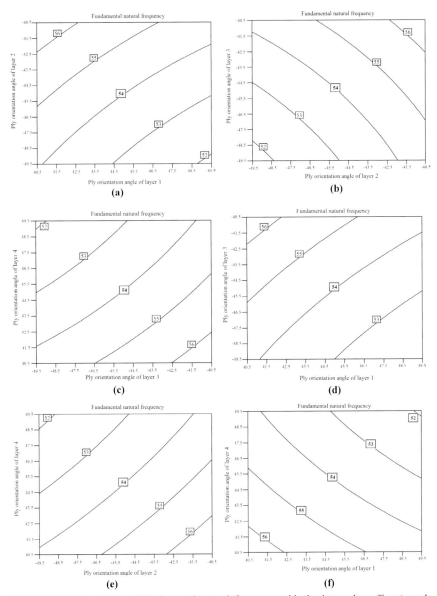

Fig. 4.6 Typical variation of fundamental natural frequency with the interaction effect (sample size = 29) due to change in ply orientation angle of angle-ply (45°/–45°/–45°/45°) composite conical shells.

all tested input parameters while transverse shear modulus is identified as least sensitive to the first three stochastic natural frequencies. As the stochastic twist angle increases, the stochastic first three natural frequencies are found to decrease

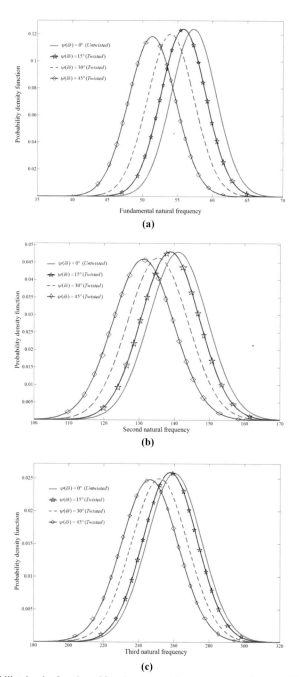

Fig. 4.7 Probability density function of first three natural frequencies (sample size = 261) of angle-ply (45°/–45°/45°/–45°) composite conical shells considering variation in angle of twist [$\psi(\bar{\omega})$] (with ± 5° variation).

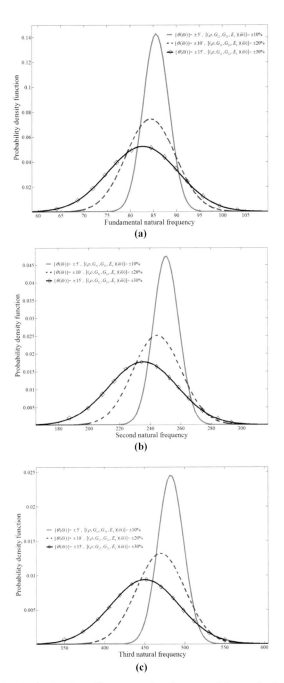

Fig. 4.8 Probability density function with respect to first three natural frequencies (sample size = 261) due to combined variation for cross-ply (0°/90°/90°/0°) conical shells.

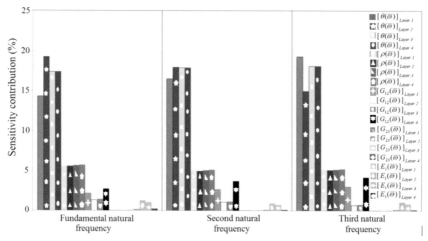

Fig. 4.9 Sensitivity contribution in percentage for combined variation in $[\theta(\bar{\omega}), \rho(\bar{\omega}), G_{12}(\bar{\omega}), G_{23}(\bar{\omega}),$ $E_1(\bar{\omega})]$ (sample size = 261) for angle-ply (45°/–45°/–45°/45°) composite conical shells.

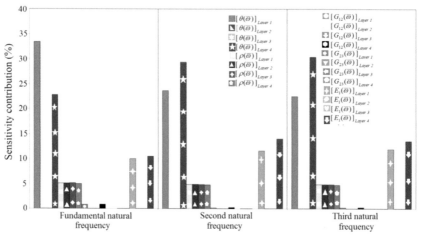

Fig. 4.10 Sensitivity contribution in percentage for combined variation in $[\theta(\bar{\omega}), \rho(\bar{\omega}), G_{12}(\bar{\omega}),$ $G_{23}(\bar{\omega}), E_1(\bar{\omega})]$ (sample size = 261) for cross-ply (0°/90°/90°/0°) composite conical shells.

subsequently due to reduction of stochastic elastic stiffness. From the sensitivity analysis of combined input variation case, the ply-orientation angles of two outer surface layers are found to dominate the variation of natural frequencies. For the combined input variation case of cross-ply laminate, null sensitivity effect is identified at two inner layers for ply orientation angle and elastic modulus while in contrast, the sensitivity of ply orientation angle for angle-ply laminate at two inner layers is found to be predominant. As the degree of variation of input parameters increases, the stochastic mean natural frequency values are found to reduce but the volatility in stochastic natural frequencies are found to increase for both angle-ply and cross-ply laminates. More complex built-up structural systems of laminated composites can be investigated in the future following the efficient D-optimal design based approach.

CHAPTER 5

Stochastic Dynamic Analysis of Doubly Curved Composite Shells

5.1 Introduction

The previous two chapters in this book deal with the effect of uncertainty in composite plates and singly curved composite shells. This chapter presents an efficient stochastic free vibration analysis of composite doubly curved shells. The stochastic finite element formulation is carried out considering rotary inertia and transverse shear deformation based on Mindlin's theory. The sampling size and computational cost in the probabilistic analysis is reduced by employing a Kriging model based approach compared to direct Monte Carlo simulation. Besides detail investigation on the stochastic natural frequencies corresponding to low frequency vibration modes, the stochastic mode shapes and frequency response functions are also presented for a typical laminate configuration. The effect of noise on the kriging based uncertainty propagation algorithm is addressed. Results are presented for different levels of noise in a probabilistic framework to provide a comprehensive idea about stochastic structural responses under the influence of simulated noise.

Composite doubly curved shells are extensively used in aerospace, marine, automotive, construction and many other industries due to its high specific stiffness and strength along with weight sensitivity and cost-effectiveness. Different forms of material and geometric uncertainties are inevitable in composite structures. In general, Monte Carlo simulation (MCS) technique is employed to generate the probabilistic description of output responses (say natural frequency) considering a large sample size. Although the uncertainty in material and geometric properties can be computed by the direct MCS method, it is inefficient and incurs high computational cost. To mitigate this lacuna, the present analysis employs the Kriging based metamodeling technique to propagate the uncertainties efficiently. Its purpose is to map response quantities (such as natural frequency) based on randomness considered in input parameters (ply orientation angle, elastic modulus, mass density, shear modulus and Poisson's ratio). The pioneering specific model and related algorithm was developed by using mathematical and statistical methods to obtain Kriging solutions (Cressie 1990, 1993, Matheron 1963). Of late, the basis of design and analysis of experiments have also been investigated by Montgomery (1991) and Michael and Norman (1974). Kriging model is being employed in geostatistics and optimization (Matheron 1963, Martin and Simpson 2005, Sakata et al. 2004, Ryu et al. 2002, Jeong et al. 2005). A Kriging based approach has been studied for locating a sampling site in the assessment of air quality by Bayraktar and Turalioglu (2005) while the correct Kriging variance was estimated using bootstrapping by Hertog et al. (2006). Further studies were also extended for estimating recoverable reserves by ordinary multi-gaussian Kriging (Emery 2005) and as probabilistic models in design (Martin and Simpson 2004). Later on, the uncertainty quantification of dynamic adaptive sampling is studied based on Kriging surrogate models by Shimoyama et al. (2013). From the open literature, it is found that the Kriging criterion provides a rational means for creating experimental designs for surrogate modeling. Stochastic analyses of composite structures have received considerable attention (Fang and Springer 1993, Carrere et al. 2009, Sarangapani and Ganguli 2013, Goyal and Kapania 2008, Afeefa et al. 2008, Ghavanloo and Fazelzadeh 2013, Fazzolari 2014, Tornabene et al. 2014a, Mantari 2012, Hu and Peng 2013, Tornabene et al. 2014b, Sriramula and Chryssanthopoulos 2009, Shaw et al. 2010, Alkhateb et al. 2009). Dey et al. (2015b) have presented a Kriging based uncertainty quantification algorithm for composite structures. A polynomial chaos expansion based probabilistic approach is presented by Scarth et al. (2014) for design of composite wings including uncertainties.

This chapter investigates the stochastic free vibration characteristics of graphite–epoxy composite doubly curved shells by using the finite element model coupled with the Kriging based metamodeling approach. In this analysis, selective representative samples are drawn using the Latin hypercube sampling algorithm over the entire domain ensuring good prediction capability of the constructed metamodels. An eight noded isoparametric quadratic element with five degrees of freedom at each node is considered in the stochastic finite element formulation. This chapter hereafter is organized as follows, Section 5.2: theoretical formulation for the dynamic analysis of doubly curved composite shells; Section 5.3: formulation of kriging based metamodeling approach; Section 5.4: kriging based stochastic approach including the effect of noise; Section 5.5: results and discussion; and Section 5.6: summary.

5.2 Governing equations for the dynamics of doubly curved composite shells

In present study, a composite cantilever shallow doubly curved shell with uniform thickness 't' and principal radii of curvature R_x and R_y along x- and y-direction respectively is considered as furnished in Fig. 5.1. Based on the first-order shear deformation theory, the displacement field of the shells may be described as

$$u(x, y, z) = u^0(x, y) - z\,\theta_x(x, y)$$
$$v(x, y, z) = v^0(x, y) - z\,\theta_y(x, y) \tag{5.1}$$
$$w(x, y, z) = w^0(x, y) = w(x, y)$$

where, u^0, v^0, and w^0 are displacements of the reference plane and θ_x and θ_y are rotations of the cross section relative to x and y axes, respectively. Each of the thin fibre of laminae can be oriented at an arbitrary angle 'θ' with reference to the x-axis. The constitutive equations for the shell can be expressed as (Chakravorty et al. 1995)

$$\{F\} = [D(\bar{\omega})]\,\{\varepsilon\} \tag{5.2}$$

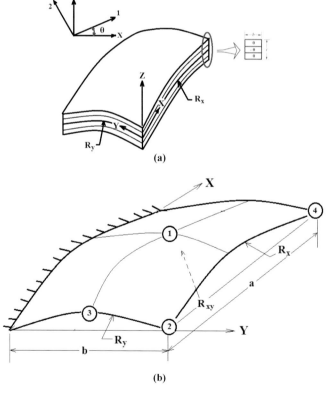

Fig. 5.1 Laminated composite shallow doubly curved shell model.

where Force resultant

$$\{F\} = \{N_x, N_y, N_{xy}, M_x, M_y, M_{xy}, Q_x, Q_y\}^T$$

$$\{F\} = \left[\int_{-h/2}^{h/2} \{\sigma_x, \quad \sigma_y, \quad \tau_{xy}, \quad \sigma_{xz}, \quad \sigma_{yz}, \quad \tau_{xyz}, \quad \tau_{xz}, \quad \tau_{yz}\} \, dz \right]^T$$

and strain $\{\varepsilon\} = \{\varepsilon_x, \varepsilon_y, \varepsilon_{xy}, k_x, k_y, k_{xy}, \gamma_{xz}, \gamma_{yz}\}^T$

$$\text{where } [D(\bar\omega)] = \begin{bmatrix} A_{11} & A_{12} & A_{16} & B_{11} & B_{12} & B_{16} & 0 & 0 \\ A_{12} & A_{22} & A_{26} & B_{12} & B_{22} & B_{26} & 0 & 0 \\ A_{16} & A_{26} & A_{66} & B_{16} & B_{26} & B_{66} & 0 & 0 \\ B_{11} & B_{12} & B_{16} & D_{11} & D_{12} & D_{16} & 0 & 0 \\ B_{12} & B_{22} & B_{26} & D_{12} & D_{22} & D_{26} & 0 & 0 \\ B_{16} & B_{26} & B_{66} & D_{16} & D_{26} & D_{66} & 0 & 0 \\ 0 & 0 & 0 & 0 & 0 & 0 & S_{44} & S_{45} \\ 0 & 0 & 0 & 0 & 0 & 0 & S_{45} & S_{55} \end{bmatrix}$$

(5.3)

The elements of elastic stiffness matrix $[D(\bar\omega)]$ can be expressed as

$$[A_{ij}(\bar\omega), \quad B_{ij}(\bar\omega), \quad D_{ij}(\bar\omega)] = \sum_{k=1}^{n} \int_{z_{k-1}}^{z_k} [\{\bar{Q}_{ij}(\bar\omega)\}_{on}]_k \; [1, \quad z, \quad z^2] \, dz \qquad i, j = 1, 2, 6$$

$$[S_{ij}(\bar\omega)] = \sum_{k=1}^{n} \int_{z_{k-1}}^{z_k} \alpha_s \, [\bar{Q}_{ij}]_k \, dz \qquad i, j = 4, 5$$

(5.4)

where $\bar\omega$ indicates the stochastic representation and α_s is the shear correction factor $(= 5/6)$ and $[\bar{Q}_{ij}]$ are elements of the off-axis elastic constant matrix which is given by

$$[\bar{Q}_{ij}]_{off} = [T_1(\bar\omega)]^{-1} \, [\bar{Q}_{ij}]_{on} \, [T_1(\bar\omega)]^{-T} \quad \text{for } i, j = 1, 2, 6$$

$$[\bar{Q}_{ij}]_{off} = [T_2(\bar\omega)]^{-1} \, [\bar{Q}_{ij}]_{on} \, [T_2(\bar\omega)]^{-T} \quad \text{for } i, j = 4, 5$$

(5.5)

$$[T_1(\bar\omega)] = \begin{bmatrix} m^2 & n^2 & 2mn \\ n^2 & m^2 & -2mn \\ -mn & mn & m^2 - n^2 \end{bmatrix} \text{ and } [T_2(\bar\omega)] = \begin{bmatrix} m & -n \\ n & m \end{bmatrix}$$

(5.6)

in which $m = Sin\theta(\bar\omega)$ and $n = Cos\theta(\bar\omega)$, wherein $\theta(\bar\omega)$ is random fibre orientation angle.

$$[Q_{ij}(\bar\omega)]_{on} = \begin{bmatrix} Q_{11} & Q_{12} & 0 \\ Q_{12} & Q_{12} & 0 \\ 0 & 0 & Q_{66} \end{bmatrix} \text{ for } i, j = 1, 2, 6 \quad [\bar{Q}_{ij}(\bar\omega)]_{on} = \begin{bmatrix} Q_{44} & Q_{45} \\ Q_{45} & Q_{55} \end{bmatrix}$$

$$\text{for } i, j = 4, 5$$

(5.7)

where

$$Q_{11} = \frac{E_1}{1-v_{12}v_{21}} \qquad Q_{22} = \frac{E_2}{1-v_{12}v_{21}} \qquad \text{and} \qquad Q_{12} = \frac{v_{12}E_2}{1-v_{12}v_{21}}$$

$$Q_{66} = G_{12} \qquad Q_{44} = G_{23} \qquad \text{and} \qquad Q_{55} = G_{13}$$

The strain-displacement relations for shallow doubly curved shells can be expressed as

$$\begin{Bmatrix} \varepsilon_x \\ \varepsilon_y \\ \gamma_{xy} \\ \gamma_{xz} \\ \gamma_{yz} \end{Bmatrix} = \begin{Bmatrix} \dfrac{\partial u}{\partial x} - \dfrac{w}{R_x} \\[6pt] \dfrac{\partial v}{\partial y} - \dfrac{w}{R_y} \\[6pt] \dfrac{\partial u}{\partial y} + \dfrac{\partial v}{\partial x} - \dfrac{2w}{R_{xy}} \\[6pt] \theta_x + \dfrac{\partial w}{\partial x} \\[6pt] \theta_y + \dfrac{\partial w}{\partial y} \end{Bmatrix}, \quad \begin{Bmatrix} k_x \\ k_y \\ k_{xy} \\ k_{xz} \\ k_{yz} \end{Bmatrix} = \begin{Bmatrix} \dfrac{\partial \theta_x}{\partial x} \\[6pt] \dfrac{\partial \theta_y}{\partial y} \\[6pt] \dfrac{\partial \theta_x}{\partial y} + \dfrac{\partial \theta_y}{\partial x} \\[6pt] 0 \\[6pt] 0 \end{Bmatrix} \qquad (5.8)$$

and $\{\varepsilon\} = [B]\{\delta_e\}$

$$\{\delta_e\} = \{u_1, \ v_1, \ w_1, \ \theta_{x1}, \ \theta_{y1}, \dots \dots u_8, \ v_8, \ w_8, \ \theta_{x8}, \ \theta_{y8}\}^T$$

$$\text{where } [B] = \sum_{i=1}^{8} \begin{bmatrix} N_{i,x} & 0 & -\dfrac{N_i}{R_x} & 0 & 0 \\[6pt] 0 & N_{i,y} & -\dfrac{N_i}{R_y} & 0 & 0 \\[6pt] N_{i,y} & N_{i,x} & -\dfrac{2N_i}{R_{xy}} & 0 & 0 \\[6pt] 0 & 0 & 0 & N_{i,x} & 0 \\[6pt] 0 & 0 & 0 & 0 & N_{i,y} \\[6pt] 0 & 0 & 0 & N_{i,y} & N_{i,x} \\[6pt] 0 & 0 & N_{i,x} & N_i & 0 \\[6pt] 0 & 0 & N_{i,y} & 0 & N_i \end{bmatrix}$$

where R_{xy} are the radii of curvatures of shallow doubly curved shells and k_x, k_y, k_{xy}, k_{xz}, k_{yz} are curvatures of the shell and u, v, w are the displacements of the mid-plane along x, y and z axes, respectively. In the case of doubly curved shallow shell, elements are employed to model the middle-surface geometry more accurately.

An eight noded isoparametric quadratic element with five degrees of freedom at each node (three translations and two rotations) is considered wherein the shape functions (N_i) (Bathe 1990) can be expressed as

$$N_i = (1 + \chi \chi_i)(1 + \varsigma \varsigma_i)(\chi \chi_i + \varsigma \varsigma_i - 1)/4 \text{ (for i = 1, 2, 3, 4)} \quad (5.9)$$

$$N_i = (1 - \chi^2)(1 + \varsigma \varsigma_i)/2 \qquad \text{(for i = 5, 7)} \quad (5.10)$$

$$N_i = (1 - \varsigma^2)(1 + \chi \chi_i)/2 \qquad \text{(for i = 6, 8)} \quad (5.11)$$

where ς and χ are the local natural coordinates of the element. The mass per unit area for doubly curved shell is expressed as

$$P(\bar{\omega}) = \sum_{k=1}^{n} \int_{z_{k-1}}^{z_k} \rho(\bar{\omega}) \, dz \quad (5.12)$$

Mass matrix is expressed as

$$[M(\bar{\omega})] = \int_{Vol} [N][P(\bar{\omega})][N] \, d(vol) \quad (5.13)$$

The stiffness matrix is given by

$$[K(\bar{\omega})] = \int_{-1}^{1} \int_{-1}^{1} [B(\bar{\omega})]^T \ [D(\bar{\omega})] \ [B(\bar{\omega})] \, d\xi \, d\eta \quad (5.14)$$

The Hamilton's principle is employed to study the dynamic nature of the composite structure (Meirovitch 1992). The principle used for the Lagrangian which is defined as

$$L_f = T - U - W \quad (5.15)$$

where T, U and W are total kinetic energy, total strain energy and total potential of the applied load, respectively. The Hamilton's principle applicable to non-conservative system is expressed as,

$$\delta H = \int_{p_i}^{p_f} [\delta T - \delta U - \delta W] \, dp = 0 \quad (5.16)$$

Hamilton's principle applied to dynamic analysis of elastic bodies states that among all admissible displacements which satisfy the specific boundary conditions, the actual solution makes the functional $\int (T + V) \, dp$ stationary, where T and W are the kinetic energy and the work done by conservative and non-conservative forces, respectively. For free vibration analysis (i.e., $\delta W = 0$), the stationary value is actually a minimum. In case of a dynamic problem without damping the conservative forces are the elastic forces developed within a deformed body and the non-conservative forces are the external force functions. The energy functional for Hamilton's principle is the Lagrangian (L_f) which includes kinetic energy (T) in addition to potential strain energy (U) of an elastic body. The expression for kinetic energy of an element is expressed as

$$T = \frac{1}{2} \{\dot{\delta}_e\}^T [M_e(\bar{\omega})]\{\dot{\delta}_e\} \tag{5.17}$$

The potential strain energy for an element of a plate can be expressed as,

$$U = U_1 + U_2 = \frac{1}{2} \{\delta_e\}^T [K_e(\bar{\omega})]\{\delta_e\} + \frac{1}{2} \{\delta_e\}^T [K_{\sigma e}(\bar{\omega})]\{\delta_e\} \tag{5.18}$$

The Langrange's equation of motion is given by

$$\frac{d}{dt}\left[\frac{\partial L f}{\partial \dot{\delta}_e}\right] - \left[\frac{\partial L f}{\partial \delta_e}\right] = \{F_e\} \tag{5.19}$$

where $\{F_e\}$ is the applied external element force vector of an element and L_f is the Lagrangian function. Substituting $L_f = T - U$, and the corresponding expressions for T and U in Lagrange's equation, the dynamic equilibrium equation for each element (Dey and Karmakar 2012b) can be expressed as

$$[M][(\bar{\omega})]\{\ddot{\delta}_e\} + ([K_e(\bar{\omega})] + [K_{\sigma e}(\bar{\omega})])\{\delta_e\} = \{F_e\} \tag{5.20}$$

After assembling all the element matrices and the force vectors with respect to the common global coordinates, the equation of motion of a free vibration system with *n* degrees of freedom can expressed as

$$[M(\bar{\omega})][\ddot{\delta}] + [K(\bar{\omega})]\{\delta\} = \{F_L\} \tag{5.21}$$

In the above equation, $\mathbf{M}(\bar{\omega}) \in \mathrm{R}^{n \times n}$ is the mass matrix, $[K(\bar{\omega})]$ is the stiffness matrix wherein $[K(\bar{\omega})] = [K_e(\bar{\omega})] + [K_{\sigma e}(\bar{\omega})]$ in which $\mathbf{K}_e(\bar{\omega}) \in \mathrm{R}^{n \times n}$ is the elastic stiffness matrix, $\mathbf{K}_{\sigma e}(\bar{\omega}) \in \mathrm{R}^{n \times n}$ is the geometric stiffness matrix (depends on initial stress distribution) while $\{\boldsymbol{\delta}\} \in \mathrm{R}^n$ is the vector of generalized coordinates and $\{\boldsymbol{F}_L\} \in \mathrm{R}^n$ is the force vector. The governing equations are derived based on Mindlin's theory incorporating rotary inertia, transverse shear deformation. For free vibration, the random natural frequencies $[\boldsymbol{\omega}_n(\bar{\omega})]$ are determined from the standard eigenvalue problem (Bathe 1990) which is represented below and is solved by the QR iteration algorithm,

$$[A(\bar{\omega})]\{\delta\} = \lambda(\bar{\omega})\{\delta\} \tag{5.22}$$

where

$$[A(\bar{\omega})] = ([K_e(\bar{\omega})] + [K_{\sigma e}(\bar{\omega})])^{-1} [M(\bar{\omega})]$$

$$\lambda(\bar{\omega}) = \frac{1}{\{\omega_n(\bar{\omega})\}^2} \tag{5.23}$$

It can be also shown that the eigenvalues and eigenvectors satisfy the orthogonality relationship that is

$$\mathbf{x}_i^T [\mathbf{M}(\bar{\omega})] \mathbf{x}_j = \lambda_{1j} \text{ and } \mathbf{x}_i^T [\mathbf{K}(\bar{\omega})] \mathbf{x}_j = \omega_j^2 \lambda_{1j} \text{ where i, j} = 1, 2\ldots\ldots.n \tag{5.24}$$

Note that the Kroneker delta functions is given by $\lambda_{lj} = 1$ for $l = j$ and $\lambda_{lj} = 0$ otherwise. The property of the eigenvectors in equation (5.24) is also known as the mass orthonormality relationship. The solution of the undamped eigenvalue problem is now standard in many finite element packages. This orthogonality property of the undamped modes is very powerful as it allows to transform a set of coupled differential equations to a set of independent equations. For convenience, the matrices are expressed as

$$\Omega(\bar{\omega}) = \text{diag} \, [\omega_1, \omega_2, \omega_3 \ldots\ldots\ldots \omega_n] \in R^{n \times n}$$
$$\text{and } X(\bar{\omega}) = [x_1, x_2 \ldots\ldots\ldots x_n] \in R^{n \times n} \tag{5.25}$$

where the eigenvalues are arranged such that $\omega_1 < \omega_2$, $\omega_2 < \omega_3$, $\ldots\ldots\ldots \omega_k <$ ω_{k+1}. The matrix X is known as the undamped modal matrix. Using these matrix notations, the orthogonality relationships (24) can be rewritten as

$$x_i^T \, [M(\bar{\omega})] \, x_j = \lambda_{lj} \text{ and } x_i^T \, [K(\bar{\omega})] \, x_j = \omega_j^2 \, \lambda_{lj} \quad \text{where i, j} = 1, 2\ldots\ldots.n \tag{5.26}$$

$$X^T [M(\bar{\omega})] \, X = I \quad \text{and} \quad X^T [K(\bar{\omega})] \, X = \Omega^2 \tag{5.27}$$

where I is a (n x n) identity matrix. We use the following coordinate transformation (as the modal transformation)

$$\delta \, (\bar{\omega}) \, (t) = X \, y(t) \tag{5.28}$$

Using the modal transformation in equation (5.28), pre-multiplying equation (5.21) by X^T and using the orthogonality relationships in (5.27), equation of motion of a damped system in the modal coordinates may be obtained as

$$\ddot{y}(t) + X^T \, C \, X \, \dot{y}(t) + \Omega^2 \, y(t) = \tilde{f}(t) \tag{5.29}$$

Clearly, unless $X^T [C] \, X$ is a diagonal matrix, no advantage can be gained by employing modal analysis because the equations of motion will still be coupled. To solve this problem, it is common to assume proportional damping. With the proportional damping assumption, the damping matrix $[C]$ is simultaneously diagonalisable with $[M]$ and $[K]$. This implies that the damping matrix in the modal coordinate is

$$[C'] = X^T [C] \, X \tag{5.30}$$

where $[C']$ is a diagonal matrix. This matrix is also known as the modal damping matrix. The damping factors ζ_j are defined from the diagonal elements of modal damping matrix as

$$C'_{jj} = 2 \, \zeta_j \, \omega_j \quad \text{where } j = 1,2,\ldots\ldots\ldots.n \tag{5.31}$$

Such a damping model, introduced by Lord Rayleigh (1877) is employed to analyse damped systems in the same manner as undamped systems since the equation of motion in the modal coordinates can be decoupled as

$$\ddot{y}_j(t) + 2\zeta_j \, \omega_j \, \dot{y}_j(t) + \omega_j^2 \, y(t) = \tilde{f}_j(t) \quad \text{where } j = 1,2,\ldots\ldots\ldots.n \tag{5.32}$$

The generalized proportional damping model expresses the damping matrix as a linear combination of the mass and stiffness matrices, that is

$$[\mathbf{C}(\overline{\omega})] = \alpha_1 [\mathbf{M}(\overline{\omega})] \mathbf{f} ([\mathbf{M}^{-1}(\overline{\omega})][\mathbf{K}(\overline{\omega})]) \tag{5.33}$$

where $\alpha_1 = 0.005$ is constant damping factor.

The transfer function matrix of the system can be obtained as

$$H(i\omega)(\overline{\omega}) = X[-\omega^2 I + 2i\omega\zeta\Omega + \Omega^2]^{-1} X^T = \sum_{j=1}^{n} \frac{X_j X_j^T}{-\omega^2 + 2i\omega\zeta_j\omega_j + \omega_j^2} \tag{5.34}$$

Using this, the dynamic response in the frequency domain with zero initial conditions can be conveniently represented as

$$\overline{\delta}(i\omega)(\overline{\omega}) = H(i\omega)\overline{f}(i\omega) = \sum_{j=1}^{n} \frac{X_j^T \overline{f}(i\omega)}{-\omega^2 + 2i\omega\zeta_j\omega_j + \omega_j^2} X_j \tag{5.35}$$

Therefore, the dynamic response of a proportionally damped system can be expressed as a linear combination of the undamped mode shapes.

5.3 Kriging based metamodeling approach

In general, a surrogate is an approximation of the Input/Output (I/O) function that is implied by the underlying simulation model (refer to Fig. 5.2). Surrogate models are fitted to the I/O data produced by the experiment with the simulation model. This simulation model may either be deterministic or random (stochastic). The Kriging model was initially developed in spatial statistics by Danie Gerhardus Krige and subsequently extended by Matheron (1963) and Cressie (1993). Kriging is a Gaussian process based modelling method, which is compact and cost effective for computation. Kriging surrogate models are employed to fit the data that are obtained for larger experimental areas rather than the areas used in low order polynomial regression. Hence Kriging models are global rather than local wherein such models are used for prediction. In the present study, Kriging model for simulation of required output (say natural frequency) is expressed as,

$$y(x) = y_0(x) + Z(x) \tag{5.36}$$

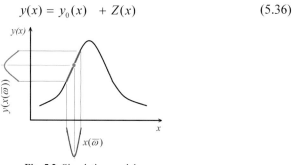

Fig. 5.2 Simulation model.

where $y(x)$ is the unknown function of interest, x is an m dimensional vector (m design variables), $y_0(x)$ is the known approximation (usually polynomial) function and $Z(x)$ represents the realization of a stochastic process with mean zero, variance, and nonzero covariance. In the model, the local deviation at an unknown point ($\mathbf{x}$) is expressed using stochastic processes. The sample points are interpolated with the Gaussian random function as the correlation function to estimate the trend of the stochastic processes. The $y_0(x)$ term is similar to a polynomial response surface, providing global model of the design space (Sacks et al. 1989b).

In the present study, $y_0(x)$ globally approximates the design space, $Z(x)$ creates the localized deviations so that the Kriging model interpolates the p-sampled data points for composite shallow doubly curved shells; however, non-interpolative Kriging models can also be created to smooth noisy data (Cressie 1993). The covariance matrix of $Z(x)$ is given as

$$Cov[Z(x^i), Z(x^j)] = \sigma^2 \, R \, [R(x^i, x^j)] \tag{5.37}$$

where $\mathbf{R}$ is a $(p \times p)$ correlation matrix and $R(\mathbf{x}^i, \mathbf{x}^j)$ is the correlation function between any two of the p-sampled data points $\mathbf{x}^i$ and $\mathbf{x}^j$. $\mathbf{R}$ is an $(p \times p)$ symmetric matrix with ones along the diagonal. The correlation function $R(\mathbf{x}^i, \mathbf{x}^j)$ is specified by the user, and a variety of correlation functions exist. Using Gaussian correlation function

$$R(x^i, x^j) = \exp\left[-\sum_{k=1}^{n} \theta_k \left| x_k^i - x_k^j \right|^2 \right] \tag{5.38}$$

where n is the number of design variables, θ_k is the unknown correlation parameters used to fit the model, and x_k^i and x_k^j are the k-th components of the sample points x^i and x_p, respectively. The predicted estimates, $\hat{y}$ of the response $y(x)$ at random values of x are defined as Kriging predictor

$$\hat{y}(x) = \hat{\beta} + r^T(x) \quad R^{-1} \quad [y - f\hat{\beta}] \tag{5.39}$$

where y is the column vector of length p that contains the sample values of the frequency responses and f is a column vector of length p that is filled with ones when $y_0(x)$ is taken as constant. Now, $r^T(x)$ is the correlation vector of length p between the random x and the sample data points $\{x^1, x^2, \ldots \ldots x^p\}$

$$r^T(x) = [\, R(x, x^1), \; R(x, x^2), \; R(x, x^3) \ldots \ldots \ldots R(x, x^p) \,]^T \tag{5.40}$$

$$\hat{\beta} = (f^T \, R^{-1} \, f)^{-1} \; f^T \, R^{-1} \; y \tag{5.41}$$

An estimate of the variance between underlying global model $\hat{\beta}$ and y is estimated by

$$\hat{\sigma}^2 = \frac{1}{p}(y - f\,\hat{\beta})^T \; R^{-1} \; (y - f\,\hat{\beta}) \tag{5.42}$$

Now the model fitting is accomplished by maximum likelihood (i.e., best guesses) for θ_k. The maximum likelihood estimates (i.e., "best guesses") for the θ_k used to fit a Kriging model are obtained as

$$Max.\Gamma(\theta_k) = -\frac{1}{2}\left[p\ln(\hat{\sigma}^2) + \ln|R|\right] \qquad (5.43)$$

where the variance σ^2 and $|R|$ are both functions of θ_k, and is solved for positive values of θ_k as optimization variables. After obtaining Kriging based surrogate, the random process $Z(x)$ provides the approximation error that can be used for improving the surrogate model. The maximum mean square error (MMSE) and maximum error (ME) are calculated as,

$$MMSE = max.\left[\frac{1}{k}\sum_{i=1}^{k}(\bar{y}_i - y_i)^2\right] \qquad (5.44)$$

$$ME\ (\%) = Max\left[\frac{y_{i,MCS} - y_{i,Kriging}}{Y_{i,MCS}}\right] \qquad (5.45)$$

where y_i and $\bar{y}_i$ are the vector of the true values and the vector corresponding to i-th prediction, respectively.

5.4 Stochastic approach using Kriging model

The stochasticity in material properties of laminated composite shallow doubly curved shells, such as longitudinal elastic modulus, transverse elastic modulus, longitudinal shear modulus, transverse shear modulus, Poisson's ratio, mass density and geometric properties such as ply-orientation angle as input parameters are considered for natural frequency analysis of composite shallow doubly curved shells. In the present study, the frequency domain feature (first three natural frequencies) is considered as output. It is assumed that the distribution of randomness of input parameters exists within a certain band of tolerance with their central deterministic mean values. Therefore the normal distribution considered is expressed as

$$f(x) = \frac{1}{\sigma\sqrt{2\pi}}\exp\left[-\frac{1}{2}\left(\frac{x-\bar{x}}{\sigma}\right)^2\right] \qquad (5.46)$$

where σ^2 is the variance of the random variable; $\bar{x}$ is the mean of the random variable. The cases wherein the random variables considered in each layer of laminate are investigated for

a) Variation of ply-orientation angle only: $\theta(\bar{\omega}) = \{\theta_1\ \theta_2\ \theta_3.........\theta_i......\theta_l\}$

b) Variation of longitudinal elastic modulus only: $E_1(\bar{\omega}) = \{E_{1(1)}\ E_{1(2)}\ E_{1(3)}.........E_{1(i)}......E_{1(l)}\}$

c) Variation of transverse elastic modulus only: $E_2(\bar{\omega}) = \{E_{2(1)}\ E_{2(2)}\ E_{2(3)}.........E_{2(i)}......E_{2(l)}\}$

d) Variation of longitudinal shear modulus only: $G_{12}(\bar{\omega}) = \{G_{12\,(1)}\ G_{12\,(2)}\ G_{12\,(3)}....G_{12\,(i)}...G_{12\,(l)}\}$

e) Variation of transverse shear modulus only: $G_{23}(\bar{\omega}) = \{G_{23\,(1)}\ G_{23\,(2)}\ G_{23(3)}....G_{23\,(i)}...G_{23\,(l)}\}$

f) Variation of Poisson's ratio only: $\mu(\bar{\omega}) = \{\mu_1\ \mu_2\ \mu_3........\mu_i......\mu_l\}$

g) Variation of mass density only: $\rho(\bar{\omega}) = \{\rho_1\ \rho_2\ \rho_3........\rho_i......\rho_l\}$

h) Combined variation of ply orientation angle, elastic modulus (longitudinal and transverse), shear modulus (longitudinal and transverse), Poisson's ratio and mass density:

$$g\{\theta(\bar{\omega}), \rho(\bar{\omega}), G_{12}(\bar{\omega}), G_{23}(\bar{\omega}), E_1(\bar{\omega})\} = \{\ \Phi_1(\theta_1...\theta_l), \Phi_2(E_{1(1)}...E_{1(l)}), \Phi_3(E_{2(1)}...E_{2(l)}),$$
$$\Phi_4(G_{12(1)}..G_{12(l)}),\ \Phi_5(G_{23(1)}..G_{23(l)}),\ \Phi_6(\mu_1...\mu_l), \Phi_7(\rho_1..\rho_l)\}$$

where θ_i, $E_{1(i)}$, $E_{2(i)}$, $G_{12(i)}$, $G_{23(i)}$, μ_i and ρ_i are the ply orientation angle, elastic modulus along longitudinal and transverse direction, shear modulus along longitudinal direction, shear modulus along transverse direction, Poisson's ratio and mass density, respectively and '*l*' denotes the number of layer in the laminate. In the present study, $\pm 10°$ for ply orientation angle with subsequent $\pm 10\%$ tolerance for material properties from deterministic mean value are considered. Latin hypercube sampling is employed for generating sample points to ensure the representation of all portions of the vector space (McKay et al. 2000). In Latin hypercube sampling, the interval of each dimension is divided into *m* non-overlapping intervals having equal probability considering a uniform distribution, so the intervals should have equal size. Moreover, the sample is chosen randomly from a uniform distribution point in each interval in each dimension and the random pair is selected considering equal likely combinations for the point from each dimension. A working flowchart for the kriging based uncertainty quantification algorithm adopted in this chapter is presented above in Fig. 5.3.

In the surrogate based approach of uncertainty quantification, a virtual mathematical model is formed for the response quantity of interest that effectively replaces the actual expensive finite element model. The surrogate model is built up on the basis of information acquired regarding the behavior of the response quantity throughout the entire design space utilizing few algorithmically chosen design points. Besides the source-uncertainties due to stochastic nature of material and geometric attributes, there remains another inevitable source for second phase of uncertainty associated with the information acquired through the design points that needs further attention (refer to Fig. 5.4). In the present chapter, this second source of uncertainty associated with the surrogate model formation has been addressed by developing an algorithm to account it in the form of random noise (Mukhopadhyay et al. 2016a). The effect of such simulated noise can be regarded as considering other sources of uncertainty besides conventional material and geometric uncertainties, such as error in measurement of responses, error in

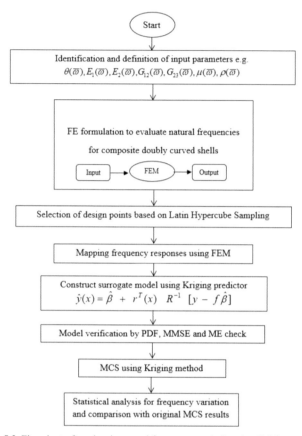

Fig. 5.3 Flowchart of stochastic natural frequency analysis using Kriging model.

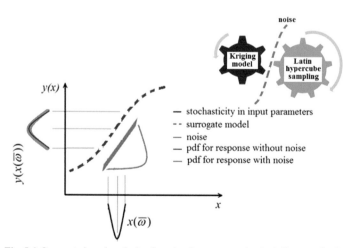

Fig. 5.4 Surrogate based analysis of stochastic system under the influence of noise.

modelling and computer simulation and various other epistemic uncertainties involved with the system. Noise effects are found to be accounted in several other studies in available literature (Nejad et al. 2005, Friswell et al. 2015, Mukhopadhyay 2018a) dealing with deterministic analysis. In the present chapter an algorithm has been presented to quantify the effect of noise for Kriging based stochastic analysis of doubly curved composite shells.

To portray the effect of noise on the surrogate based uncertainty quantification algorithm, different levels of noise have been introduced in the responses of design points while constructing the surrogate models. In the proposed approach, Gaussian white noise with a specific level (p) has been introduced in the set of output responses (natural frequencies), which is subsequently used for surrogate model formation

$$f_{ijN} = f_{ij} + p \times \xi_{ij} \qquad (9.22)$$

where, f denotes natural frequency with the subscript i and j as frequency number and sample number in the design point set, respectively. ξ_{ij} is a function that generates normally distributed random numbers. Subscript N is used here to indicate the noisy frequency. Thus a simulated noisy dataset (i.e., the sampling matrix for surrogate model formation) is formed by introducing pseudo random noise in the responses, while the input design points are kept unaltered. Subsequently for each dataset, surrogate based MCS is carried out to quantify the uncertainty of composite laminates following a non-intrusive approach as described in Fig. 5.5. From the

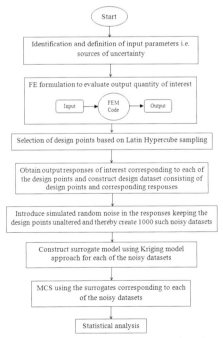

Fig. 5.5 Flowchart for analyzing the effect of noise on surrogate based uncertainty quantification.

flowchart it can be understood that quantification of the effect of noise involves carrying out 1000 surrogate based MCS requiring formation of such surrogates 1000 times for analyzing each noise level. The kind of analysis carried out here will provide a comprehensive idea about the robustness of surrogate based uncertainty quantification algorithm under noisy data.

5.5   Results and discussion

In the present study, four layered graphite–epoxy symmetric angle-ply and cross-ply laminated composite cantilever shallow doubly curved shells namely spherical $(R_x/R_y = 1)$, hyperbolic paraboloid $(R_x/R_y = -1)$ and elliptical paraboloid $(R_x/R_y \neq 1)$ are considered. The length, width and thickness of the composite laminate considered in the present analysis are 1 m, 1 mm and 5 mm, respectively. Material properties of graphite–epoxy composite as given by Qatu and Leissa (1991b) are considered with deterministic mean value as $E_1 = 138.0$ GPa, $E_2 = 8.96$ GPa, $G_{12} = 7.1$ GPa, $G_{13} = 7.1$ GPa, $G_{23} = 2.84$ GPa, $\mu = 0.3$, $\rho = 3202$ kg/m^3. A typical discretization of (6×6) mesh on plan area with 36 elements, 133 nodes with natural coordinates of an isoparametric quadratic plate bending element are considered for the present FEM approach.

For full scale MCS, number of original FE analysis is same as the sampling size. In general for complex composite structures, the performance function is not available as an explicit function of the random design variables. The random response in terms of natural frequencies of the composite structure can only be evaluated numerically at the end of a structural analysis procedure such as the finite element method which is often time-consuming. The present Kriging method is employed to find a predictive and representative surrogate model relating each natural frequency to a number of input variables. The Kriging surrogate models are used to determine the first three natural frequencies corresponding to given values of input variables, instead of time-consuming deterministic FE analysis. The surrogate model thus represents the result of the structural analysis encompassing every possible combination of all input variables. From this, thousands of combinations of all design variables can be created and performed a pseudo analysis for each variable set, by simply adopting the corresponding predictive values. The probability density function (PDF) is plotted as the benchmark of bottom line results. Due to paucity of space, only a few important representative results are furnished.

5.5.1   Validation

The present computer code is validated with the results available in the open literature. Table 5.1 presents the non-dimensional fundamental frequencies of graphite–epoxy composite twisted plates with different ply-orientation angle given by Qatu and Leissa (1991). Convergence studies are performed using uniform mesh division of (4 x 4), (6 x 6), (8 x 8), (10 x 10) and (20 x 20) wherein the (6 x 6) mesh is found to provide best results with the least difference compared to

Table 5.1 Non-dimensional fundamental natural frequencies [$\omega = \omega_n L^2 \sqrt{(\rho/E_1 h^2)}$] of three layered [$\theta, -\theta, \theta$] graphite-epoxy twisted plates, L/b = 1, b/h = 20, ψ = 30°.

Ply-orientation Angle, θ	Present FEM					Qatu and Leissa (1991a)
	4 x 4	6 x 6	8 x 8	10 x 10	20 x 20	
15°	0.8588	0.8618	0.8591	0.8543	0.8004	0.8759
30°	0.6753	0.6790	0.6752	0.6722	0.6232	0.6923
45°	0.4691	0.4732	0.4698	0.4578	0.4057	0.4831
60°	0.3189	0.3234	0.3194	0.3114	0.2769	0.3283

benchmarking results in Qatu and Leissa (1991a) and the results also corroborate monotonic downward convergence towards higher as well as lower mesh sizes. The differences between the results by Qatu and Leissa (1991) and the present finite element approach can be attributed to the consideration of transverse shear deformation and rotary inertia in the present FEM approach and also to the fact that Ritz method overestimates the structural stiffness of the composite plates. Moreover, increasing the size of matrix because of higher mesh size increases the ill-conditioning of the numerical eigenvalue problem. Hence, the lower mesh size (6 x 6) is employed in the present analysis due to computational efficiency. Further validation of the finite element code is carried out for a composite shell considering the converged mesh size. Table 5.2 presents the non-dimensional fundamental natural frequencies of isotropic, corner point-supported spherical and hyperbolic paraboloid shells obtained from the presented code along with the results reported by Chakravorty et al. (1995) and Leissa and Narita (1984).

Another convergence study is carried out for selection of sample size with respect to maximum mean square error (MMSE) and maximum error (in percentage) using Kriging method compared to original MCS (10,000 samples) with respect to variation of only the ply-orientation angle in each layer of graphite–epoxy angle-ply (45°/–45°/–45°/45°) composite cantilever spherical shells as furnished in Table 5.3. To optimize the computation time in conjunction with MMSE and ME, a sample size of 625 is chosen for combined variation of twenty eight (28) input parameters for the present study. In contrast, a sample size of 30 is chosen in the similar fashion for individual random variation with four input parameters for the present study. All these studies depict an excellent agreement and hence it demonstrates the

Table 5.2 Non-dimensional fundamental frequencies $\left[\omega = \omega_n a^2 \sqrt{\dfrac{12\rho(1-\mu^2)}{E_1 t^2}} \right]$ of isotropic, corner point-supported spherical and hyperbolic paraboloidal shells considering a/b = 1, a'/a = 1, a/t = 100, a/R = 0.5, μ = 0.3.

R_x/R_y	Shell type	Present FEM	Leissa and Narita (1984)	Chakravorty et al. (1995)
1	Spherical	50.74	50.68	50.76
–1	Hyperbolic paraboloid	17.22	17.16	17.25

Table 5.3 Convergence study for maximum mean square error (MMSE) and maximum error (in percentage) using Kriging model compared to original MCS with different sample sizes for combined variation of 28 nos input parameters of graphite-epoxy angle-ply (45°/–45°/–45°/45°) composite cantilever spherical shells, considering $E_1 = 138$ GPa, $E_2 = 8.9$ GPa, $G_{12} = G_{13} = 7.1$ GPa, $G_{23} = 2.84$ GPa, $t = 0.005$ m, $\mu = 0.3$.

Sample size	Parameter	Fundamental natural frequency	Second natural frequency	Third natural frequency
450	MMSE	0.0289	0.1968	0.2312
	Max Error (%)	2.4804	7.6361	6.5505
500	MMSE	0.0178	0.1466	0.2320
	Max Error (%)	1.6045	2.6552	3.0361
550	MMSE	0.0213	0.1460	0.2400
	Max Error (%)	1.2345	2.0287	1.8922
575	MMSE	0.0207	0.1233	0.2262
	Max Error (%)	1.1470	1.8461	1.7785
600	MMSE	0.0177	0.1035	0.2071
	Max Error (%)	1.1360	1.7208	1.7820
625	MMSE	0.0158	0.0986	0.1801
	Max Error (%)	1.0530	1.7301	1.6121
650	MMSE	0.0153	0.0966	0.1755
	Max Error (%)	0.9965	1.8332	1.6475

capability of the computer codes developed and insures the accuracy of analyses. Figure 5.6 depicts a representative plot describing the relationship between the original FE model and the constructed Kriging model for fundamental natural frequencies signifying the accuracy of the present Kriging model. A comparative study addressing the probability density functions (PDF) for both original MCS and Kriging model using sample size of 10,000 is carried out corresponding to first three natural frequencies with typical laminated shell configuration as furnished in Fig. 5.7 (individual variation of ply orientation angle) and Fig. 5.8 (combined random variation of parameters). The low scatterness of the points found around the diagonal line in Fig. 5.6 and the low deviation obtained between the probability distribution function estimations of original MCS and Kriging responses in Fig. 5.7 and Fig. 5.8 corroborate the fact that the present surrogate models are accurately formed. These two plots are checked for all other cases and are found in good agreement ensuring the efficiency and accuracy of the constructed Kriging models. While evaluating the statistics of responses through full scale MCS, computational time is exorbitently high because it involves number of repeated FE analysis. However, in the present method, MCS is conducted in conjunction with Kriging model. Here, although the same sampling size as in direct MCS (with sample size of 10,000) is considered, the number of original FE analysis is much less compared to direct MCS and is equal to number of samples required to construct the Kriging

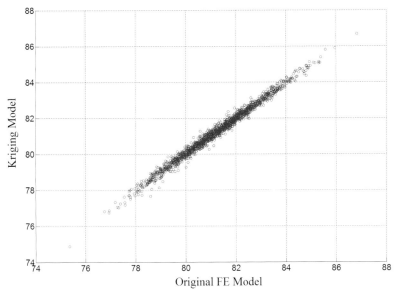

Fig. 5.6 Scatter plot for Kriging model with respect to original FE model of fundamental natural frequency for combined variation of ply orientation angle $[\theta(\bar{\omega})]$, longitudinal elastic modulus $[E_1(\bar{\omega})]$, transverse elastic modulus $[E_2(\bar{\omega})]$, longitudinal shear modulus $[G_{12}(\bar{\omega})]$, Transverse shear modulus $[G_{23}(\bar{\omega})]$, poisson's ratio $[\mu(\bar{\omega})]$ and mass density $[\rho(\bar{\omega})]$ for composite catilevered spherical shells, considering sample size = 10,000, E_1 = 138 GPa, E_2 = 8.9 GPa, $G_{12} = G_{13}$ = 7.1 GPa, G_{23} = 2.84 GPa, ρ = 3202 kg/m³, t = 0.005 m, μ = 0.3.

surrogate model. Hence, the computational time and effort expressed in terms of FE calculation is reduced compared to full scale direct MCS. This provides an efficient affordable way for simulating the uncertainties in natural frequency.

5.5.2 Statistical analysis

The variation of material properties like elastic modulus (logitudinal and transverse) $[E_1(\bar{\omega}), E_2(\bar{\omega})]$, shear modulus (logitudinal and transverse) $[G_{12}(\bar{\omega}), G_{23}(\bar{\omega})]$, poisson's ratio $[\mu(\bar{\omega})]$ and mass density $[\rho(\bar{\omega})]$ considered in each layer of the composite shallow doubly curved shells are scaled randomly in the range having the lower and the upper limit as $\pm 10\%$ variability with respective mean values, while for ply orientation angle $[\theta(\bar{\omega})]$ the bound is considered as within $\pm 10°$ fluctuation with respective mean values of each layer of laminates. The Kriging surrogate models are formed to generate first three natural frequencies for angle-ply and cross-ply composite shallow doubly curved shells. The natural frequencies of tested angle-ply (45°/–45°/–45°/45°) laminate are found to be lower than that of the same for cross-ply (0°/90°/90°/0°) laminate irrespective of geometry of the shells and stochasticity considered in random input parameters. Table 5.4 and Table 5.5 present the maximum values, minimum values and standard deviation using the Kriging model for first three natural frequencies obtained due to individual stochasticity of ply-orientation

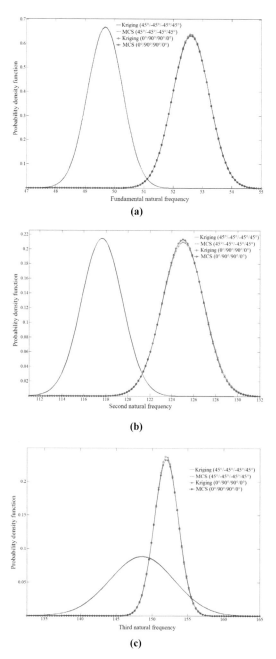

Fig. 5.7 Probability density function obtained by original MCS and Kriging model with respect to first three natural frequencies for individual variation of ply orientation angle $[\theta(\bar{\omega})]$ for composite elliptical paraboloid shells, considering sample size = 10,000, $R_x \neq R_y$, $R_{xy} = \alpha$, $E_1 = 138$ GPa, $E_2 = 8.9$ GPa, $G_{12} = G_{13} = 7.1$ GPa, $G_{23} = 2.84$ GPa, $\rho = 3202$ kg/m^3, $t = 0.005$ m, $\mu = 0.3$.

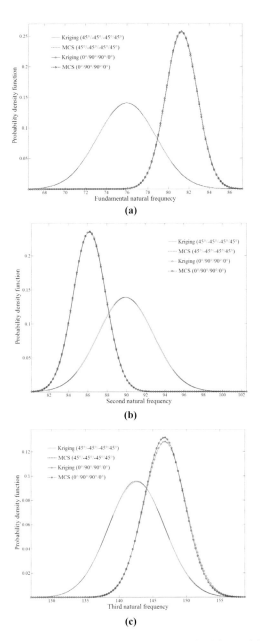

Fig. 5.8 Probability density function obtained by original MCS and Kriging model with respect to first three natural frequencies for combined variation of ply orientation angle $[\theta(\bar{\omega})]$, mass density $[\rho(\bar{\omega})]$, longitudinal shear modulus $[G_{12}(\bar{\omega})]$, transverse shear modulus $[G_{23}(\bar{\omega})]$ and longitudinal elastic modulus $[E_1(\bar{\omega})]$ and transverse elastic modulus $[E_2(\bar{\omega})]$ for composite spherical shells, considering sample size = 10,000, E_1=138 GPa, E_2=8.9 GPa, $G_{12}=G_{13}$=7.1 GPa, G_{23}=2.84 GPa, ρ=3202 kg/m^3, t=0.005 m, μ=0.3.

Table 5.4 Individual stochasticity of $[\theta(\bar\omega), E_1(\bar\omega), E_2(\bar\omega), G_{12}(\bar\omega), G_{23}(\bar\omega), \mu(\bar\omega), \rho(\bar\omega)]$ for angle-ply $(45°/-45°/-45°/45°)$ composite shells.

Type	Value	HP				SP				EP			
		SD	Mean	Max	Min	SD	Mean	Max	Min	SD	Mean	Max	Min
$\theta(\bar\omega)$	f_3^s	0.21	28.40	29.28	27.79	2.57	76.44	84.04	69.76	0.60	49.68	51.93	47.94
	f_2^s	1.26	108.68	113.02	105.54	2.32	90.55	97.52	85.06	1.85	117.65	123.91	113.17
	f_1^s	2.79	164.52	171.42	156.30	3.53	144.10	154.33	134.58	4.52	148.61	160.73	136.05
$E_1(\bar\omega)$	f_3^s	0.22	28.33	28.98	27.66	0.46	75.35	76.69	73.94	0.33	49.32	50.32	48.29
	f_2^s	0.86	108.53	111.10	105.87	0.86	90.71	93.03	88.29	0.94	116.81	119.42	114.07
	f_1^s	1.19	163.77	167.29	160.10	1.01	142.26	145.01	139.32	0.87	146.75	149.16	144.13
$E_2(\bar\omega)$	f_3^s	0.07	28.34	28.52	28.16	0.17	69.21	69.64	68.75	0.08	49.34	49.54	49.13
	f_2^s	0.28	108.58	109.27	107.87	0.28	78.97	79.68	78.21	0.14	116.85	117.22	116.48
	f_1^s	0.22	163.84	164.40	163.27	0.30	128.22	128.99	127.40	0.11	146.80	147.10	146.49
$G_{12}(\bar\omega)$	f_3^s	0.15	29.41	29.84	28.95	0.62	80.06	81.89	78.12	0.32	49.33	50.25	48.35
	f_2^s	0.51	112.32	113.80	110.74	0.50	94.45	95.89	92.90	0.72	116.84	118.94	114.61
	f_1^s	1.02	171.49	174.47	168.32	1.09	150.51	153.71	147.12	1.20	146.77	150.28	143.00
$G_{23}(\bar\omega)$	f_3^s	0.00	28.34	28.35	28.34	0.00	75.38	75.39	75.37	0.00	49.34	49.34	49.33
	f_2^s	0.01	108.58	108.60	108.55	0.00	90.76	90.77	90.75	0.00	116.85	116.86	116.84
	f_1^s	0.02	163.84	163.89	163.78	0.01	142.30	142.33	142.28	0.01	146.80	146.82	146.77
$\mu(\bar\omega)$	f_3^s	0.00	28.34	28.35	28.33	0.01	75.38	75.40	75.36	0.01	49.34	49.36	49.31
	f_2^s	0.01	108.58	108.62	108.54	0.02	90.76	90.81	90.71	0.03	116.85	116.93	116.78
	f_1^s	0.02	163.84	163.89	163.79	0.03	142.30	142.38	142.23	0.02	146.80	146.87	146.73
$\rho(\bar\omega)$	f_3^s	0.41	28.35	29.68	27.19	1.09	75.41	78.95	72.31	0.71	49.35	51.67	47.33
	f_2^s	1.57	108.62	113.72	104.15	1.31	90.79	95.06	87.06	1.69	116.89	122.38	112.09
	f_1^s	2.37	163.89	171.60	157.16	2.06	142.35	149.04	136.50	2.12	146.85	153.75	140.81

Table 5.5 Individual stochasticity of [$\theta(\overline{\omega})$, $E_1(\overline{\omega})$, $E_2(\overline{\omega})$], $G_{12}(\overline{\omega})$, $G_{23}(\overline{\omega})$, $\mu(\overline{\omega})$, $\rho(\overline{\omega})$] for cross-ply (0°/90°/90°/0°) composite shells.

Type	Value	HP				SP				EP			
		SD	Mean	Max	Min	SD	Mean	Max	Min	SD	Mean	Max	Min
$\theta(\omega)$	f_1^c	0.49	30.03	31.31	28.72	0.57	81.76	82.53	78.96	0.62	52.60	54.25	50.66
	f_2^c	1.56	110.29	114.45	106.04	0.64	86.82	89.20	84.43	1.90	125.03	129.67	117.30
	f_3^c	3.14	173.55	179.15	159.86	1.63	148.23	150.16	139.11	1.67	151.99	155.85	145.55
$E_1(\omega)$	f_1^c	0.28	30.15	30.98	29.30	0.79	82.59	84.93	80.16	0.49	52.90	54.45	51.36
	f_2^c	0.93	110.32	113.12	107.46	0.89	86.94	89.37	84.35	1.22	125.48	129.08	121.84
	f_3^c	1.71	174.71	179.77	169.47	1.56	150.23	154.42	145.68	1.43	152.96	157.49	148.60
$E_2(\omega)$	f_1^c	0.08	30.16	30.37	29.94	0.06	75.38	75.55	75.20	0.14	52.92	53.29	52.53
	f_2^c	0.25	110.37	111.01	109.69	0.10	90.76	91.02	90.49	0.25	125.52	126.19	124.79
	f_3^c	0.38	174.78	175.77	173.71	0.11	142.30	142.59	142.01	0.44	153.07	154.14	151.88
$G_{12}(\omega)$	f_1^c	0.11	30.89	31.20	30.58	0.23	84.13	84.76	83.47	0.16	52.91	53.38	52.43
	f_2^c	0.53	114.02	115.55	112.47	0.34	89.50	90.47	88.44	0.43	125.51	126.80	124.16
	f_3^c	0.54	178.47	179.98	176.93	0.23	152.08	152.73	151.38	0.43	153.06	154.25	151.73
$G_{23}(\omega)$	f_1^c	0.00	30.16	30.17	30.16	0.00	82.62	82.63	82.61	0.00	52.92	52.92	52.91
	f_2^c	0.01	110.37	110.38	110.35	0.00	86.97	86.98	86.96	0.01	125.52	125.54	125.50
	f_3^c	0.02	174.78	174.84	174.71	0.01	150.44	150.47	150.42	0.02	153.07	153.12	153.03
$\mu(\omega)$	f_1^c	0.00	30.16	30.17	30.15	0.01	82.62	82.65	82.59	0.00	52.92	52.93	52.91
	f_2^c	0.02	110.37	110.42	110.31	0.01	86.97	87.00	86.94	0.03	125.52	125.58	125.46
	f_3^c	0.03	174.78	174.87	174.69	0.02	150.44	150.50	150.40	0.02	153.07	153.13	153.02
$\rho(\omega)$	f_1^c	0.44	30.17	31.59	28.93	1.19	82.65	86.53	79.25	0.77	52.94	55.42	50.76
	f_2^c	1.60	110.40	115.59	105.87	1.26	87.00	91.09	83.42	1.82	125.56	131.47	120.41
	f_3^c	2.53	174.84	183.06	167.66	2.18	150.49	157.57	144.31	2.21	153.13	160.32	146.84

angle $[\theta(\bar\omega)]$, elastic modulus (logitudinal and transverse) $[E_1(\bar\omega), E_2(\bar\omega)]$, shear modulus (logitudinal and transverse) $[G_{12}(\bar\omega), G_{23}(\bar\omega)]$, and poisson's ratio $[\mu(\bar\omega)]$, mass density $[\rho(\bar\omega)]$ in each layer due to randomness considered for angle-ply and cross-ply composite cantilever shells, respectively. In general, for both angle-ply and cross-ply laminates with respect to individual variation of input parameters, the maximum and minimum of mean value for fundametal natural frequency is observed for spherical shells and hyperbolic paraboloid shells, respectively while the same is found to be intermediate for elliptical paraboloid shells. In contrast, the maximum and minimum of mean value for second natural frequency is observed for elliptical paraboloid shells and spherical shells, respectively while the same is found to be intermediate for hyperbolic paraboloid shells. Interestingly, the maximum of mean value for third natural frequency is observed for hyperbolic paraboloid shells, trailed by elliptical paraboloid shells and the minimum mean value of third natural frequency is identified for spherical shells. Such variation in mean value of first three natural frequencies for three different types of shells can be attributed to the fact of predominance of the geometric curvature effect which influences the elastic stiffness of the structure differently. Table 5.6 presents the maximum

Table 5.6 Maximum value, minimum value, mean value and standard deviation (SD) using Kriging model for first three natural frequencies obtained due to combined stochasticity $[\theta(\bar\omega), E_1(\bar\omega), E_2(\bar\omega)], G_{12}(\bar\omega), G_{23}(\bar\omega), \mu(\bar\omega), \rho(\bar\omega)]$ for graphite-epoxy angle-ply (45°/–45°/–45°/45°) and cross-ply (0°/90°/90°/0°) composite shallow doubly curved shells considering $E_1 = 138$ GPa, $E_2 = 8.9$ GPa, $G_{12} = G_{13} = 7.1$ GPa, $G_{23} = 2.84$ GPa, $\rho = 3202$ kg/m³, $t = 0.005$ m, $\mu = 0.3$.

Shell type	Values	Angle-ply (45°/–45°/–45°/45°)			Cross-ply (0°/90°/90°/0°)		
		Fundamental natural frequency	Second natural frequency	Third natural frequency	Fundamental natural frequency	Second natural frequency	Third natural frequency
Spherical (SP)	Min value	67.81	81.14	129.79	75.46	80.55	133.98
	Max value	86.15	101.39	156.77	86.99	91.71	158.45
	Mean value	75.98	89.96	142.61	81.31	86.21	146.79
	Standard Deviation	2.82	2.86	4.15	1.55	1.69	3.11
Hyperbolic paraboloid (HP)	Min value	26.46	100.86	149.01	27.38	100.54	151.89
	Max value	30.53	117.66	174.55	32.26	118.04	184.67
	Mean value	28.16	107.62	162.00	29.76	109.37	170.69
	Standard Deviation	0.54	2.26	3.77	0.71	2.47	4.29
Elliptical paraboloid (EP)	Min value	45.97	108.11	128.37	48.20	112.48	138.22
	Max value	53.22	127.70	163.86	56.46	135.33	160.86
	Mean value	49.34	116.89	147.24	52.35	123.97	149.81
	Standard Deviation	1.04	2.79	4.91	1.14	2.93	3.14

value, minimum value, mean value and standard deviation (SD) using the Kriging model (625 samples) for first three natural frequencies obtained due to combined stochasticity $[\theta(\bar{\omega}), E_1(\bar{\omega}), E_2(\bar{\omega}), G_{12}(\bar{\omega}), G_{23}(\bar{\omega}), \mu(\bar{\omega}), \rho(\bar{\omega})]$ for graphite–epoxy angle-ply (45°/–45°/–45°/45°) and cross-ply (0°/90°/90°/0°) composite shallow doubly curved shells which as expected shows the maximum volatility compared to individual random variation in input parameter.

Figure 5.9 presents the [*SD/Mean*] of first three natural frequencies for individual variation of input parameters and combined variation for angle-ply (45°/–45°/–45°/45°) and cross-ply (0°/90°/90°/0°) composite shallow doubly curved shells. The degree of influence of the ply-orientation angle $[\theta(\bar{\omega})]$ on tested

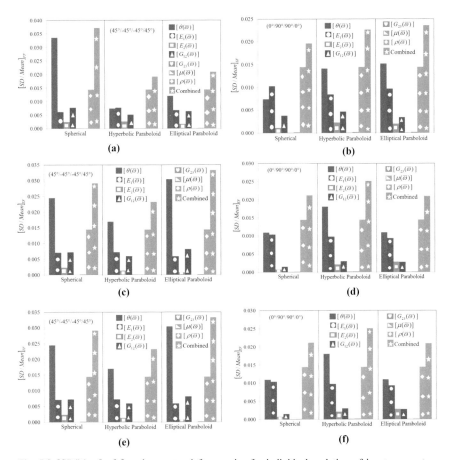

Fig. 5.9 [*SD/Mean*] of first three natural frequencies for individual variation of input parameters $[\theta(\bar{\omega})]$, $[E_1(\bar{\omega})]$, $[E_2(\bar{\omega})]$, $[G_{12}(\bar{\omega})]$, $[G_{23}(\bar{\omega})]$, $[\mu(\bar{\omega})]$, $[\rho(\bar{\omega})]$, and combined variation $[\theta(\bar{\omega}), E_1(\bar{\omega}), E_2(\bar{\omega}), G_{12}(\bar{\omega}), G_{23}(\bar{\omega}), \mu(\bar{\omega}), \rho(\bar{\omega})]$ for angle-ply (45°/–45°/–45°/45°) and cross-ply (0°/90°/90°/0°) composite shallow doubly curved shells (spherical, hyperboilic paraboloid and elliptical paraboloid), considering deterministic values as E_1 = 138 GPa, E_2 = 8.9 GPa, $G_{12} = G_{13}$ = 7.1 GPa, G_{23} = 2.84 GPa, ρ = 3202 kg/m³, t = 0.005 m, μ = 0.3.

natural frequencies is found to be maximum for spherical, hyperbolic paraboloid and elliptical paraboloid angle-ply composite laminate considering individual stochaticity of input parameters. In contrast, mass density $[\rho(\bar{\omega})]$ of spherical, hyperbolic paraboloid and elliptical paraboloid cross-ply composite laminate is identified to be most sensitive for first three natural frequencies while individual stochaticity of input parameters are considered. In both the cases of individual stochasticity, shear modulus along transverse direction $[G_{23}(\bar{\omega})]$ is observed to be least sensitive irrespective of type of shells. Due to the cascading effect of variation of inputs in each layer of laminate, the combined variation of all input parameters is found to be most volatile for tested natural frequencies compared to individual input variations. The probability density function with respect to first three natural frequencies of combined variation for cross-ply $(0°/90°/90°/0°)$ composite hyperbolic paraboloid shallow doubly curved shells considering $\pm 5°$, $\pm 10°$ and $\pm 15°$ for ply orientation angle with subsequent $\pm 5\%$, $\pm 10\%$ and $\pm 15\%$ tolerance for material properties respectively from deterministic mean value are considered as furnished in Fig. 5.10. Representative stochastic modeshapes for first three natural frequencies of angle-ply composite cantilever elliptical paraboloid shells are furnished in Fig. 5.11 wherein the first chordwise bending mode is observed corresponding to its fundamental natural frequency while the dominance of combined bending and torsion mode is found for second and third natural frequencies. It is also identified that the symmetry modes are absent and the nodal lines indicate with zero displacement amplitude. In the present study, a driving point and three cross points are considered to ascertain the corresponding amplitude (in dB) of the frequency response function (FRF) wherein point 2 is considered as driving point while points 1, 3 and 4 are considered as the three cross response points as indicated in Fig. 5.1b assuming 0.5% damping factor for all the modes. A typical frequency response function for combined variation $[\theta(\bar{\omega}),$ $E_1(\bar{\omega}), E_2(\bar{\omega})], G_{12}(\bar{\omega}), G_{23}(\bar{\omega}), \mu(\bar{\omega}), \rho(\bar{\omega})]$ of both angle-ply $(45°/-45°/-45°/45°)$ and cross-ply $(0°/90°/90°/0°)$ composite cantilever elliptical paraboloid shells are furnished in the Fig. 5.12. For a small amount of variability in input parameters, higher frequency shows wider volatility in simulation bounds of FRF compared to the same at lower frequency ranges. This can be attributed to the fact that due to multiplier effect at higher frequencies, a wider volatility of frequency response function is occurred.

The results presented so far in this chapter are without consideration of the effect of noise (i.e., $p = 0$). Effect of noise on the kriging based uncertainty quantification algorithm is presented considering spherical cantilever composite shells. Scatter plot bounds depicting the effect of simulated random noise on the prediction of Kriging based surrogates are presented in Fig. 5.13 for different levels of noise using 1000 simulations for each case. From the figure, it is evident that as the level of noise increases, the deviation of the points from diagonal line also becomes more, indicating higher influence of noise on surrogate predictions. Figure 5.14 presents probability density function plots for first three natural frequencies showing the effect of noise with different levels. It can be noticed from the figures that

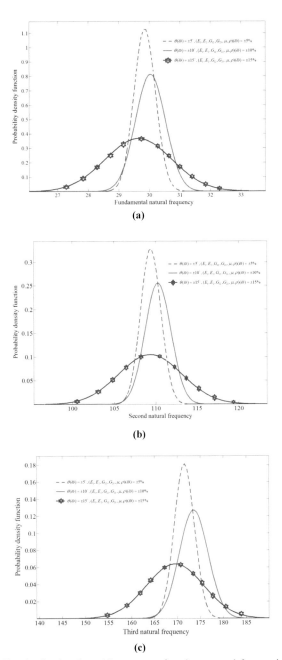

Fig. 5.10 Probability density function with respect to first three natural frequencies with different combined variation for cross-ply $(0°/90°/90°/0°)$ composite hyperbolic paraboloid shallow doubly curved shells considering $E_1 = 138$ GPa, $E_2 = 8.9$ GPa, $G_{12} = G_{13} = 7.1$ GPa, $G_{23} = 2.84$ GPa, $\rho = 3202$ kg/m³, t = 0.005 m, $\mu = 0.3$.

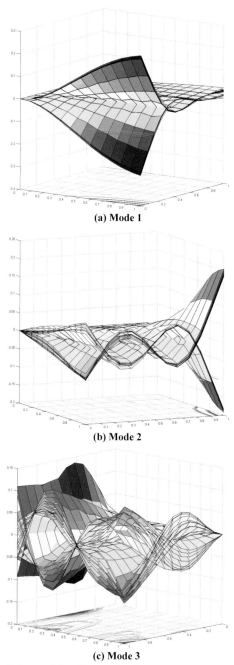

(a) Mode 1

(b) Mode 2

(c) Mode 3

Fig. 5.11 Effect on modeshapes of first three modes due to combined stochasticity for four layered angle-ply ($45°/–45°/–45°/45°$) composite cantilever elliptical paraboloid shells considering $E_1 = 138$ GPa, $E_2 = 8.9$ GPa, $G_{12} = G_{13} = 7.1$ GPa, $G_{23} = 2.84$ GPa, $\rho = 3202$ Kg/m³, $t = 0.005$ m, $\mu = 0.3$.

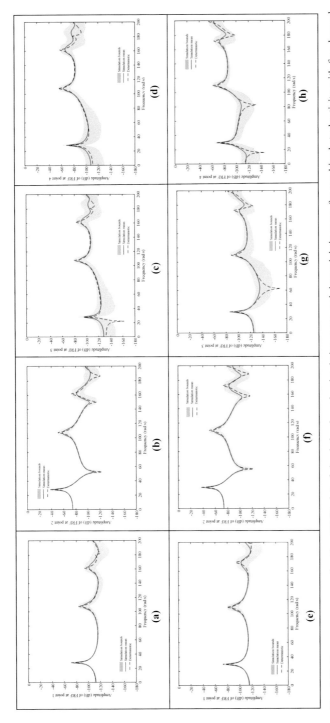

Fig. 5.12 Frequency response function (FRF) plot of simulation bounds, simulation mean and deterministic mean for combined stochasticity with four layered graphite epoxy composite cantilever elliptical paraboloid shells considering $E_1 = 138$ GPa, $E_2 = 8.9$ GPa, $G_{12} = G_{13} = 7.1$ GPa, $G_{23} = 2.84$ GPa, $\rho = 3202$ Kg/m³, $t = 0.005$ m, $\mu = 0.3$.

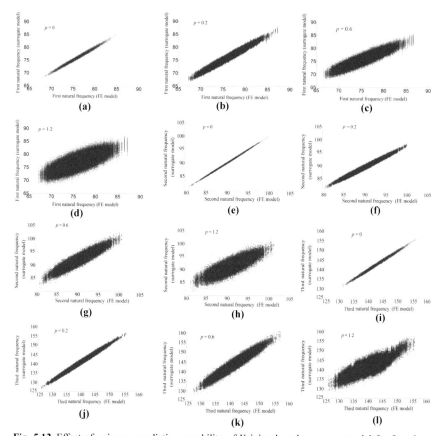

Fig. 5.13 Effect of noise on prediction capability of Kriging based surrogate model for first three natural frequencies for noise level $(p) = 0$, 0.2, 0.6 and 1.2.

the response bounds increase with increasing level of noise for all three natural frequencies. Normalised standard deviations (with respect to deterministic values of respective natural frequencies) for the first three natural are presented in Fig. 5.15. The figure shows that the bounds of normalized standard deviation decreases for higher modes of frequencies indicating subsequent reduction in sensitivity of noise.

5.6 Summary

This chapter presents a layer-wise random variable approach for the stochastic dynamic responses of angle-ply and cross-ply composite doubly curved shells. By employing the kriging based metamodelling approach, actual number of finite element simulation is significantly reduced compared to direct Monte Carlo simulation without compromising the accuracy of results. It is observed that kriging model can handle large number of input parameters (i.e., high dimensional systems). The ply-orientation angle and mass density are found to be most sensitive

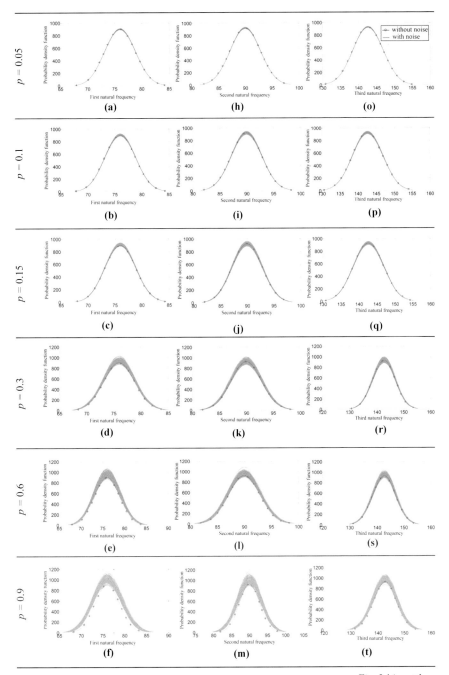

Fig. 5.14 contd....

...Fig. 5.14 contd.

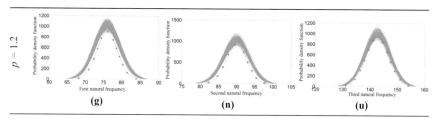

Fig. 5.14 Probability density function of first three natural frequencies for different levels of noise.

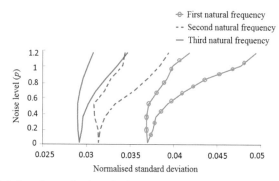

Fig. 5.15 Variation of normalised standard deviation with noise level for natural frequencies.

among the tested input parameters while transverse shear modulus is identified as the least sensitive for first three natural frequencies. In general, for both angle-ply and cross-ply laminates with respect to individual variation of input parameters, the maximum and minimum of mean value for fundametal natural frequency is observed for spherical shells and hyperbolic paraboloid shells, respectively while the same is found to be intermediate for elliptical paraboloid shells. The first chordwise bending mode is observed corresponding to its fundamental frequency while the dominance of combined bending and torsion mode is found for second and third natural frequencies, respectively. For a given amount of variability in the random input parameters, more volatility in the output quantities are observed in the higher frequency ranges compared to lower frequency ranges irrespective of frequency response function points.

In this chapter, another prospective source of uncertainty has been identified in the surrogate based uncertainty propagation approaches besides the conventionally considered uncertainties and the effect of same has been analyzed through introducing simulated noise in the system. The effect of such simulated noise can be regarded as inclusion of other sources of uncertainty beside the conventionally considered stochastic material and geometric parameters, such as error in measurement of responses, error in modelling and computer simulation and various other epistemic uncertainties involved with the system. The kind of analysis presented in this chapter provides a thorough insight on the stochastic responses under investigation. The representative results are presented for laminated composite spherical shallow shell based on Kriging approach considering different levels of noise, wherein it is evident that the simulated noise has considerable effect on stochastic dynamics of the system. Consideration of the effect of such noise is thus an important criterion for robust and comprehensive analysis of stochastic systems. More complex forms of composite structures and various other global responses can be investigated in future following the efficient kriging based uncertainty propagation approach presented in this chapter.

CHAPTER 6

Effect of Operational Uncertainties on the Stochastic Dynamics of Composite Laminates

6.1 Introduction

The previous chapters in this book deal with various material and structural uncertainties in composite plates and shells. The effect of noise on the surrogate based uncertainty quantification algorithms is also addressed. The present chapter investigates the effect of rotational uncertainty under operating condition in the dynamic responses of composite shells. A response surface method based on the central composite design algorithm is used for the quantification of rotational and ply-level uncertainties. The stochastic eigenvalue problem is solved by using the QR iteration algorithm. An eight noded isoparametric quadratic element with five degrees of freedom at each node is considered in the finite element formulation. Sensitivity analysis is carried out to address the influence of different input parameters on the output natural frequencies. The sampling size and computational cost is reduced by employing the present surrogate based approach compared to direct Monte Carlo simulation. The stochastic mode shapes are also depicted for a typical laminate configuration.

Composite shell structures are extensively used in aerospace, marine, automobile and civil industries due to their high strength and stiffness to weight ratios. Numerical modeling of realistic composite structures is a demanding task. However, turbomachinary blades such as wind turbine blades can be, as a first approximation, idealized as shallow conical shells in order to simplify the numerical simulation process. The production of shell-like composite structures is always subjected to large variability due to manufacturing imperfection and operational factors. Composite structures are difficult to manufacture accurately according to the exact design specifications resulting in unavoidable uncertainties (material and structural variability). The rotational speed of turbomachinary blades itself might be uncertain, varying in a range around its working speeds due to fluctuation of payload during operation. Even the lower ranges of rotational speeds for wind turbine blades are always subjected to large variability due to uncertain air velocity. In reality wind turbine blades of do not rotate at constant speed rather they rotate about their axis at variable rotational speeds which may lead to structural damage of the blades. Hence it results in unavoidable dynamic instability of the system caused due to variations of rotational speeds coupled with other sources of uncertainties. Moreover, the variability in one input parameter may propagate and influence another and the final output quantity (say random natural frequency) of the system may have a significant cascading effect due to accumulation of risk. Therefore, it is essential to quantify the effect of such uncertainties in the response quantities of interest for composite structures. In this chapter a probabilistic approach of uncertainty quantification is presented based the Monte Carlo simulation, which is carried out using polynomial regression method to achieve computational efficiency. The polynomial regression model is developed by using mathematical and statistical methods based on central composite design (CCD) as a sampling tool (Montgomery 1991, Borkowski 1995). The basis of design and analysis of experiments (Myers and Montgomery 2002, Khuri and Mukhopadhyay 2010) have been investigated extensively in the last two decades. The CCD algorithm is found to be widely employed in various fields of engineering including the stochastic analysis of composites (Aslan 2008, Fang and Perera 2009, Zhu and Chen 2014, Dey et al. 2015f). In the present analysis, selective representative samples are drawn uniformly over the entire parameter domain using the CCD algorithm, ensuring good prediction capability of the constructed response surface model in the whole design space including the tail regions. The response surfaces of the output quantities of interest, formed on the basis of polynomial regression, are used for quantifying the rotational and ply-level uncertainties of the composite shells. A sensitivity analysis is carried out to investigate the relative importance of the individual input parameters by making use of the constructed response surface models. This chapter hereafter is organized as follows, Section 6.2: theoretical formulation for the dynamic analysis of composite shells including the effect of rotation; Section 6.3: formulation of response surface method using central composite design and the algorithm for response surface based stochastic dynamic analysis; Section 6.4: results and discussion; and Section 6.5: summary.

6.2 Governing equations for the dynamic analysis of composite shells including the effect of rotation

The shallow conical shell with length Lo, reference width b, thickness t, vertex angle ϕ_{ve} and base subtended angle of cone ϕ_o is depicted in Fig. 6.1. The component of

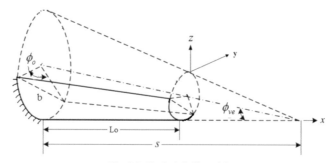

Fig. 6.1 Conical shell model.

radius of curvature in the chordwise direction $r_y(x,y)$ is a parameter varying both in the x- and y-directions. The variation in x-direction is linear. There is no curvature along the spanwise direction ($r_x = \infty$). The cantilever shell is clamped along x=0 with radius of twist r_{xy}. Thus a shallow conical shell of uniform thickness, made of laminated composite is considered. A shallow shell is characterized by its middle surface and is defined by the equation,

$$z = -0.5\,[\,(x^2/r_x) + (2xy/r_{xy}) + (y^2/r_y)\,] \tag{6.1}$$

The radius of twist (r_{xy}), length (Lo) of the shell and twist angle (Ψ) are expressed as (Liew et al. 1994)

$$r_{xy} = -\,Lo/\tan\psi \tag{6.2}$$

The constitutive equations for composite conical shell are given by

$$\{F\} = [D(\bar\omega)]\,\{\varepsilon\} \tag{6.3}$$

where   force resultant $\{F\} = \{N_x, N_y, N_{xy}, M_x, M_y, M_{xy}, Q_x, Q_y\}^{\mathrm{T}}$ (6.4)

$$\{F\} = \left[\int_{-t/2}^{t/2} \{\sigma_x,\quad \sigma_y,\quad \tau_{xy},\quad \sigma_x z,\quad \sigma_y z,\quad \tau_{xy} z,\quad \tau_{xz},\quad \tau_{yz}\}\, z\,dz\right]^{\mathrm{T}}$$

$$\{\varepsilon\} = \{\,\varepsilon_x, \varepsilon_y, \varepsilon_{xy}, k_x, k_y, k_{xy}, \gamma_{xz}, \gamma_{yz}\,\}^{\mathrm{T}}$$

$$\text{and}\ \ [D(\bar\omega)] = \begin{bmatrix} A_{ij}(\bar\omega) & B_{ij}(\bar\omega) & 0 \\ B_{ij}(\bar\omega) & D_{ij}(\bar\omega) & 0 \\ 0 & 0 & S_{ij}(\bar\omega) \end{bmatrix} \tag{6.5}$$

The elements of elastic stiffness matrix $[D(\bar{\omega})]$ is expressed as

$$[A_{ij}(\bar{\omega}), \quad B_{ij}(\bar{\omega}), \quad D_{ij}(\bar{\omega})] = \sum_{k=1}^{n} \int_{z_{k-1}}^{z_k} [\{\bar{Q}_{ij}(\bar{\omega})\}_{on}]_k \; [1, \; z, \; z^2] \, dz \qquad i, j = 1, 2, 6$$

(6.6)

$$[S_{ij}(\bar{\omega})] = \sum_{k=1}^{n} \int_{z_{k-1}}^{z_k} \alpha_s \; [\{\bar{Q}_{ij}(\bar{\omega})\}_{on}]_k \; dz \qquad i, j = 4, 5$$

where α_s is the shear correction factor and is assumed as 5/6. $[\bar{Q}_{ij}]$ is the off-axis elastic constant matrix which is given by

$$[\bar{Q}_{ij}(\bar{\omega})]_{off} = [T_1(\bar{\omega})]^{-1} \; [\bar{Q}_{ij}]_{on} \; [T_1(\bar{\omega})]^{-T} \qquad \text{for } i, j = 1,2,6$$

$$[\bar{Q}_{ij}(\bar{\omega})]_{off} = [T_2(\bar{\omega})]^{-1} \; [\bar{Q}_{ij}]_{on} \; [T_2(\bar{\omega})]^{-T} \qquad \text{for } i, j = 4,5$$

(6.7)

$$\text{where} \quad [T_1(\bar{\omega})] = \begin{bmatrix} m^2 & n^2 & 2mn \\ n^2 & m^2 & -2mn \\ -mn & mn & m^2 - n^2 \end{bmatrix} \quad \text{and}$$

(6.8)

$$[T_2(\bar{\omega})] = \begin{bmatrix} m & -n \\ n & m \end{bmatrix}$$

in which $m = Sin\theta(\bar{\omega})$ and $n = Cos\theta(\bar{\omega})$, wherein $\theta(\bar{\omega})$ is random fibre orientation angle.

$$[Q_{ij}(\bar{\omega})]_{on} = \begin{bmatrix} Q_{11} & Q_{12} & 0 \\ Q_{12} & Q_{12} & 0 \\ 0 & 0 & Q_{66} \end{bmatrix} \qquad \text{for } i, j = 1, 2, 6 \qquad (6.9)$$

$$[\bar{Q}_{ij}(\bar{\omega})]_{on} = \begin{bmatrix} Q_{44} & Q_{45} \\ Q_{45} & Q_{55} \end{bmatrix} \qquad \text{for } i, j = 4, 5 \qquad (6.10)$$

where

$$Q_{11} = \frac{E_1}{1 - \nu_{12}\nu_{21}} \qquad Q_{22} = \frac{E_2}{1 - \nu_{12}\nu_{21}} \quad \text{and} \quad Q_{12} = \frac{\nu_{12}E_2}{1 - \nu_{12}\nu_{21}}$$

$$Q_{66} = G_{12} \qquad Q_{44} = G_{23} \quad \text{and} \quad Q_{55} = G_{13}$$

The mass per unit area for conical shell is expressed as

$$P(\bar{\omega}) = \sum_{k=1}^{n} \int_{z_{k-1}}^{z_k} \rho(\bar{\omega}) \, dz \qquad (6.11)$$

Mass matrix is expressed as

$$[M(\overline{\omega})] = \int_{Vol} [N][P(\overline{\omega})][N]d(vol) \qquad (6.12)$$

The stiffness matrix is given by

$$[K(\overline{\omega})] = \int_{-1}^{1}\int_{-1}^{1}[B(\overline{\omega})]^{T} \quad [D(\overline{\omega})] \quad [B(\overline{\omega})] \, d\xi \, d\eta \qquad (6.13)$$

From Hamilton's principle (Meirovitch 1992), the dynamic equilibrium equation for rotational speeds is derived employing Lagrange's equation of motion and neglecting the Coriolis effect, the equation in global form is expressed as (Dey and Karmakar 2012a)

$$[M(\overline{\omega})]\{\ddot{\delta}_e\} \; + \; ([K_e(\overline{\omega})] \; + [K_{ce}(\overline{\omega})]\{\delta_e\} = \{F_e(\overline{\omega})\} + \{F_{ce}(\overline{\omega})\} \qquad (6.14)$$

After assembling all the element matrices and the force vectors with respect to the common global coordinates, the equation of motion of a free vibration system with n degrees of freedom can be expressed as

$$[M(\overline{\omega})][\ddot{\delta}] \; + [K(\overline{\omega}) \; + \; K_\sigma(\overline{\omega})]\{\delta\} = \{F\} + \{F(\Omega^2)\} \qquad (6.15)$$

where $\{F(\Omega^2)\}$ is the vector of nodal equivalent centrifugal forces $\{F\}$, is the global vector of externally applied force and $\{\delta\}$ is the global displacement vector. $[K_\sigma]$ depends on the initial stress distribution and is obtained by the iterative procedure upon solving

$$(K_e(\overline{\omega}) \; + \; K_\sigma(\overline{\omega}))\{\delta\} = \{F(\Omega^2)\} \qquad (6.16)$$

Angular velocity matrix components contributing towards acceleration vector is given as (Sreenivasamurthy and Ramamurti 1981)

$$[A_o(\overline{\omega})] \; = \; \begin{bmatrix} \{\sigma_y^2(\overline{\omega}) + \sigma_z^2(\overline{\omega})\} & -\sigma_x(\overline{\omega})\,\sigma_y(\overline{\omega}) & -\sigma_x(\overline{\omega})\,\sigma_z(\overline{\omega}) \\ -\sigma_x(\overline{\omega})\,\sigma_y(\overline{\omega}) & \{\sigma_x^2(\overline{\omega}) + \sigma_z^2(\overline{\omega})\} & -\sigma_y(\overline{\omega})\sigma_z(\overline{\omega}) \\ -\sigma_x(\overline{\omega})\,\sigma_z(\overline{\omega}) & -\sigma_y(\overline{\omega})\,\sigma_z\overline{\omega}) & \{\sigma_x^2(\overline{\omega}) + \sigma_y^2(\overline{\omega})\} \end{bmatrix} \qquad (6.17)$$

The element centrifugal force is given by

$$[F_{ce}(\overline{\omega})] = \rho \int_{V} [N]^{T}[A_o(\overline{\omega})]\begin{bmatrix} t_x + x \\ t_y + y \\ t_z + z \end{bmatrix} dV \qquad (6.18)$$

where $\{\rho\}$ is the mass density, $[N]$ stands for the shape function matrix and $\{t_x, t_y, t_z\}$ are the fixed translational offsets expressed with respect to the plate coordinate system. The element geometric stiffness matrix due to rotation is given by (Cook et al. 1989)

$$[K_{ce}(\overline{\omega}) = \int_V [G]^T [M_\sigma(\overline{\omega})][G]\,dV \qquad (6.19)$$

where the matrix [G] consists of derivatives of shape functions and $[M_\sigma(\overline{\omega})]$ is the matrix of initial in-plane stress resultants caused by rotation. Considering randomness of input parameters like ply-orientation angle, mass density, rotational speeds, the equation of motion of free vibration system with n degrees of freedom can be expressed as

$$[M(\overline{\omega})][\ddot{\delta}] + [K(\overline{\omega})]\{\delta\} = \{F_L\} \qquad (6.20)$$

In the above equation, $\mathbf{M}(\overline{\omega}) \in R^{n \times n}$ is the mass matrix, $[K(\overline{\omega})]$ is the stiffness matrix wherein $[K(\overline{\omega})] = [K_e(\overline{\omega})] + [K_{\sigma e}(\overline{\omega})]$ in which $\mathbf{K}_e(\overline{\omega}) \in R^{n \times n}$ is the elastic stiffness matrix, $\mathbf{K}_{\sigma e}(\overline{\omega}) \in R^{n \times n}$ is the geometric stiffness matrix (depends on initial stress distribution) while $\{\delta\} \in R^n$ is the vector of generalized coordinates and $\{F_L\} \in R^n$ is the force vector. The governing equations are derived based on Mindlin's theory incorporating rotary inertia, transverse shear deformation. For free vibration, the random natural frequencies $[\omega_n(\overline{\omega})]$ are determined from the standard eigenvalue problem (Bathe 1990) which is represented below and is solved by the QR iteration algorithm,

$$[A(\overline{\omega})]\{\delta\} = \lambda(\overline{\omega})\{\delta\} \qquad (6.21)$$

where $[A(\overline{\omega})] = ([K_e(\overline{\omega})] + [K_{\sigma e}(\overline{\omega})])^{-1} [M(\overline{\omega})]$

$$\lambda(\overline{\omega}) = \frac{1}{\{\omega_n(\overline{\omega})\}^2} \qquad (6.22)$$

6.3 Surrogate model: central composite design

A central composite design is an experimental design, used in response surface methodology to construct a second order (quadratic) model for the response variable without three-level factorial experiment. The statistical measure of goodness of a model is obtained by least squares regression analysis for the minimum generalized variance of the estimates of the model coefficients. CCD is employed to provide a mathematical and statistical approach in portraying the input-output mapping by construction of meta-model by using the algorithmically obtained sample set (design points). Considering the problem of estimating the coefficients of a linear approximation model by least squares regression analysis

$$y = \beta_o + \sum_{i=1}^k \beta_i x_i + \varepsilon \qquad (6.23)$$

In matrix form the above equation can be expressed as

$$Y = X\beta + \varepsilon \qquad (6.24)$$

where 'Y' is a vector of observations of sample size, 'ε' is the vector of errors having normal distribution with zero mean, 'X' is the design matrix and 'β' is a vector of unknown model coefficients (β_0 and β_i). The design matrix is a set of combinations of the values of the coded variables, which specifies the settings of the design parameters to be performed during data observation. The representative two factor central composite design is furnished in Fig. 6.2. In CCD, second order design is used consisting of the following three portions for obtaining design points: a complete (or a fraction of) 2^k factorial design coded as ± 1 (corresponding to the lower and upper value bound of the design space) consisting of 2^k design points, $2k$ axial points coded as $\pm \alpha$ ($\alpha \geq 1$) and n_0 centre points as shown in Fig. 6.2. Here 'k' is the number of input variables. Thus the total number of design points in CCD model, $n = 2^k + 2k + n_0$ where $n_0 = 1$ for present numerical study. CCD possesses the following properties according to the chosen values of α and n_0:

a) Rotatable (used for up to 5 factors to creates a design with standard error of predictions equal at points equidistant from the centre of the design).
b) Face-centred (the axial points into the faces of the cube at ± 1 levels to produce a design with each factor having 3 levels).
c) Spherical (all factorial and axial points on the surface of a sphere of radius equals to square root of the number of factors).
d) Orthogonal quadratic (α values allowing the quadratic terms to be independently estimated from the other terms).
e) Practical (used for 6 or more factors wherein α value is the fourth root of the number of factors).

If the number of input parameters (k) is more than 5, the 2^k design requires a large number of design points which is encountered by employing either one-half fraction design (consisting of one-half the number of points) or one-fourth fraction design (consisting of one-fourth the number of points). In general, a 2^{-m} th fraction of a 2^k design consists of 2^{k-m} points from a full 2^k design wherein m is chosen in such a way that $2^{k-m} \geq$ number of unknowns in the response surface equation.

In this study, the constructed CCD models provide an approximate meta-model equation which relates the input random parameters 'x_i' (say ply orientation angle of each layer of laminate, rotational speed) and output 'y' (say natural frequency)

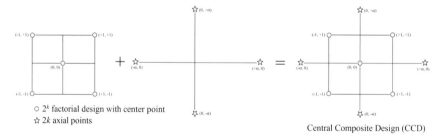

Fig. 6.2 Central composite design with two factors.

for a particular system. The CCD model developed is actually an approximate mathematical and statistical model representing a certain inherent property of a physical system and it maps the input parameters 'x_i' to the corresponding output responses 'y' by an explicit function

$$y = f(x_1, x_2, x_3, x_4, \ldots\ldots x_i \ldots\ldots x_k) + \varepsilon \qquad (6.25)$$

where 'f' denotes the approximate response function, 'ε' is the statistical error term having a normal distribution with null mean value. The input variables are usually coded as dimensionless variables with zero as mean value and a standard deviation of 'x_i'. The first order and second order polynomials are expressed as

First order model (interaction), $y = \beta_o + \sum_{i=1}^{k} \beta_i x_i + \sum_{i=1}^{k}\sum_{j>i}^{k} \beta_{ij} x_i x_j + \varepsilon$
$$(6.26)$$

Second order model,

$$y = \beta_o + \sum_{i=1}^{k} \beta_i x_i + \sum_{i=1}^{k}\sum_{j>i}^{k} \beta_{ij} x_i x_j + \sum_{i=1}^{k} \beta_{ii} x_i^2 + \varepsilon \qquad (6.27)$$

The surrogate model is used to fit approximately for a set of points in the design space using a multiple regression fitting scheme. While forming CCD models, insignificant input features are screened out and not considered in the final meta-model. A quantitative evaluation for effect of each parameter on the total model variance is carried out using analysis of variance (ANOVA) method according to its F-test value

$$F_p = \frac{n-k-1}{k}\left(\frac{SS_R}{SS_E}\right) \qquad (6.28)$$

where F_p denotes the *F*-test value of any input parameter 'p' while n, SS_E and SS_R are the number of samples used in the design procedure, sum of squares due to the model and the residual error, respectively. If F_p exceeds the selected criterion value, the input parameter 'p' is considered to be significantly influencial factor with respect to the chosen output feature. The percentage contribution of each input parameter (including the contribution of the interaction terms) to the total model variance is obtained for each input random variables for each layer of laminated composite conical shells. An optimized surrogate model is formed by adding or deleting input factors through backward elimination, forward addition or stepwise elimination or addition. It involves the calculation of the P-value (probability value, gives the risk of falsely rejecting a given hypothesis) and *Prob.* > *F* value (gives the proportion of time one would expect to get the stated *F*-value if no factor effects are significant). The meta-model constructed is checked by two basic criteria such as coefficient of determination or R^2 term (measure of the amount of variation around the mean explained by the model) and R^2_{adj} term (measure of the amount of variation with respect to mean value explained by the model, adjusted for the number of terms in the model). The adjusted R^2 decreases as the number of terms in the model increases if those additional terms don't add value to the model) and R^2_{pred} (measure of the prediction capability of the response surface model) expressed as follows.

$$R^2 = \left(\frac{SS_R}{SS_T} \right) = 1 - \left(\frac{SS_E}{SS_T} \right) \quad \text{where } 0 \le R^2 \le 1 \qquad (6.29)$$

$$R^2_{adj} = 1 - \frac{\left(\dfrac{SS_E}{n-k-1} \right)}{\left(\dfrac{SS_T}{n-1} \right)} = 1 - \left(\frac{n-1}{n-k-1} \right)\left(1 - R^2 \right) \quad \text{where } 0 \le R^2_{adj} \le 1 \qquad (6.30)$$

$$R^2_{pred} = 1 - \left(\frac{PRESS}{SS_T} \right) \quad \text{where } 0 \le R^2_{pred} \le 1 \qquad (6.31)$$

where $SS_T = SS_E + SS_R$ is the total sum of square and *PRESS* (predicted residual error sum of squares) measures the goodness of fit of the model corresponding to chosen samples in the design space. The stochasticity of ply-orientation angle in each layer of laminate and rotational speeds as input parameters are considered for composite cantilever conical shells. Figure 6.3 presents the flowchart of stochastic

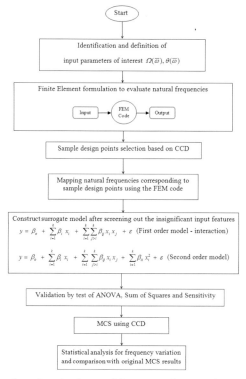

Fig. 6.3 Flowchart of stochastic natural frequency using central composite design.

natural frequency by using central composite design. In this study, frequency domain feature (first three natural frequencies) is considered as output. It is assumed that the distribution of randomness of input parameters exists within a certain band of tolerance with their central deterministic mean values. The cases wherein the random variables considered in each layer of laminate are investigated for:

a) Variation of ply-orientation angle only:
$$\theta(\bar{\omega}) = \{\theta_1\ \theta_2\ \theta_3.........\theta_i......\theta_l\}$$

b) Variation of mass density only: $\Omega(\bar{\omega})$

c) Combined variation of ply orientation angle and rotational speeds:
$$g\{\theta(\bar{\omega}),\Omega(\bar{\omega})\} = \{\ \Phi_1(\theta_1..............\theta_l), \Phi_2(\Omega)\}$$

where θ_i and Ω_i are the ply orientation angle and rotational speed, respectively and '*l*' denotes the number of layer in the laminate. In this study, $\pm 5°$ for ply orientation angle with subsequent $\pm 10\%$ tolerance for rotational speeds respectively from deterministic mean value are considered.

6.4 Results and discussion

This study considers an eight layered graphite–epoxy symmetric angle-ply [$(\theta°/–\theta°/\theta°/–\theta°)$s] composite cantilever shallow conical shells with a square plan-form (L/b$_0$ = 1), curvature ratio (b$_0$/R$_y$) of 0.5 and thickness ratio of 1000. Material properties of graphite–epoxy composite (Quatu and Leissa 1991) is considered with mean value as E_1 = 138 GPa, E_2 = 8.96 GPa, v = 0.3, G_{12} = G_{13} = 7.1 GPa, G_{23} = 2.84 GPa, ρ = 3202 kg/m^3. A typical discretization of (6×6) mesh on plan area with 36 elements, 133 nodes with natural coordinates of an isoparametric quadratic plate bending element are considered for the present FEM. For full scale MCS, number of original FE analysis is same as the sampling size. In general for complex composite structures, the performance function is not available as an explicit function of the random design variables. The random response in terms of natural frequencies of the composite structure can only be evaluated numerically at the end of a structural analysis procedure such as the finite element method which is often time-consuming. The present CCD is employed to find a predictive and representative meta-model equation by using one-half fraction design. The meta-models are used to determine the first three natural frequencies corresponding to given values of input variables, instead of time-consuming deterministic FE analysis. The response surface thus represents the result (or output) of the structural analysis encompassing (in theory) every reasonable combination of all input variables. Due to paucity of space, only a few important representative results are furnished.

6.4.1 Validation

The present computer code is validated with the results available in the open literature. Table 6.1 presents the non-dimensional fundamental frequencies of graphite–epoxy composite twisted plates with different ply-orientation angle. The convergence study is carried out to determine the natural frequencies for composite conical shells as furnished in Table 6.2. Table 6.3 presents the non-dimensional fundamental frequencies of graphite–epoxy composite rotating cantilever plate. Convergence studies are performed using uniform mesh division of (6 x 6) and (8 x 8) and the results are found to be nearly equal, with the difference being around 1% and the results also corroborate monotonic downward convergence. The differences between the results by Liew et al. (1994) and the present FEM approach can be attributed to consideration of rotary inertia and transverse shear deformation in the present FEM approach and also to the fact that Ritz method overestimates the structural stiffness. Moreover, increasing the size of matrix because of higher mesh size increases the ill-conditioning of the numerical eigenvalue problem. Hence, the lower mesh size (6 x 6) is employed in the present analysis due to computational efficiency. The comparative study depicts an excellent agreement with the previously published results and hence it demonstrates the capability of the computer codes developed and insures the accuracy of analyses.

In CCD, a representative sample size of 150 is considered for each layer of individual variation of ply-orientation angle and rotational speeds, respectively. Due to increase in number of input variables considered for combined random variation of ply-orientation angle and rotational speeds, the subsequent sample size of 280 is adopted to meet the convergence criteria. Figure 6.4 depicts a representative

Table 6.1 Non-dimensional fundamental natural frequencies [$\omega = \omega_n L^2 \sqrt{(\rho/E_1 t^2)}$] of three layered [$\theta, -\theta, \theta$] graphite–epoxy twisted plates, L/b = 1, b/t = 20, $\psi = 30°$.

Fibre Orientation Angle, θ (deterministic)	Present FEM (deterministic mean)	Qatu and Leissa (1991)
15°	0.8618	0.8759
30°	0.6790	0.6923
45°	0.4732	0.4831
60°	0.3234	0.3283

Table 6.2 Non-dimensional fundamental frequencies [$\omega = \omega_n b^2 \sqrt{(\rho t/D)}$, $D = E t^3/12(1-\nu^2)$] for the untwisted shallow conical shells with $\nu = 0.3$, s/t = 1000, $\phi_o = 30°$, $\phi_{ve} = 15°$.

Aspect ratio (L/s)	Present FEM (8 x 8) (deterministic mean)	Present FEM (6 x 6) (deterministic mean)	Liew et al. (1994)
0.6	0.3524	0.3552	0.3599
0.7	0.2991	0.3013	0.3060
0.8	0.2715	0.2741	0.2783

Table 6.3 Non-dimensional fundamental frequencies [$\omega = \omega_n L^2 \sqrt{(\rho t/D)}$] of graphite-epoxy composite rotating cantilever plate, L/b$_o$ = 1, t/L = 0.12, $D = E t^3/\{12(1-\nu^2)\}$, $\nu = 0.3$

Ω	Present FEM	Sreenivasamurthy and Ramamurti (1981)
0.0	3.4174	3.4368
1.0	4.9549	5.0916

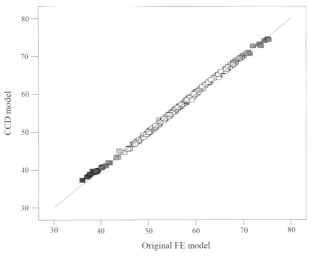

Fig. 6.4 Central composite design (CCD) model with respect to Original FE model of fundamental natural frequencies for combined variation of rotational speed and ply-orientation angle of angle-ply [(45°/–45°/45°/–45°)s] composite cantilever conical shells, considering $E_1 = 138$ GPa, $E_2 = 8.9$ GPa, $G_{12} = G_{13} = 7.1$ GPa, $G_{23} = 2.84$ GPa, $\rho = 3202$ kg/m³, $t = 0.003$ m, $\nu = 0.3$, Lo/s=0.7, $\phi_o = 45°$, $\phi_{ve} = 20°$.

plot describing the relationship between the original finite element model and the constructed CCD meta-model for fundamental natural frequencies signifying the accuracy of the present meta-model. Figure 6.5 illustrates the comparison of the probability density functions (PDF) for both original MCS and CCD using a sample size of 10,000 corresponding to first three natural frequencies considering individual as well as combined variation of ply orientation angle, rotational speed. The low scatterness of the points found around the diagonal line in Fig. 6.4 and the low deviation obtained between the pdf estimations of original MCS and CCD responses in Fig. 6.5 corroborates the fact that CCD meta-models are formed accurately. These two plots are checked and are found in good agreement ensuring the efficiency and accuracy of the present constructed CCD. While evaluating the statistics of responses through full scale MCS, computational time is exorbitently high because it involves number of repeated FE analysis. However, in the present method, MCS is conducted in conjunction with CCD model. Here, although the same sampling size as in direct MCS (with sample size of 10,000) is considered, the number of FE analysis is much less compared to original MCS and is equal to number of representative sample required to construct the CCD meta-model. The representative CCD equation is formed on which the full sample size of direct MCS is conducted. Hence, the computational time and effort expressed in terms of FE calculation is reduced compared to full scale direct MCS. Hence, in order to save computational time, the present constructed CCD methodology is employed instead of traditional Monte Carlo simulation. This provides an efficient affordable way for simulating the uncertainties in natural frequency. The sensitivity of a given material or geometric property to each random variable is also quantified in the

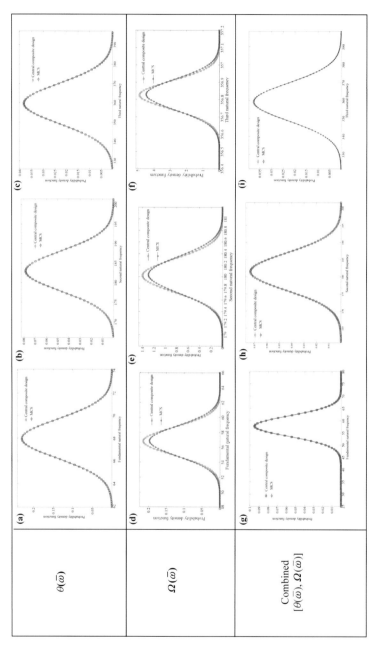

Fig. 6.5 Probability density function obtained by original MCS and Central composite design (CCD) with respect to first three natural frequencies (Hz) indicating for variation of only ply orientation angle [$\theta(\bar{\omega})$], only rotational speeds [$\Omega(\bar{\omega})$] and combined [$\theta(\bar{\omega})$, $\Omega(\bar{\omega})$] for graphite–epoxy angle-ply [($\theta°/-\theta°/\theta°/-\theta°$)s] composite conical shells, considering sample size = 10,000, E_1 = 138 GPa, E_2 = 8.9 GPa, G_{12} = G_{13} = 7.1 GPa, G_{23} = 2.84 GPa, ρ = 3202 kg/m³, θ = 45° (±5° variation), Ω = 100 rpm (±10% variation), t = 0.003 m, v = 0.3, Lo/s = 0.7, ϕ_o = 45°, ϕ_{ve} = 20°.

present meta-model context. The representative probability density function for angle-ply laminate with respect to stochastic input parameters are compared with the results predicted by a Monte Carlo simulation (10,000 samples) wherein a good agreement is observed as furnished in Fig. 6.5. In the present analysis, the values of R^2, R^2_{adj} and R^2_{pred} are found to be close to the one ensuring the best fit. The difference between R^2_{adj} and R^2_{pred} is found less than 0.2 which indicates that the model can be used for further prediction. In addition to above, another check is carried out namely, adequate precision which compares the range of the predicted values at the design points to the average prediction error. For all cases of the present CCD meta-model, its value is consistently found greater than four which indicates the present model is adequate to navigate the design space. The computational time required in the present CCD approach is observed to be around (1/67) times (for individual variation of inputs) and (1/35) times (for combined variation of inputs) of direct Monte Carlo simulation.

6.4.2 Statistical analysis

The variation of rotational speeds $[\Omega(\bar{\omega})]$ ranging from 25 rpm to 125 rpm in step of 25 rpm are scaled randomly in the range having the lower and the upper limit as ±10% variability (as per industry standard manufacturing tolerance) with respective mean values, while for ply orientation angle $[\theta(\bar{\omega})]$ ranging from 0° to 90° in step of 15° in each layer of the composite laminate the bound is considered as within ±5° fluctuation (as per industry standard manufacturing tolerance) with respective mean deterministic values. The CCD meta-models are formed to generate first three natural frequencies for graphite–epoxy composite cantilever conical shells. The natural frequencies of tested angle-ply laminate are found to be reduced as the ply orientation angle increases irrespective of rotational speeds. Table 6.4 indicates the maximum values, minimum values, mean values and standard deviation for first three natural frequencies obtained due to individual and combined stochasticity in ply-orientation angle and rotational speeds for eight layered graphite–epoxy angle-ply composite conical shells wherein the results obtained using original MCS and CCD are observed to be in good agreeement. Due to the cascading effect of variability resulting from combined stochasticity considered in nine input parameters in each layer, the bandwidth of variation of natural frequency is found to be higher than the stochasticity considered for variation of any individual input parameter. Considering the individual stochasticity in ply orientation angle $[\theta(\bar{\omega})]$ and rotational speed $[\Omega(\bar{\omega})]$, a parametric study is carried out for maximum values, minimum values, means and standard deviations of first three natural frequencies as furnished in Table 6.5 and Table 6.6, respectively. In general, the stochastic natural frequencies and its volatility are found to decrease with increase of fiber orientation angle and rotational speeds.

Figure 6.6 presents the sensitivity (in percentage) of natural frequenices with rescpect to the ply orientation angle for eight layered graphite–epoxy angle-ply composite conical shells. Based on the sensitivity analysis using CCD the significant

Table 6.4 Comparative study between MCS (10,000 samples) and central composite design results for maximum values, minimum values and percentage of deviation for first three natural frequencies (Hz) obtained for four layered graphite-epoxy angle-ply [(θ°/−θ°/θ°/−θ°)s] composite conical shells considering $E_1 = 138$ GPa, $E_2 = 8.9$ GPa, $G_{12} = G_{13} = 7.1$ GPa, $G_{23} = 2.84$ GPa, $\rho = 3202$ kg/m³, $t = 0.003$ m, $v = 0.3$, $\theta = 45°$ (± 5° variation), $\Omega = 100$ rpm (± 10% variation), $Lo/s = 0.7$, $\phi_o = 45°$, $\phi_{ve} = 20°$.

Parameter	Type of analysis	$[\theta(\overline{\omega})]$			$[\Omega(\overline{\omega})]$			$[\theta(\overline{\omega})].[\Omega(\overline{\omega})]$		
		FNF	SNF	TNF	FNF	SNF	TNF	FNF	SNF	TNF
Max value	MCS	73.9836	201.0528	396.413	59.92618	180.5032	356.9618	69.34976	201.2482	397.9168
	CCD	73.9588	201.003	396.294	59.62804	180.442	356.9445	68.2373	200.4119	396.5191
	Deviation (%)	0.033	0.025	0.030	0.498	0.034	0.005	1.604	0.416	0.351
Min value	MCS	62.0268	166.3178	326.3826	52.83288	179.4259	356.6496	32.09585	163.0662	326.0832
	CCD	62.0353	166.3642	326.4586	53.14518	179.4583	356.663	32.41708	161.1799	322.6536
	Deviation (%)	−0.014	−0.028	−0.023	−0.591	−0.018	−0.004	−1.001	1.157	1.052
Mean value	MCS	67.9652	182.9898	359.6682	56.86125	180.0369	356.8293	57.79991	181.5352	359.3835
	CCD	67.9545	182.9665	359.6045	56.82357	180.0214	356.826	57.74899	181.4775	359.2861
	Deviation (%)	0.016	0.013	0.018	0.066	0.009	0.001	0.088	0.032	0.027
Standard Deviation	MCS	1.7401	5.0895	10.4315	2.0162	0.3067	0.0888	4.206369	5.435023	10.6258
	CCD	1.7386	5.0742	10.3890	1.8555	0.2843	0.0814	4.178373	5.452521	10.54578
	Deviation (%)	0.089	0.299	0.407	7.973	7.292	8.231	0.666	−0.322	0.753

Table 6.5 Maximum value, minimum value, mean value and standard deviation (SD) of first three natural frequencies (Hz) obtained by central composite design method (10,000 samples) due to individual stochasticity of ply-orientation angle in each layer for graphite-epoxy angle-ply $[(\theta°/{-}\theta°/\theta°/{-}\theta°)s]$ ($\pm\,5°$ variation) composite conical shells, considering $E_1 = 138$ GPa, $E_2 = 8.9$ GPa, $G_{12} = G_{13} = 7.1$ GPa, $G_{23} = 2.84$ GPa, $\rho = 3202$ kg/m³, $t = 0.003$ m, $v = 0.3$, $Lo/s = 0.7$, $\phi_o = 45°$, $\phi_{ve} = 20°$.

$\theta(\bar{\omega})$	Fundamental natural frequency (FNF)				Second natural frequency (SNF)				Third natural frequency (TNF)			
	Max	Min	Mean	SD	Max	Min	Mean	SD	Max	Min	Mean	SD
0°	134.841	120.913	128.9257	2.2486	399.0981	366.6211	389.0579	5.2026	766.6234	706.74	750.1372	9.7164
15°	125.885	116.052	121.2905	1.5321	367.722	333.234	351.1758	6.2208	711.5098	642.9308	678.387	12.4653
30°	104.396	90.0987	97.2914	2.1967	290.5747	243.7575	267.4248	7.2909	568.6029	476.5848	523.2368	14.4622
45°	73.9588	62.0353	67.95457	1.7386	201.003	166.3642	182.9665	5.0742	396.294	326.4586	359.6045	10.3890
60°	53.1915	47.4424	50.3265	0.9014	142.902	127.1338	134.8277	2.4918	280.4437	248.5910	263.7762	5.1014
75°	43.6186	41.2282	42.3783	0.3816	117.1097	111.0909	113.8941	0.9490	228.7734	216.8533	222.3898	1.8849
90°	40.3862	39.8526	40.0995	0.0874	108.9529	107.9854	108.3816	0.1629	212.6769	210.8513	211.5691	0.3058

Table 6.6 Central composite design results (10,000 samples) for first three natural frequencies (Hz) due to stochasticity of rotational speeds $[\Omega(\bar{\omega})]$ (in rpm with $\pm 10\%$ variation) for angle-ply $[(45°/-45°/45°/-45°)s]$ conical shells considering $E_1 = 138$ GPa, $E_2 = 8.9$ GPa, $G_{12} = G_{13} = 7.1$ GPa, $G_{23} = 2.84$ GPa, $\rho = 3202$ kg/m^3, $t = 0.003$ m, $v = 0.3$, $Lo/s = 0.7$, $\phi_o = 45°$, $\phi_{ve} = 20°$.

$[\Omega(\bar{\omega})]$	Fundamental natural frequency				Second natural frequency				Third natural frequency			
	Max	Min	Mean	SD	Max	Min	Mean	SD	Max	Min	Mean	SD
0	-	-	67.4010 (determ-inistic)	-	-	-	181.5862 (deterministic)	-	-	-	357.2181 (determ-inistic)	-
25	67.2203	66.5726	66.9462	0.1887	181.5603	181.4726	181.5238	0.0255	357.2043	357.1882	357.1963	0.0046
50	66.2054	64.7281	65.5373	0.4263	181.4266	181.2223	181.3356	0.0591	357.1790	357.1367	357.1592	0.0122
75	64.0176	61.1246	62.7071	0.8332	181.1034	180.6880	180.9182	0.1200	357.1081	357.0063	357.0626	0.0294
100	59.6280	53.1451	56.8235	1.8555	180.4420	179.4583	180.0214	0.2844	356.9445	356.6630	356.8260	0.0815
125	52.6861	39.8621	46.2973	3.6973	178.4963	177.1830	177.6214	0.3919	356.3245	355.9771	356.0645	0.0969

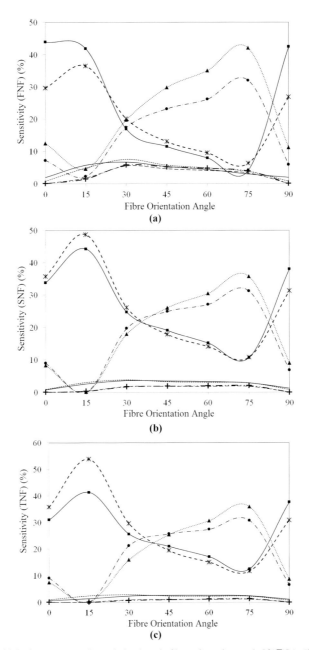

Fig. 6.6 Sensitivity in percentage for variation in only fibre orientation angle $[\theta(\bar{\omega})]$ ($\pm 5°$ variation) for eight layered graphite-epoxy angle-ply $[(\theta°/-\theta°/\theta°/-\theta°)s]$ composite conical shells, considering $E_1 = 138$ GPa, $E_2 = 8.9$ GPa, $G_{12} = G_{13} = 7.1$ GPa, $G_{23} = 2.84$ GPa, $\rho = 3202$ kg/m³, $t = 0.003$ m, $v = 0.3$, $Lo/s = 0.7$, $\phi_o = 45°$, $\phi_{ve} = 20°$ (FNF – fundamental natural frequency, SNF – second natural frequency and TNF – Third natural frequency).

input parameter are screened. The effect of individual stochasticity of ply orientation angle of each and individual layer separately for first three natural frequencies of angle-ply [(θ°/–θ°/θ°/–θ°)s] composite conical shells, wherein the sensitivity of two outermost layers (i.e., layer 1 and layer 8) are found to be maximum for θ = 0°, 15°, 30° and 90° while immediate underneath layers (i.e., layer 2 and layer 7) of the two outermost layers are observed to be maximum sensitive for θ = 45°, 60° and 75°. This is due the fact that only bending modes are predominant which results in maximum strains at outer layers. This consequently make the sensitively of natural frequencies with respect to fiber angles more sensitive at outmost layers. A complementary effect in sensitivity is identified between ultimate and penultimate outer layers corresponding to all ply orientation angles. In contrast, the two middle layers (i.e., layer 4 and layer 5) are found to be least sensitive for first three natural frequencies considering individual stochasticity of ply-orientation angle. Hence the variation in elastic stiffness due to randomness of ply orientation angle predominetly influence the first three natural frequencies. The relative sensitivity (ratio of sensitivity at a particular speed and total sensitivity) and relative frequencies (ratio of deterministic mean of rotating frequency to stationary frequency) with respect to rotational speeds for first three natural frequencies of eight layered graphite–epoxy angle-ply [(45°/–45°/45°/–45°)s] composite conical shells are furnished in Fig. 6.7 and Fig. 6.8 respectively. The relative sensitivity of rotational speeds for the first three naural frequencies are observed to increase with the increase of rotational speeds. A threshold rotational speed is observed below which the relative sensitivity of rotational speeds is found to decrease with increase in number of modes while the reservse trend is identified for the same

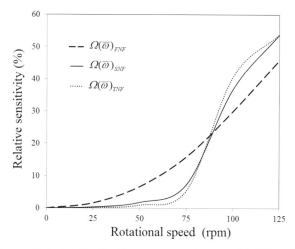

Fig. 6.7 Relative sensitivity (rotational) with respect to rotational speed (±10% variation) for eight layered graphite–epoxy angle-ply [(45°/–45°/45°/–45°)s] composite conical shells, considering $E_1 = 138$ GPa, $E_2 = 8.9$ GPa, $G_{12} = G_{13} = 7.1$ GPa, $G_{23} = 2.84$ GPa, $\rho = 3202$ kg/m³, $t = 0.003$ m, $v = 0.3$, Lo/s = 0.7, $\phi_o = 45°$, $\phi_{ve} = 20°$ (FNF – fundamental natural frequency, SNF – second natural frequency and TNF – Third natural frequency).

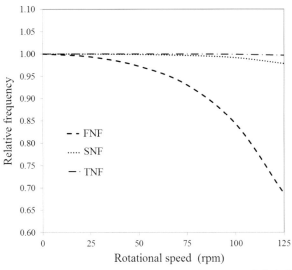

Fig. 6.8 Relative frequency with respect to rotational speed ($\pm 10\%$ variation) for eight layered graphite-epoxy angle-ply [(45°/–45°/45°/–45°)s] composite conical shells, considering $E_1 = 138$ GPa, $E_2 = 8.9$ GPa, $G_{12}=G_{13} = 7.1$ GPa, $G_{23} = 2.84$ GPa, $\rho = 3202$ kg/m³, $t = 0.003$ m, $v=0.3$, $Lo/s = 0.7$, $\phi_o = 45°$, $\phi_{ve} = 20°$ (FNF – fundamental natural frequency, SNF – second natural frequency and TNF – third natural frequency).

beyond that threshold rotational speed. This is mainly due to the fact that before the threshold speed, the contribution of the first mode is more than other modes. However, the second and third modes contribution becomes more significant after the threshold speed. The relative frequencies are found to decrease with increase of rotational speeds. As the number of modes increases the rate of decrease of relative frequencies are observed to decrease with increase of rotational speeds. The trend of random fluctuation of rotational speeds are portrayed by probability the density function corresponding to first three natural frequencies as furnished in Fig. 6.9. As the rotational speed increases the stochastic mean of first three natural frequencies are found to decrease whereas the standard deviation of the same are observed to increase with increase of rotational speed. This can be attributed as the coupling effect between geometric stiffness due to rotation and elastic stiffness. Figure 6.10 presents the representative difference in deterministic (stationary) and stochastic (rotational) modeshapes considering individual stochasticity in rotational speed wherein the spanwise bending is predominently observed for first three modes. The significant effect in modeshapes for variation of rotational speeds may occur during real-time operation of such composite structures which is predominantly caused due to its random change in geometric stiffness in addition to its elastic stiffness. The first three natural modes shows the basic bending modeshapes for deterministic plot without considering any rotation (i.e., stationary case) while stochastic modeshapes with random spanwise bending modes are obtained for random variation of rotational speeds. It is also to be noted that in the first mode, a

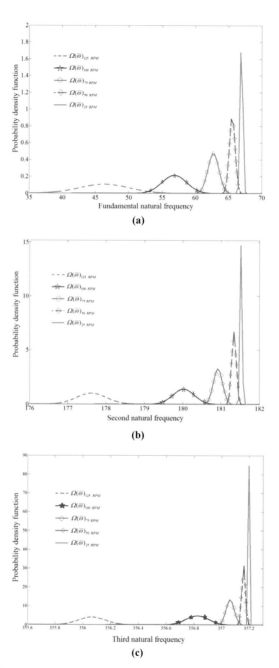

Fig. 6.9 Probability density function with respect to first three natural frequencies due to only variation in rotational speeds ($\Omega(\bar{\omega})$ = 25, 50,75,100,125 rpm with ± 10% variation) for angle-ply [(45°/–45°/45°/–45°)s] composite conical shells considering E_1 = 138 GPa, E_2 = 8.9 GPa, G_{12} = G_{13} = 7.1 GPa, G_{23} = 2.84 GPa, ρ = 3202 kg/m³, t = 0.003 m, v = 0.3, Lo/s = 0.7, ϕ_o = 45°, ϕ_{ve} = 20°.

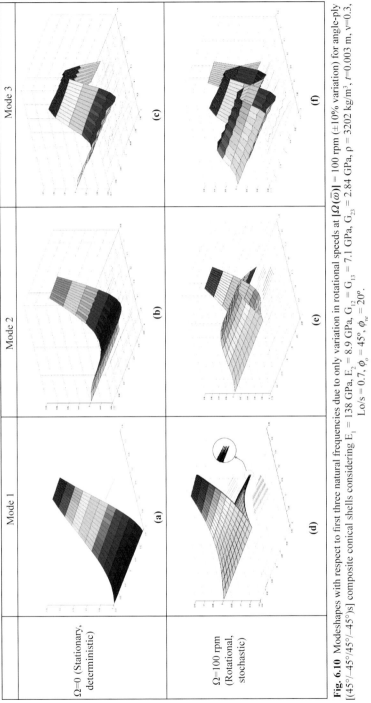

Fig. 6.10 Modeshapes with respect to first three natural frequencies due to only variation in rotational speeds at $[\Omega(\bar{\omega})] = 100$ rpm ($\pm 10\%$ variation) for angle-ply $[(45°/-45°/45°/-45°)s]$ composite conical shells considering $E_1 = 138$ GPa, $E_2 = 8.9$ GPa, $G_{12} = G_{13} = 7.1$ GPa, $G_{23} = 2.84$ GPa, $\rho = 3202$ kg/m^3, $r = 0.003$ m, $v = 0.3$, Lo/s $= 0.7$, $\phi_o^2 = 45°$, $\phi_{ve} = 20°$.

smooth surface is obtained while in case of the second and third modes, the wrinkles show on the represented surface. These undulations on the surfaces of second and third mode shapes are caused due to consideration of lower mesh size in the present study (6 × 6 mesh). This problem can be mitigated by using a higher mesh size which will lead to higher cost of computation. The present stochastic analysis is carried out with 6 × 6 mesh size to achieve computational efficiency and accuracy.

6.5 Summary

This chapter investigates the individual and combined effect of ply level and rotational uncertainty on the first three natural frequencies of composite cantilever conical shells. The number of actual finite element simulation is reduced in the present analysis using central composite design compared to direct Monte Carlo simulation and is equal to the number of samples required to construct the CCD model. The results obtained employing CCD are compared with the results of the direct Monte Carlo Simulation method and are found to establish the accuracy and computational efficiency. The computational time required in the present CCD approach is observed to be around (1/67) times (for individual variation of inputs) and (1/35) times (for combined variation of inputs) of direct Monte Carlo simulation. The first three natural frequencies are found to reduce as the ply orientation angle increases for a particular rotational speed. The ply orientation angles of ultimate and penultimate outer layers are complementarily found to be most sensitive while the middle layers are observed to be least sensitive for first three natural frequencies. The stochastic means of first three natural frequencies are found to decrease as the rotational speed increases while the standard deviations of the same are observed to increase with increase of rotational speed. The relative sensitivity of rotational speeds for first three naural frequencies are observed to increase with the increase of rotational speeds. The relative frequencies are found to decrease with increase of rotational speeds. As the number of modes increases the rate of decrease of relative frequencies are observed to decrease with increase of rotational speeds. The spanwise bending mode is observed to be predominent for first three modes.

CHAPTER 7

Effect of Environmental Uncertainties on the Free Vibration Analysis of Composite Laminates

7.1 Introduction

Various forms of material and geometric uncertainties for composite structures have been discussed in the previous chapters including the effect of operational conditions such as rotational uncertainty. The present chapter deals with the uncertainty caused by inevitable random variation of environmental factors such as temperature. The propagation of thermal uncertainty in composite structures has significant computational challenges. This chapter presents the thermal, ply-level and material uncertainty propagation in frequency responses of laminated composite plates by employing a surrogate model which is capable of dealing with both correlated and uncorrelated input parameters. In the present generalized high dimensional model representation (GHDMR) based approach, diffeomorphic modulation under observable response preserving homotopy (D-MORPH) regression is utilized to ensure the hierarchical orthogonality of high dimensional model representation component functions. The stochastic range of thermal field includes elevated temperatures up to 375 K and sub-zero temperatures up to cryogenic range of 125 K.

Composite structures may be exposed to variation in environmental (hot or cold) conditions during the service life such as an aircraft wing made of composite materials experiences a wide range of temperature variation from take-off to level-flight depending on the altitude of flight. The stochasticity in frequency responses due to thermal effects on composites is an important criterion for the overall design. In specific applications, the thermal effect acts as a fundamental factor for design consideration. These changes of thermal environments produce uncertain responses in the laminated composite structures. Hence the thermal condition has significant effect on the frequency characteristics and overall dynamic performance of composites. The natural frequency of composite structure under different thermo-mechanical loading conditions relies on its system parameters. The uncertainties of input parameters such as temperature, ply orientation angle and material properties lead to uncertainties in the natural frequency of the composite structures. The free vibration of laminated plates as an effect of environment has been considered earlier by Whitney and Ashton (1971). The effect of environments on the material properties of composites has been studied by many researchers, for example, Ishikawa et al. (1978), Strife and Prewo (1979) and Bowles and Tompkins (1989). Ample published work is found on frequency and buckling analysis in conjunction to thermal and hygrothermal behavior (Chen and Chen 1989, Nakagiri et al. 1990, Sairam and Sinha 1992, Shen 2001, Srikanth and Kumar 2003, Shariyat 2007, Pandey et al. 2008, Lal et al. 2011, Dey et al. 2015e). The concept of random vibration is exhaustively utilised in many engineering application (Marano and Greco 2011, Mahmoudkhani et al. 2013, Guan et al. 2014). Most of the literatures are deterministic in nature, which lacks in portraying the probable deviation caused by random input parameters. Due to the presence of a large number of inter-dependent factors in production of composites, the system input parameters are generally random in nature. The allowable responses for conventional material is expected to be close to their mean values as fewer parameters are involved in their production process; while in contrast for composites, a range of random fluctuation in system parameter may occur due to a large number of system properties. Even after ensuring the effective quality control of production process in sensitive applications, the allowable responses are normally found to scatter widely with respect to the mean values. The knowledge of input variabilities and corresponding range of stochastic responses may serve to control the purposes of lightweight design, which is one of the important characteristics for composites. Therefore efficient computational modelling and analysis is needed considering randomness in material properties and ply orientation angle including the effect of thermal uncertainty to ensure optimization, operational safety and reliability. Such issues can be addressed by employing the probabilistic method, which quantifies the uncertainties in frequency responses.

The focus of the present chapter includes the stochastic analysis of natural frequencies for laminated composite plates subjected to uncertain thermo-mechanical loading. A surrogate model is used based on the generalized high dimensional model representation (GHDMR) approach wherein D-MORPH (Diffeomorphic Modulation under Observable Response Preserving Homotopy) regression is employed to ensure the hierarchical orthogonality of HDMR component functions (Li and Rabitz 2012). Random sampling high dimensional model representation (RS-HDMR) is found to be employed for uncertainty quantification of natural frequency in composite plates considering three input parameters namely the fibre-orientation angle, elastic modulus and mass density (Dey et al. 2015a), wherein the input parameters are independent to each other. In the present study, the sources of uncertainty for natural frequency are considered as layer wise variation of material properties, ply orientation angle and temperature. Material properties of fibre reinforced composites are temperature dependent. Due to this reason, the input parameters become co-related to each other in case of layer wise combined variation of material properties, ply orientation angle and temperature. The co-related input parameters cannot be mapped for the corresponding output response using conventional high dimensional model representation (HDMR) approach (different variants of HDMR can be found in available literature such as Cut-HDMR (Rabitz and Alis 1999, Rabitz et al. 1999, Shorter et al. 1999), RS-HDMR (Li et al. 2006), mp-Cut-HDMR (Li et al. 2001), Multicut-HDMR Li et al. (2004), lp-RS-HDMR Li et al. (2003) depending primarily on sampling scheme (Li and Rabitz 2012). The present GHDMR can efficiently take care of both independent as well as co-related input parameters under a relaxed vanishing condition. The extended bases are used as basis functions to approximated HDMR component functions and D-MORPH regression is used to determine the coefficients in the GHDMR algorithm. In the present chapter, a layer-wise random variable approach is employed in conjunction with finite element formulation to figure out the random eigenvalue problem. The numerical results are shown for the first three natural frequencies with individual and combined variation of the stochastic input parameters. The present probabilistic approach is validated with Monte Carlo simulation wherein a small random variation is considered as tolerance zone. This chapter hereafter is organized as follows, Section 7.2: theoretical formulation for the dynamic analysis of composite laminates under thermo-mechanical loading; Section 7.3: surrogate based stochastic approach (the detail formulation of GHDMR based surrogate modelling can be found in Chapter 3); Section 7.4: results and discussion; and Section 7.5: summary.

7.2 Governing equations for composite laminates under thermo-mechanical loading

Consider a laminated composite cantilever plate (Fig. 7.1) of thickness '*t*' consisting of *n* number of thin lamina, the stress strain relations in the presence of temperature can be represented as (Kollar and Springer 2009)

$$\begin{Bmatrix} \sigma_x(\tilde{\omega}) \\ \sigma_y(\tilde{\omega}) \\ \tau_{xy}(\tilde{\omega}) \end{Bmatrix} = \begin{bmatrix} \bar{Q}_{11}(\tilde{\omega}) & \bar{Q}_{12}(\tilde{\omega}) & \bar{Q}_{16}(\tilde{\omega}) \\ \bar{Q}_{12}(\tilde{\omega}) & \bar{Q}_{22}(\tilde{\omega}) & \bar{Q}_{26}(\tilde{\omega}) \\ \bar{Q}_{16}(\tilde{\omega}) & \bar{Q}_{26}(\tilde{\omega}) & \bar{Q}_{66}(\tilde{\omega}) \end{bmatrix} \begin{Bmatrix} \varepsilon_x^o - e_x^T(\tilde{\omega}) \\ \varepsilon_y^o - e_y^T(\tilde{\omega}) \\ \gamma_{xy}^o - e_{xy}^T(\tilde{\omega}) \end{Bmatrix} \qquad (7.1)$$

and

$$\begin{Bmatrix} \tau_{xz}(\tilde{\omega}) \\ \tau_{yz}(\tilde{\omega}) \end{Bmatrix} = \begin{bmatrix} \bar{Q}_{44}(\tilde{\omega}) & \bar{Q}_{45}(\tilde{\omega}) \\ \bar{Q}_{45}(\tilde{\omega}) & \bar{Q}_{55}(\tilde{\omega}) \end{bmatrix} \begin{Bmatrix} \gamma_{xz}^o \\ \gamma_{yz}^o \end{Bmatrix} \qquad (7.2)$$

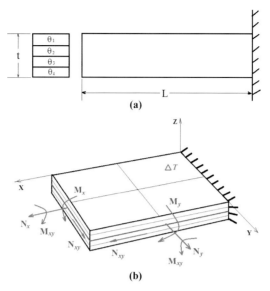

Fig. 7.1 Laminated composite cantilever plate.

where, σ_x, σ_y, τ_{xy}, τ_{xz}, τ_{yz} are normal and shear stresses; ε_x^o, ε_y^o, γ_{xy}^o, γ_{xz}^o, γ_{yz}^o are normal and shear strains. The e_x^T, e_y^T, e_{xy}^T values are the thermal strain components due to temperature in x-y reference axes, which are derived from the corresponding values in the fiber axes after applying the transformations expressed as

$$\{e^T(\tilde{\omega})\} = \begin{Bmatrix} e_x^T(\tilde{\omega}) \\ e_y^T(\tilde{\omega}) \\ e_{xy}^T(\tilde{\omega}) \end{Bmatrix} = \begin{bmatrix} Cos^2\theta(\tilde{\omega}) & Sin^2\theta(\tilde{\omega}) \\ Sin^2\theta(\tilde{\omega}) & Cos^2\theta(\tilde{\omega}) \\ -2Cos\theta(\tilde{\omega})Sin\theta(\tilde{\omega}) & 2Sin\theta(\tilde{\omega})Cos\theta(\tilde{\omega}) \end{bmatrix} \begin{Bmatrix} \alpha_1 \\ \alpha_2 \end{Bmatrix} \Delta T(\tilde{\omega})$$

(7.3)

where, in this chapter, $(\tilde{\omega})$ indicates the randomness of the corresponding variables and α_1, α_2 are the thermal expansion coefficients of lamina in longitudinal and lateral directions and their values are considered as -0.3×10^6/K and 28.1×10^6/K, respectively. $\Delta T(\tilde{\omega}) = T(\tilde{\omega}) - T_o(\tilde{\omega})$, where $T_o(\tilde{\omega})$ is the reference variable temperature in Kelvin. T is exposed random temperature in Kelvin. Here, $\theta(\tilde{\omega})$ is the random ply orientation angle of the lamina with reference to x-axis. The non-mechanical in-plane stress and moment resultants due to thermal environment are expressed as

$$\begin{Bmatrix} N_x^t(\tilde{\omega}) \\ N_y^t(\tilde{\omega}) \\ N_{xy}^t(\tilde{\omega}) \end{Bmatrix} = \sum_{k=1}^{n} \int_{Z_{k-1}}^{Z_k} [\bar{Q}_{ij}(\tilde{\omega})]_k \ [e^T(\tilde{\omega})]_k \ dz \quad \text{where } i, j = 1, 2, 6 \qquad (7.4)$$

$$\text{and} \quad \begin{Bmatrix} M_x^t(\tilde{\omega}) \\ M_y^t(\tilde{\omega}) \\ M_{xy}^t(\tilde{\omega}) \end{Bmatrix} = \sum_{k=1}^{n} \int_{Z_{k-1}}^{Z_k} [\bar{Q}_{ij}(\tilde{\omega})]_k \ [e^T(\tilde{\omega})]_k \ z \, dz \quad \text{where } i, j = 1, 2, 6 \quad (7.5)$$

The force and moment resultants are modified to include the thermal field by the constitutive equations (Yang et al. 1966, Jones 1975) for the composite plate are given by

$$\{F(\tilde{\omega})\} = [D(\tilde{\omega})]\{\varepsilon\} \ - \ \{F^t(\tilde{\omega})\} \tag{7.6}$$

where

$$\{F(\tilde{\omega})\} = [N_x(\tilde{\omega}) \ \ N_y(\tilde{\omega}) \ \ N_{xy}(\tilde{\omega}) \ \ M_x(\tilde{\omega}) \ \ M_y(\tilde{\omega}) \ \ M_{xy}(\tilde{\omega}) \ Q_x(\tilde{\omega}) \ Q_y(\tilde{\omega})]^T \tag{7.7}$$

$$[D(\tilde{\omega})] = \begin{bmatrix} A_{11}(\tilde{\omega}) & A_{12}(\tilde{\omega}) & A_{16}(\tilde{\omega}) & B_{11}(\tilde{\omega}) & B_{12}(\tilde{\omega}) & B_{16}(\tilde{\omega}) & 0 & 0 \\ A_{12}(\tilde{\omega}) & A_{22}(\tilde{\omega}) & A_{26}(\tilde{\omega}) & B_{12}(\tilde{\omega}) & B_{22}(\tilde{\omega}) & B_{26}(\tilde{\omega}) & 0 & 0 \\ A_{16}(\tilde{\omega}) & A_{26}(\tilde{\omega}) & A_{66}(\tilde{\omega}) & B_{16}(\tilde{\omega}) & B_{26}(\tilde{\omega}) & B_{66}(\tilde{\omega}) & 0 & 0 \\ B_{11}(\tilde{\omega}) & B_{12}(\tilde{\omega}) & B_{16}(\tilde{\omega}) & D_{11}(\tilde{\omega}) & D_{12}(\tilde{\omega}) & D_{16}(\tilde{\omega}) & 0 & 0 \\ B_{12}(\tilde{\omega}) & B_{22}(\tilde{\omega}) & B_{26}(\tilde{\omega}) & D_{12}(\tilde{\omega}) & D_{22}(\tilde{\omega}) & D_{26}(\tilde{\omega}) & 0 & 0 \\ B_{16}(\tilde{\omega}) & B_{26}(\tilde{\omega}) & B_{66}(\tilde{\omega}) & D_{16}(\tilde{\omega}) & D_{26}(\tilde{\omega}) & D_{66}(\tilde{\omega}) & 0 & 0 \\ 0 & 0 & 0 & 0 & 0 & 0 & S_{44}(\tilde{\omega}) & S_{45}(\tilde{\omega}) \\ 0 & 0 & 0 & 0 & 0 & 0 & S_{45}(\tilde{\omega}) & S_{55}(\tilde{\omega}) \end{bmatrix} \tag{7.8}$$

The non-mechanical loads due to uncertain thermal condition

$$\{F^t(\tilde{\omega})\} = [N_x^t(\tilde{\omega}) \ \ N_y^t(\tilde{\omega}) \ \ N_{xy}^t(\tilde{\omega}) \ \ M_x^t(\tilde{\omega}) \ \ M_y^t(\tilde{\omega}) \ \ M_{xy}^t(\tilde{\omega}) \ \ 0 \ \ 0 \ ]^T \tag{7.9}$$

$$\{\varepsilon\} = [\varepsilon_x^o \quad \varepsilon_y^o \quad \gamma_{xy}^o \quad k_x \quad k_y \quad k_{xy} \quad \gamma_{xz}^o \quad \gamma_{yz}^o \ ]^T \tag{7.10}$$

The stiffness coefficients are defined as

$$[A_{ij}(\tilde{\omega}), \ B_{ij}(\tilde{\omega}), \ D_{ij}(\tilde{\omega})] = \sum_{k=1}^{n} \int_{Z_{k-1}}^{Z_k} [\bar{Q}_{ij}(\tilde{\omega})]_k \ [1, \quad z, \quad z^2] \, dz \quad \text{where } i, j = 1, 2, 6 \tag{7.11}$$

$$\text{and} \ \ [S_{ij}(\tilde{\omega})] = \alpha_{scf} \sum_{k=1}^{n} \int_{Z_{k-1}}^{Z_k} [\bar{Q}_{ij}(\tilde{\omega})]_k \ dz \quad \text{where } i, j = 4, 5 \tag{7.12}$$

where α_{scf} is the shear correction factor and is assumed as 5/6. $[\bar{Q}_{ij}(\tilde{\omega})]$ in above equation (7.11) and (7.12) is defined as

$$[\bar{Q}_{ij}(\tilde{\omega})] = [T_1(\tilde{\omega})]^{-1} \ [Q_{ij}(\tilde{\omega})] \ [T_1(\tilde{\omega})]^{-T} \ \text{ for } i, j = 1, 2, 6 \tag{7.13}$$

$$[\bar{Q}_{ij}(\tilde{\omega})] = [T_2(\tilde{\omega})]^{-1} [Q_{ij}(\tilde{\omega})] [T_2(\tilde{\omega})]^{-T} \quad \text{for } i, j = 4, 5 \tag{7.14}$$

where

$$[T_1(\tilde{\omega})] = \begin{bmatrix} Cos^2\theta(\tilde{\omega}) & Sin^2\theta(\tilde{\omega}) & 2Sin\theta(\tilde{\omega})Cos\theta(\tilde{\omega}) \\ Sin^2\theta(\tilde{\omega}) & Cos^2\theta(\tilde{\omega}) & -2Cos\theta(\tilde{\omega})Sin\theta(\tilde{\omega}) \\ -Cos\theta(\tilde{\omega})Sin\theta(\tilde{\omega}) & Sin\theta(\tilde{\omega})Cos\theta(\tilde{\omega}) & Cos^2\theta(\tilde{\omega}) - Sin^2\theta(\tilde{\omega}) \end{bmatrix}$$

$$\text{and} \quad [T_2(\tilde{\omega})] = \begin{bmatrix} Cos\theta(\tilde{\omega}) & -Sin\theta(\tilde{\omega}) \\ Sin\theta(\tilde{\omega}) & Cos\theta(\tilde{\omega}) \end{bmatrix}$$

$$\text{where} \quad [Q_{ij}(\tilde{\omega})]_k = \begin{bmatrix} Q_{11}(\tilde{\omega}) & Q_{12}(\tilde{\omega}) & 0 \\ Q_{12}(\tilde{\omega}) & Q_{22}(\tilde{\omega}) & 0 \\ 0 & 0 & Q_{66}(\tilde{\omega}) \end{bmatrix} \quad \text{for } i, j = 1, 2, 6$$

$$\text{and} \quad [Q_{ij}(\tilde{\omega})]_k = \begin{bmatrix} Q_{44}(\tilde{\omega}) & 0 \\ 0 & Q_{55}(\tilde{\omega}) \end{bmatrix} \quad \text{for } i, j = 4, 5$$

where Q_{ij} are the in-plane element of the stiffness matrix. From Hamilton's principle (Meirovitch 1992), the dynamic equilibrium equation can be expressed as Dey and Karmakar (2012).

$$[M(\tilde{\omega})]\{\ddot{\delta}_e\} + [K_e(\tilde{\omega})] \{\delta_e\} = \{F_e\} \tag{7.15}$$

After assembling all the element matrices and the force vectors with respect to the common global coordinates, the resulting equilibrium equation is formulated. Considering randomness of input parameters like temperature, ply-orientation angle, elastic modulus, etc., the equation of motion of free vibration system with n degrees of freedom can expressed as

$$[M(\tilde{\omega})][\ddot{\delta}] + [K(\tilde{\omega})]\{\delta\} = 0 \tag{7.16}$$

where $\{\delta\} \in R^n$ is the vector of generalized coordinates and $\{F_L\} \in R^n$ is the force vector. The governing equations are derived based on Mindlin's Theory (Cook et al. 1989) incorporating rotary inertia, and transverse shear deformation. The random natural frequencies $[\omega_n(\tilde{\omega})]$ are determined from the standard eigenvalue problem (Bathe 1990) and uncertain random frequencies are calculated by solving the QR iteration algorithm. In the present study, an eight noded isoparametric quadratic element with five degrees of freedom at each node (three translations and two rotations) is considered for finite element formulation with respect to laminated composite cantilever plate wherein the shape functions (N_i) are as follows

$$N_i = (1 + \xi \xi_i)(1 + \varsigma \varsigma_i)(\xi \xi_i + \varsigma \varsigma_i - 1)/4 \quad \text{(for } i = 1, 2, 3, 4) \tag{7.17}$$

$$N_i = (1 - \xi^2)(1 + \varsigma \varsigma_i)/2 \quad \text{(for } i = 5, 7) \tag{7.18}$$

$$N_i = (1 - \varsigma^2)(1 + \xi \xi_i)/2 \quad \text{(for } i = 6, 8) \tag{7.19}$$

where ζ and ξ are the local natural coordinates of the element. The element stiffness matrix is given by

$$[K_e(\tilde{\omega})] = \int_{-1}^{+1}\int_{-1}^{+1}[B]^T[D(\tilde{\omega})][B][J_c]d\xi d\varsigma \qquad (7.20)$$

where, $[B]$ is the strain displacement matrix and $[D(\tilde{\omega})]$ is the random stress-strain matrix. The strain displacement matrix, $[B] = [[B_1], [B_2], \ldots\ldots\ldots[B_8]]$

$$[B_i] = \begin{bmatrix} N_{i,x} & 0 & 0 & 0 & 0 \\ 0 & N_{i,y} & 0 & 0 & 0 \\ N_{i,y} & N_{i,x} & 0 & 0 & 0 \\ 0 & 0 & 0 & N_{i,x} & 0 \\ 0 & 0 & 0 & 0 & N_{i,y} \\ N_{i,y} & -N_{i,x} & 0 & N_{i,y} & N_{i,x} \\ 0 & 0 & N_{i,x} & N_i & 0 \\ 0 & 0 & N_{i,y} & 0 & N_i \end{bmatrix} \qquad (7.21)$$

The element mass matrix is obtained from the Integral

$$[M_e(\tilde{\omega})] = \int_{-1}^{+1}\int_{-1}^{+1}[N]^T[\rho(\tilde{\omega})][N][J_c]d\xi d\varsigma \qquad (7.22)$$

where, $[N]$ is the shape function matrix and $[\rho(\tilde{\omega})]$ is the random inertia matrix. The derivatives of the shape function, N_i with respect to x, y are expressed in term of their derivatives with respect to ξ and ζ by the following relationship

$$\begin{bmatrix} N_{i,x} \\ N_{i,y} \end{bmatrix} = [J_c]^{-1}\begin{bmatrix} N_{i,\xi} \\ N_{i,\varsigma} \end{bmatrix} \quad \text{where} \quad \text{Jacobian } [J_c] = \begin{bmatrix} \dfrac{\partial x}{\partial \xi} & \dfrac{\partial y}{\partial \xi} \\ \dfrac{\partial x}{\partial \varsigma} & \dfrac{\partial x}{\partial \varsigma} \end{bmatrix} \qquad (7.23)$$

7.3 Random input representation and surrogate based uncertainty propagation scheme

The random input parameters such as ply-orientation angle and temparature in each layer of laminate are considered for composite cantilever plates. It is assumed that the distribution of random input parameters exists within a certain band of tolerance with their crisp values. The cases wherein the input variables considered in each layer of laminate are as follows:

a) Variation of ply-orientation angle $\theta(\tilde{\omega}) = \{\theta_1\ \theta_2\ \theta_3\ldots\ldots\theta_i\ldots\ldots\theta_l\}$
   only:

b) Variation of longitudinal elastic modulus only:

$$E_1(\tilde{\omega}) = \{E_{1(1)} \quad E_{1(2)} \quad E_{1(3)} ....E_{1(i)} ......E_{1(l)}\}$$

c) Variation of shear modulus only:

$$G_{12}(\tilde{\omega}) = \{G_{12(1)} \quad G_{12(2)} \quad G_{12(3)} ...G_{12(i)} ...G_{12(l)}\}$$

d) Variation of temperature only:

$$T(\tilde{\omega}) = \{T_{(1)} \quad T_{(2)} \quad T_{(3)} ....T_{(i)} ......T_{(l)}\}$$

e) Combined variation of ply orientation angle, elastic modulus, shear modulus and temperature:

$$[\theta, E_1, G_{12}, T(\tilde{\omega})] = [(\theta_1...\theta_l), (E_{1(1)}...E_{1(l)}), (G_{12(1)}...G_{12(l)}), (T_{(1)}...T_{(l)})]$$

where θ_i, $E_{1(i)}$, $G_{12(i)}$ and $T_{(i)}$ are the ply orientation angle, and temperature, respectively and 'l' denotes the number of layer in the laminate. In the present study, $\pm 5°$ variations for ply orientation angle, $\pm 10\%$ volatility in material properties and $\pm$ 25 K tolerance for temperature, respectively are considered from their respective deterministic values. Figure 7.2 presents the flowchart of dynamic analysis using GHDMR with D-MORPH. It is worth mentioning that material properties such as E_1 and G_{12} are considered as temperature dependant in the present study. Thus in case of the combined variation of ply orientation angle, elastic modulus, shear

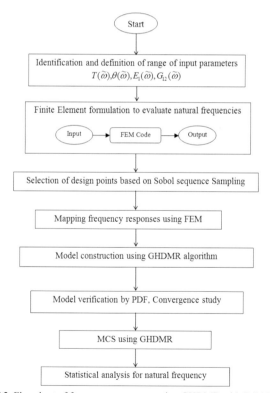

Fig. 7.2 Flowchart of frequency responses using GHDMR with D-MORPH.

modulus and temperature, correlated input variables are needed to be mapped for natural frequencies as discussed in the introduction section.

7.4 Results and discussion

The present study considers four layered graphite–epoxy angle-ply $[(\theta°/–\theta°/\theta°/\theta°)]$ and cross-ply $(0°/90°/0°/90°)$ composite cantilever plates. An eight noded isoparametric plate bending element is considered for finite element formulation. Due to paucity of space, only a few important representative results are furnished. Table 7.1 presents the convergence study of non-dimensional fundamental natural frequencies of three layered graphite–epoxy untwisted composite plates (Qatu and Leissa 1991a). Table 7.2 presents the non-dimensional natural frequencies for simply-supported graphite–epoxy symmetric cross-ply $(0°/90°/90°/0°)$ composite plates (Sai and Sinha 1992). In both the cases, close agreement with benchmarking results are obtained at (6×6) mesh size. Material properties and their variation with temperature (Panda et al. 2014) are furnished in Table 7.3. Considering mean temperature as $T = 300$ K and thickness $t = 0.004$ m, the deterministic values of material properties are considered as $E_1 = E_2 = 15.4$ GPa, $\nu = 0.43$, $G_{12} = G_{13} = G_{23} = 3.56$ GPa, $\rho = 1660$ kg/m³. The present GHDMR methodology is employed to find a predictive and representative surrogate model

Table 7.1 Convergence study for non-dimensional fundamental natural frequencies $[\omega = \omega_n \, L^2 \, \sqrt{(\rho/E_1 t^2)}]$ of three layered $(\theta°/–\theta°/\theta°)$ graphite-epoxy untwisted composite plates, a/b = 1, b/t = 100, considering $E_1 = 138$ GPa, $E_2 = 8.96$ GPa, $G_{12} = 7.1$ GPa, $\nu_{12} = 0.3$.

Ply angle, θ	Present FEM (4 × 4)	Present FEM (6 × 6)	Present FEM (8 × 8)	Present FEM (10 × 10)	Qatu and Leissa (1991)
0°	1.0112	1.0133	1.0107	1.004	1.0175
45°	0.4556	0.4577	0.4553	0.4549	0.4613
90°	0.2553	0.2567	0.2547	0.2542	0.2590

Table 7.2 Non-dimensional natural frequencies $[\omega = \omega_n \, a^2 \, \sqrt{(\rho/E_2 t^2)}]$ for simply-supported graphite-epoxy symmetric cross-ply $(0°/90°/90°/0°)$ composite plates considering a/b = 1, $T = 325$K, a/t = 100.

Frequency	Present FEM (4 × 4)	Present FEM (6 × 6)	Present FEM (8 × 8)	Present FEM (10 × 10)	Sai Ram and Sinha [40]
1	8.041	8.061	8.023	8.001	8.088
2	18.772	19.008	18.684	18.552	19.196
3	38.701	38.981	38.597	38.443	39.324

Table 7.3 Material properties of glass/epoxy lamina at different temperatures, $E_1 = E_2$, $G_{12} = G_{13} = G_{23}$, mass density $(\rho) = 1660$ Kg/m³, $\nu = 0.43$, Panda et al. (2014).

Material properties (GPa)	Temperature (K)						
	123	150	200	250	300	350	400
E_1	15.4	15.4	15.4	15.4	15.4	14.93	14.7
G_{12}	3.56	3.56	3.56	3.56	3.56	3.51	3.48

relating each natural frequency to a number of input variables. The present surrogate models are used to determine the first three natural frequencies corresponding to given values of input variables, instead of time-consuming deterministic finite element analysis. The probability density function is plotted as the benchmark of bottom line results. Due to paucity of space, only a few important representative results are furnished. The variation of temperature is scaled in the range with the lower and the upper limit (tolerance limit) as ± 25 K with respective mean values while for ply orientation angle as within ± 5° fluctuation (as per standard of composite manufacturing industry) with respective deterministic values. Both angle-ply and cross-ply laminated composite cantilever plates are considered for the present analysis.

Figure 7.3 presents the scatter plot which establishes the accuracy of the present model with respect to the original finite element model corresponding to random fundamental natural frequencies for combined variation of temperature and ply orientation angle. Table 7.4 presents the convergence study of the present method compared to direct Monte Carlo simulation (MCS) for first three natural frequencies due to individual variation of ply-orientation angle and temperature of angle-ply (45°/–45°/45°/–45°) composite cantilever plate while Table 7.5 represents the convergence study of the present method with direct MCS for first three natural frequencies due to combined variation of temparature, ply-orientation angle, elastic modulus and shear modulus of angle-ply (45°/–45°/45°/–45°) composite cantilever plate. Figure 7.4 presents the comparative probability density function plot with respect to first three natural frequencies due to individual and combined variation of

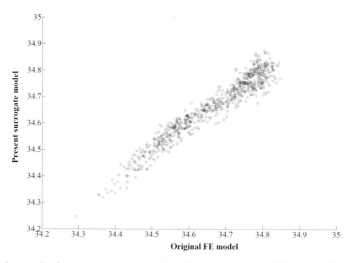

Fig. 7.3 Scatter plot for present surrogate model with respect to original FE model of fundamental natural frequencies for combined variation of ply-orientation angle and temperature of angle-ply (45°/–45°/45°/–45°) composite cantilever plate, considering $E_1 = E_2 = 15.4$ GPa, $G_{12} = G_{13} = G_{23} = 3.56$ GPa, $T = 300$K, $\rho = 1660$ Kg/m³, $t = 0.004$ m, $v = 0.43$.

Table 7.4 Convergence study of first three natural frequencies (Hz) due to individual variation of ply-orientation angle and temperature of angle-ply (45°/−45°/45°/−45°) composite cantilever plate considering $E_1 = E_2 = 15.4$ GPa, $G_{12} = G_{13} = G_{23} = 3.56$ GPa, $\rho = 1660$ Kg/m³, $t = 0.004$ m, $\nu = 0.43$, $T_{mean} = 300$ K.

Parameter	Values	f_1				f_2				f_3			
		MCS (10,000)	Present method (sample run)			MCS (10,000)	Present method (sample run)			MCS (10,000)	Present method (sample run)		
			32	64	128		32	64	128		32	64	128
$\theta(\widetilde{\omega})$	Max	34.8601	34.8999	34.8664	34.8783	98.1667	98.4748	98.6176	98.5928	216.7606	217.8250	216.7751	217.1431
	Min	34.2870	34.2531	34.2767	34.2941	84.9534	84.8309	84.8548	84.9552	205.8846	204.4143	205.7557	206.1325
	Mean	34.6546	34.6468	34.6509	34.6561	92.0607	92.0099	92.0172	92.0485	213.6560	213.4836	213.5772	213.6981
	SD	0.1061	0.1095	0.1068	0.1068	2.4501	2.4564	2.4550	2.4673	2.0449	2.1606	2.0635	2.0517
$T(\widetilde{\omega})$	Max	34.6904	34.6996	34.6932	34.6922	93.0976	93.1461	93.1307	93.1468	214.4427	214.6334	214.5056	214.4707
	Min	34.4488	34.4591	34.4536	34.4554	88.24453	88.3375	88.3098	88.3482	209.4478	209.5902	209.5364	209.5452
	Mean	34.5872	34.5879	34.5877	34.5872	90.7581	90.7659	90.7653	90.7634	212.3069	212.3222	212.3173	212.3069
	SD	0.0422	0.0428	0.0429	0.04287	0.8561	0.8707	0.8710	0.8686	0.8882	0.9011	0.9034	0.9016

Table 7.5 Convergence study of the present method with direct Monte Carlo simulation (MCS) for first three natural frequencies (Hz) due to combined variation of temperature, ply-orientation angle, elastic modulus and shear modulus of angle-ply (45°/–45°/45°/–45°) composite cantilever plate.

Frequency	Method	Sample size	Parameters			
			Max	**Min**	**Mean**	**Standard Deviation**
f_1	MCS	10,000	34.8508	34.2855	34.5996	0.1187
	Present method	32	34.9831	34.0428	34.5836	0.1019
		64	34.9002	34.2454	34.5978	0.0985
		128	35.0770	34.0130	34.6000	0.1403
		256	34.8492	34.2701	34.5999	0.1187
		512	34.8511	34.2638	34.5997	0.1195
		1024	34.8521	34.2628	34.5996	0.1198
f_2	MCS	10,000	96.4222	84.7779	90.7088	2.4025
	Present method	32	99.0099	79.0589	90.4359	2.0766
		64	96.9819	84.2358	90.6973	1.9971
		128	97.0225	83.4915	90.6970	2.4065
		256	96.3939	84.6852	90.7179	2.4068
		512	96.4816	84.4608	90.7116	2.4215
		1024	96.4883	84.4624	90.7110	2.4225
f_3	MCS	10,000	216.6953	205.2277	212.2366	2.4307
	Present method	32	219.6954	200.6440	211.9285	2.1035
		64	218.1019	204.7512	212.1888	1.9985
		128	224.9086	197.1597	212.2513	3.3649
		256	216.7612	205.1236	212.2363	2.4305
		512	216.7192	204.7821	212.2321	2.4455
		1024	216.7172	204.7808	212.2319	2.4414

stochastic input parameters of angle-ply (45°/–45°/45°/–45°) composite cantilever plate for both the MCS as well as the present method.

In the present analysis, a sample size of 64 is considered for layerwise individual variation of stochastic input parameters, while due to increment of number of input variables for combined random variation, the subsequent sample size of 512 is adopted to meet the convergence criteria. Here, although the same sampling size as in direct MCS (10,000 samples) is considered, the number of actual FE analysis is much less compared to original MCS and is equal to the number representative sample required to construct the surrogate model. The surrogate

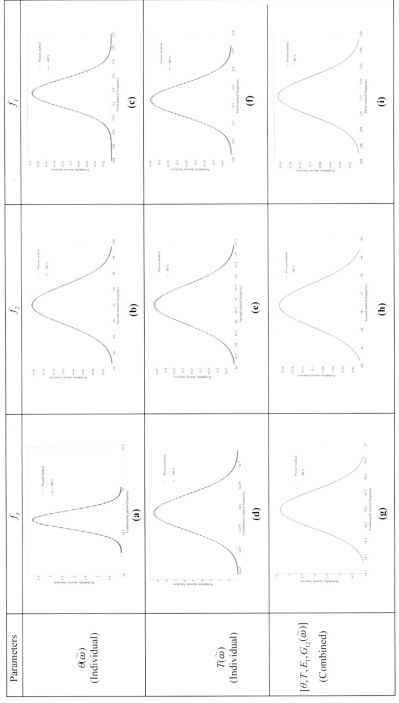

Fig. 7.4 (a–i) Probability density function with respect to first three natural frequencies (Hz) due to individual and combined variation of ply orientation angle, elastic modulus, shear modulus and temperature of angle-ply (45°/−45°/45°/−45°) composite cantilever plate at mean temperatute (T_{mean}) = 300 K.

model is formed on which the full sample size of direct MCS is conducted. Hence, the computational time and effort expressed in terms of FE calculation is reduced compared to full scale direct MCS. This provides an efficient affordable way for simulating the uncertainties in natural frequency.

A comparative study on variation of stochastic natural frequencies is carried out for angle-ply (45°/–45°/45°/–45°) and cross-ply (0°/90°/0°/90°) composite cantilever plate due to individual variation of elastic modulus and shear modulus as furnished in Fig. 7.5. From Fig. 7.5, it is observed that the mean fundamental natural frequency for angle-ply laminate is found to be slightly lower than that of the same for cross-ply laminate while a significant higher mean values are obtained at second and third modes for angle-ply compared to cross-ply. Considering only variation of temperature of angle-ply (45°/–45°/45°/–45°) and cross-ply (0°/90°/0°/90°) composite cantilever plate, the probability density function (PDF) with respect to first three natural frequencies are plotted in Fig. 7.6 wherein it is found that as the temperature increases the variabilities of first three natural frequencies of angle-ply laminate are increased and the probability density function curves become more steeper as the temperature increases. This can be attributted to the fact that the rise in temeprature influences the thermo-mechanical loading due to random variation leading to change in the system properties. In contrast, the reverse trend is identified for cross-ply laminated composite plates due to the reduction effect of 0° and 90° on effective stiffness of the laminate. On the other hand, Fig. 7.7 presents the ply level quantification of uncertainty in first three natural frequencies in terms of the Probability density function for angle-ply [(θ°/–θ°/ θ°/–θ°) where θ = Ply orientation angle] and cross-ply (0°/90°/0°/90°) composite cantilever plate. Due to random variation of ply orientation angle, the elastic stiffness of the laminated composite plate is found to be varied which in turn influence the frequency responses irrespective of laminate configuration. The effect of combined variation of input parameters is also carried out in addition to individual variation of inputs in conjunction to stochastic natural frequencies for composite laminated plates as furnished in Fig. 7.8. The ply orientation angle, elastic modulus, shear modulus and temperature of angle-ply (45°/–45°/45°/–45°) and cross-ply (0°/90°/0°/90°) are considered as random input variables wherein the upper and lower bounds of volatility in natural frequencies are found to be wider than that of individual variation of inputs irrespective of laminate configuration. This corroborates with the fact that the combined effect of random input parameters leads to increase the variation in outputs compared to individual cases.

In the present study, the relative coefficient of variance (RCV) (normalized mean to standard deviation ratio) due to variation of temperature is also quantified for each layer for angle-ply and cross-ply laminate as furnished in Fig. 7.9. The two outer-most layers of the angle-ply laminate are found to be most sensitive to temperature variation for first three modes while the maximum sensitiveness of temperature is observed only at the bottom layer of the cross-ply laminate. In contrast, the least sensitivity to temperature variation is identified at the third intermediate layer for first three modes irrespective of laminate configuration. The

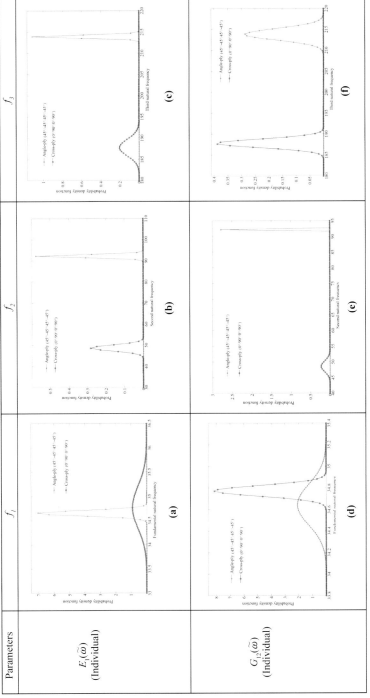

Fig. 7.5 (a–f) Probability density function with respect to first three natural frequencies (Hz) due to individual variation of elastic modulus, shear modulus of angleply $(45°/−45°/45°/−45°)$ and cross-ply $(0°/90°/0°/90°)$ composite cantilever plate at mean temperature $(T_{mean}) = 300$ K.

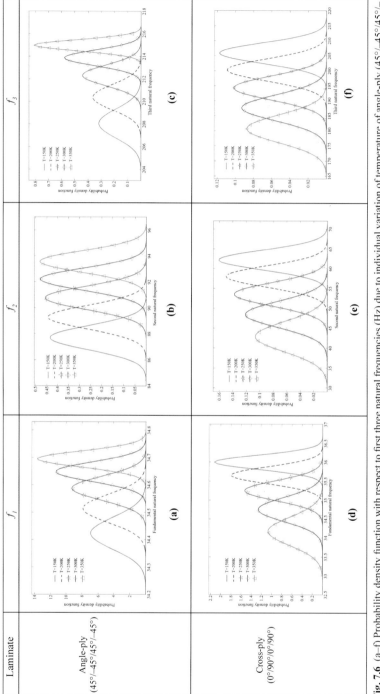

Fig. 7.6 (a–f) Probability density function with respect to first three natural frequencies (Hz) due to individual variation of temperature of angle-ply (45°/−45°/45°/−45°) and cross-ply (0°/90°/0°/90°) composite cantilever plate.

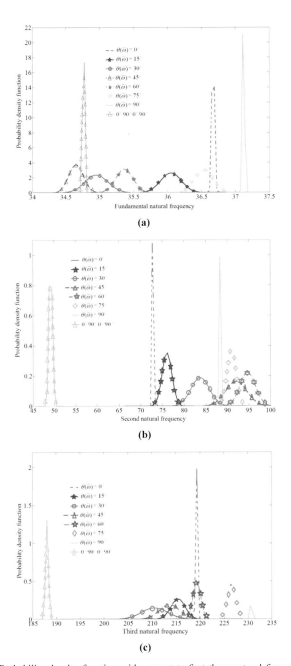

Fig. 7.7 (a–c) Probability density function with respect to first three natural frequencies (Hz) due to individual variation of ply orientation angle of angle-ply [$(\theta°/-\theta°/\theta°/-\theta°)$ where $\theta = 0°$, 15°, 30°, 45°, 60°, 75° and 90°] and cross-ply (0°/90°/0°/90°) composite cantilever plate at mean temperature $(T_{mean}) = 300$ K.

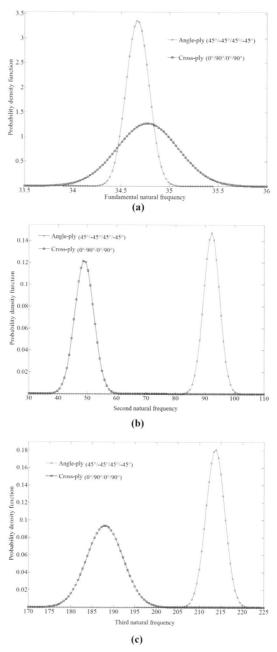

Fig. 7.8 (a–c) Probability density function with respect to first three natural frequencies (Hz) due to combined variation of ply orientation angle, elastic modulus, shear modulus and temperature of angle-ply (45°/–45°/45°/–45°) and cross-ply (0°/90°/0°/90°) composite cantilever plate at mean temperature (T_{mean}) = 300 K.

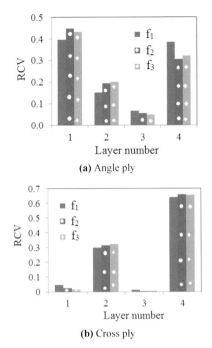

(a) Angle ply

(b) Cross ply

Fig. 7.9 Relative coefficient of variance (RCV) of first three natural frequencies due to variation of temperature (layerwise) for angle-ply (45°/–45°/45°/–45°) and cross-ply (0°/90°/0°/90°) composite cantilever plate at mean temperature (T_{mean}) = 300 K.

layerwise ply-level sensitiveness to temperature variation for fundamental mode is studied to map the sensitivity of each layer due to the influence of the ply orientation angle on variation of temperature as furnished in Fig. 7.10a–d. The least sensitivity is observed at $\theta(\widetilde{\omega}) = 45°$ for outer layers of the angle-ply laminate.

7.5 Summary

This chapter illustrates the layer-wise thermal uncertainty propagation in laminated composite plates. The ranges of variation in first three natural frequencies are analyzed considering both individual and combined stochasticity of input parameters. A generalized high dimensional model representation (GHDMR) model in conjunction with diffeomorphic modulation under observable response preserving homotopy (D-MORPH) regression is employed to map the input parameters (both correlated and uncorrelated) and natural frequencies. After utilizing the surrogate modelling approach, the number of finite element simulations is found to be exorbitently reduced compared to the original Monte Carlo simulation without compromising the accuracy of results. The computational expense is reduced by (1/156) times (individual stochasticity) and (1/19) times (combined stochasticity) of the original Monte Carlo simulation. It is observed that as the temperature increases

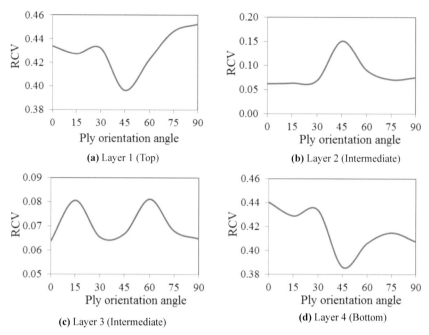

Fig. 7.10 Relative coefficient of variance (RCV) of fundamental mode due to variation of temperature (layerwise) for angle-ply ($\theta°/-\theta°/\theta°/-\theta°$) ($\theta$=Ply orientation angle) composite cantilever plate at mean temperature (T_{mean}) = 300 K.

the variabilities of first three natural frequencies of angle-ply laminate are increased and the probability density function become steeper. The two outer-most layers of the angle-ply laminate is found to be most sensitive to temperature variation for first three modes while the maximum sensitivity of temperature is observed at the bottom layer of the cross-ply laminate. It is found that stochastic variation of temperature influences the natural frequencies significantly and thus it is a crucial design parameter from the operational safety and serviceabilty point of view. The numerical results presented in this chapter provide a comprehensive idea for design and control of laminated composite structures under thermal environment.

CHAPTER **8**

Effect of Cutout on the Stochastic Free Vibration Analysis of Twisted Composite Panels

8.1 Introduction

The effect of various source-uncertainties (resulting from material and structural attributes) and the inevitable effect of uncertain environmental and operational conditions on composite laminates have been addressed in the previous chapters. However, various application-specific requirements are often needed to be met in engineering structures such as cutout in plates and shells. The present chapter deals with the effect of cutout on stochastic dynamic responses of composite laminates. Support vector regression (SVR) model in conjunction with Latin hypercube sampling is used in this investigation as a surrogate of the actual finite element model to achieve computational efficiency. The convergence of the present algorithm for laminated composite curved panels with cutout is validated with original finite element (FE) analysis along with traditional Monte Carlo simulation (MCS). Variations of input parameters (both individual and combined cases) are studied to portray their relative effect on the output quantity of interest. The layer-wise variability of structural and material properties is included considering the effect of twist angle, cutout sizes and different geometries (such as cylindrical, spherical, hyperbolic paraboloid and plate). The sensitivities of input parameters in terms of coefficient of variation are enumerated to project the relative importance of different random inputs on natural frequencies. Subsequently, the noise induced effects on the

SVR based computational algorithm are presented to map the inevitable variability in practical field of applications.

Composite panels are employed in a plenitude of shallow weight-sensitive load bearing structural components for wide range of aerospace, automotive, nuclear, marine, and civil engineering applications. The cutouts of composite panels are inevitable primarily for practical considerations. These are generally utilized not only to access ports for mechanical and electrical systems, but also to serve as doors and windows. Moreover, it is employed for inspection or maintenance purposes of the system. The dynamic behavior composite laminates may fluctuate significantly due to presence of cutout. In other words, the free vibration characteristics of composite curved panels are affected due to variability in shape, size and location of cutouts. Its effects are more difficult to quantify when such composite panels are subjected to random variability with uncertain geometric and material properties. The combined effect of cutout along with different stochastic material and geometric parameters may cause wide range of uncertainty in vibration behaviour of the structure which may lead to sudden failures due to resonance. It is essential to quantify the uncertain natural frequencies of such composite structural components accurately and thereby follow a design process accounting for all the uncertainties appropriately. Composite structures have more uncertainties and variabilities in the structural properties than conventional structures because of large number of structural parameters (inter-dependent in nature) and complex manufacturing and fabrication processes leading to less overall design control. In order to have more inclusive and realistic analysis, they can be modelled as stochastic structures, i.e., structures with uncertain system parameters (both in inputs and outputs). Beside these, the inherent errors involved in finite element modelling lead to inaccuracy in results. A brief review of the literature dealing with the effect of cutout in composite laminates and stochastic analysis of general composite structures is presented in the next paragraph.

A detail bibliography of previous investigations on dynamics of composite plates with cutouts is given in review papers (Rajamani and Prabhakaran 1977a, 1977b, Reddy 1982). The past investigations (Lee et al. 1987, Lee and Lim 1992, Beslin and Guyader 1996, Huang and Sakiyama 1999, Rezaeepazhand and Jafari 2005, Hota and Padhi 2007) incorporating cutouts are primarily confined to buckling and free vibration analysis of composite plates in a deterministic framework. Thornburgh and Hilburger (2006) carried out both experimental as well as numerical studies of composite panels with cutouts subjected to compressive load while Dimopoulos and Gantes (2015) employed the numerical methods to design the cylindrical steel shells with cutouts. The deterministic free vibration analyses of laminated composite shells with cutout have been studied by many researchers (Sai and Sreedhar 2002, Poore et al. 2008, Hu and Peng 2013, Park et al. 2009, Malekzadeh et al. 2013, Lee and Chung 2010, Parthasarthy et al. 1986). The mode shapes and natural frequencies were investigated for cross-ply laminates with square cut-outs by Jenq et al. (1993). Dey et al. (2016f) presented the stochastic dynamic responses of composite structures with cutout. Eiblmeier and Loughlan

(1997) studied the buckling analysis of composite panels with circular shaped cutouts. Sivakumar et al. (1999) considered large amplitude oscillation on frequency analyses of composite plates with cutout. Multi-dimensional deterministic studies were carried out to conduct investigations on behavior of composite and sandwich plate or shells with cutouts such as Anuja and Katukam (2015) presented parametric studies on the cutouts in heavily loaded aircraft beams, Mondal et al. (2015) studied the dynamic performance of sandwich composite plates with circular hole. Venkatachari et al. (2016) investigated the influence of environment for free vibration of composite laminates with cutouts and Yu et al. (2016) studied the buckling and free vibration analyses of laminated composite plates with complicated cutouts employing the first order shear deformation theory and level set method.

The present chapter is focused to quantify the uncertain natural frequencies for composite panels with cutouts following an efficient support vector regression based algorithm in conjunction with finite element analysis. The effect of twist angle is considered in the analysis along with cutouts. Analyzing the effect of twist angle in composite structures is immensely important for the applications in turbomachinery blades, wind turbines and various aircraft components. Composite structures are often pre-twisted due to design and operational needs such as the wing twist of aircraft is provided to maintain optimum angle of attack preventing negative lift or thrust and maximizing the aerodynamic efficiency. In general, the deterministic approach of finite element analysis concerning cutout and twist becomes computationally inefficient and costly when the input parameters considered at each nodal points of each discretized element becomes random with respect to its meshing pattern and boundary condition. The number of elements depends not only on cutout-size but also on its shape and location. Moreover, due to random variability of each input parameters at element level throughout the structure, the application of identical isotropic plate elements for computing the element mass and stiffness matrices will never match with reality. Thus uncertainty quantification for such structures following traditional Monte Carlo simulation based approaches is prohibitively expensive because of the fact that an exorbitant numbers of expensive finite element iterations are required for separate random input parameter sets. This complex problem of composites can be effectively handled by using Support Vector Regression (SVR) which is employed as an efficient surrogate of the expensive finite element model allowing rigorous occurrence of virtual iterations to be exercised with cost-effectiveness.

In stochastic structural problems, it can be restricted to consideration of the two-class problems, namely, with linear and non-linear classifier, without loss of generality. In such problems, the aim is to segregate the two classes by means of a function which is induced from known random dataset. In other words, the prime objective is to produce a classifier that will work well on unpredictable random data, i.e., it generalizes well. The conformity of such phenomenon can be efficiently dealt by the SVR model. Due to inherent complexity, composite structures have intrusive variability of geometric and material properties in both linear and nonlinear domain while analyzing the structural uncertainty. The random variation of individual parameters and combined parameters are considered in the present investigation along with the effect of cutout. This chapter is organized hereafter as, Section 8.2: stochastic finite element formulation for composite curved panels with cutout, Section 8.3: brief description of support vector regression, Section 8.4: support vector regression based uncertainty quantification algorithm for composite laminates with cutout including the effect of noise, Section 8.5: results and discussion, Section 8.6: summary.

8.2 Governing equations for composite laminates with cutout

In the present study, the composite panels with central cutout (as shown in Fig. 8.1) are considered. The stress resultants can be expressed in terms of the mid-plane strains and curvatures as

$$
\begin{Bmatrix} N_i(\bar{\omega}) \\ M_i(\bar{\omega}) \\ Q_i(\bar{\omega}) \end{Bmatrix} = \begin{bmatrix} A_{ij}(\bar{\omega}) & B_{ij}(\bar{\omega}) & 0 \\ B_{ij}(\bar{\omega}) & D_{ij}(\bar{\omega}) & 0 \\ 0 & 0 & S_{ij}(\bar{\omega}) \end{bmatrix} \begin{Bmatrix} \varepsilon_j \\ k_j \\ \gamma_j \end{Bmatrix}
\tag{8.1}
$$

The constitutive equation (Kollar and Springer 2009) is given by

$$
\{F\} = [\mathrm{D}(\bar{\omega})]\,\{\varepsilon\}
\tag{8.2}
$$

Subsequently, extension, bending-stretching coupling and bending terms can be expressed as

$$
[A_{ij}(\bar{\omega}),\ B_{ij}(\bar{\omega}),\ D_{ij}(\bar{\omega})] = \sum_{k=1}^{n} \int_{z_{k-1}}^{z_k} [\{\bar{Q}_{ij}(\bar{\omega})\}_{on}]_k\ [1,\ z,\ z^2]\, dz \qquad i, j = 1, 2, 6
\tag{8.3}
$$

The transverse shear term can be derived from

$$
[S_{ij}(\bar{\omega})] = \sum_{k=1}^{n} \int_{z_{k-1}}^{z_k} \alpha_s\ [\bar{Q}_{ij}(\bar{\omega})]_k\ dz \qquad i, j = 4, 5
\tag{8.4}
$$

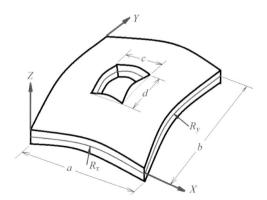

Fig. 8.1 Laminated composite curved panel with cutout.

where $\bar{\omega}$ indicates the representation of randomness while α_s denotes the shear correction factor (α_s=5/6) and $[\bar{Q}_{ij}]$ are the off-axis elastic constant matrix for elements which can be expressed as

$$\bar{Q}_{11}(\bar{\omega}) = Q_{11}(\bar{\omega})\, m^4 + 2[Q_{12}(\bar{\omega})+2Q_{66}(\bar{\omega})]\, m^2 n^2 + Q_{22}(\bar{\omega})\, n^4$$

$$\bar{Q}_{12}(\bar{\omega}) = [Q_{11}(\bar{\omega}) + Q_{22}(\bar{\omega}) - 4Q_{66}(\bar{\omega})]\, m^2 n^2 + Q_{12}(\bar{\omega})\, (m^4 + n^4)$$

$$\bar{Q}_{22}(\bar{\omega}) = Q_{11}(\bar{\omega})\, n^4 + 2[Q_{12}(\bar{\omega})+2Q_{66}(\bar{\omega})]\, m^2 n^2 + Q_{22}(\bar{\omega})\, m^4 \qquad (8.5)$$

$$\bar{Q}_{16}(\bar{\omega}) = [Q_{11}(\bar{\omega}) - Q_{12}(\bar{\omega}) - 2Q_{66}(\bar{\omega})]\, n\, m^3 + [Q_{12}(\bar{\omega}) - Q_{22}(\bar{\omega}) + 2Q_{66}(\bar{\omega})]\, m\, n^3$$

$$\bar{Q}_{26}(\bar{\omega}) = [Q_{11}(\bar{\omega}) - Q_{12}(\bar{\omega}) - 2Q_{66}(\bar{\omega})]\, m\, n^3 + [Q_{12}(\bar{\omega}) - Q_{22}(\bar{\omega}) + 2Q_{66}(\bar{\omega})]\, n\, m^3$$

$$\bar{Q}_{66}(\bar{\omega}) = [Q_{11}(\bar{\omega}) + Q_{22}(\bar{\omega}) - 2Q_{12}(\bar{\omega}) - 2Q_{66}(\bar{\omega})]\, m^2 n^2 + Q_{66}(\bar{\omega})]\,(m^4 + n^4)$$

The off-axis elastic constant matrix linked with transverse shear deformation can be expressed as

$$\bar{Q}_{44}(\bar{\omega}) = G_{13}(\bar{\omega})\, m^2 + G_{23}(\bar{\omega})n^2$$

$$\bar{Q}_{45}(\bar{\omega}) = [G_{13}(\bar{\omega}) - G_{23}(\bar{\omega})]\, m\, n \qquad (8.6)$$

$$\bar{Q}_{55}(\bar{\omega}) = G_{13}(\bar{\omega})\, n^2 + G_{23}(\bar{\omega})\, m^2$$

where $m = \sin\theta(\bar{\omega})$ and $n = \cos\theta(\bar{\omega})$, wherein $\theta(\bar{\omega})$ is random ply orientation angle. The on-axis terms can be represented as

$$[Q_{ij}(\bar{\omega})]_{on} = \begin{bmatrix} Q_{11}(\bar{\omega}) & Q_{12}(\bar{\omega}) & 0 \\ Q_{12}(\bar{\omega}) & Q_{12}(\bar{\omega}) & 0 \\ 0 & 0 & Q_{66}(\bar{\omega}) \end{bmatrix} \quad \text{for } i, j = 1, 2, 6$$

$$\qquad (8.7)$$

$$[\bar{Q}_{ij}(\bar{\omega})]_{on} = \begin{bmatrix} Q_{44}(\bar{\omega}) & Q_{45}(\bar{\omega}) \\ Q_{45}(\bar{\omega}) & Q_{55}(\bar{\omega}) \end{bmatrix} \quad \text{for } i, j = 4, 5$$

where

$$Q_{11} = \frac{E_1(\bar{\omega})}{1 - v_{12}(\bar{\omega})\, v_{21}(\bar{\omega})} \qquad Q_{22} = \frac{E_2(\bar{\omega})}{1 - v_{12}(\bar{\omega})\, v_{21}(\bar{\omega})} \qquad Q_{12} = \frac{v_{12}(\bar{\omega})\, E_2(\bar{\omega})}{1 - v_{12}(\bar{\omega})\, v_{21}(\bar{\omega})}$$

$$Q_{66} = G_{12}(\bar{\omega}) \qquad\qquad Q_{44} = G_{23}(\bar{\omega}) \qquad\qquad Q_{55} = G_{13}(\bar{\omega})$$

In present study, the uniaxial in-plane periodic loads are considered for composite panel with cutout. The differential equations of equilibrium can be expressed as (Chandrashekhara 1989)

$$\frac{\partial N_x(\overline{\omega})}{\partial x} + \frac{\partial N_{xy}(\overline{\omega})}{\partial y} - \frac{1}{2}C_2\left(\frac{1}{R_y(\overline{\omega})} - \frac{1}{R_x(\overline{\omega})}\right)\frac{\partial M_{xy}(\overline{\omega})}{\partial y}$$

$$+ C_1\left(\frac{Q_x(\overline{\omega})}{R_x(\overline{\omega})} + \frac{Q_y(\overline{\omega})}{R_{xy}(\overline{\omega})}\right) = P_1(\overline{\omega})\frac{\partial^2 u(\overline{\omega})}{\partial t^2} + P_2(\overline{\omega})\frac{\partial^2 \theta_x(\overline{\omega})}{\partial t^2}$$

(8.8)

$$\frac{\partial N_{xy}(\overline{\omega})}{\partial x} + \frac{\partial N_y(\overline{\omega})}{\partial y} + \frac{1}{2}C_2\left(\frac{1}{R_y(\overline{\omega})} - \frac{1}{R_x(\overline{\omega})}\right)\frac{\partial M_{xy}(\overline{\omega})}{\partial x}$$

$$+ C_1\left(\frac{Q_y(\overline{\omega})}{R_y(\overline{\omega})} + \frac{Q_x(\overline{\omega})}{R_{xy}(\overline{\omega})}\right) = P_1(\overline{\omega})\frac{\partial^2 v(\overline{\omega})}{\partial t^2} + P_2(\overline{\omega})\frac{\partial^2 \theta_y(\overline{\omega})}{\partial t^2}$$

(8.9)

$$\frac{\partial Q_x(\overline{\omega})}{\partial x} + \frac{\partial Q_y(\overline{\omega})}{\partial y} - \frac{N_x(\overline{\omega})}{R_x(\overline{\omega})} - \frac{N_y(\overline{\omega})}{R_y(\overline{\omega})} - 2\frac{N_{xy}(\overline{\omega})}{R_{xy}(\overline{\omega})}$$

$$+ N_x^0(\overline{\omega})\frac{\partial^2 w(\overline{\omega})}{\partial x^2} + N_y^0(\overline{\omega})\frac{\partial^2 w(\overline{\omega})}{\partial y^2} = P_1(\overline{\omega})\frac{\partial^2 w(\overline{\omega})}{\partial t^2}$$

(8.10)

$$\frac{\partial M_x(\overline{\omega})}{\partial x} + \frac{\partial M_{xy}(\overline{\omega})}{\partial y} - Q_x(\overline{\omega}) = P_3(\overline{\omega})\frac{\partial^2 \theta_x(\overline{\omega})}{\partial t^2} + P_2(\overline{\omega})\frac{\partial^2 u(\overline{\omega})}{\partial t^2} \quad (8.11)$$

$$\frac{\partial M_{xy}(\overline{\omega})}{\partial x} + \frac{\partial M_y(\overline{\omega})}{\partial y} - Q_y(\overline{\omega}) = P_3(\overline{\omega})\frac{\partial^2 \theta_y(\overline{\omega})}{\partial t^2} + P_2(\overline{\omega})\frac{\partial^2 u(\overline{\omega})}{\partial t^2} \quad (8.12)$$

wherein C_1 and C_2 are represented as tracers of shear deformable version of the theories of Sanders ($C_1 = C_2 = 1$), Love ($C_1 = 1$ and $C_2 = 0$), and Donnells ($C_1 = C_2 = 0$). $N_x(\overline{\omega})$, $N_y(\overline{\omega})$ and $N_{xy}(\overline{\omega})$ denote the stochastic in-plane stress resultants, $M_x(\overline{\omega})$, $M_y(\overline{\omega})$ and $M_{xy}(\overline{\omega})$ represents the stochastic moment resultants while $Q_x(\overline{\omega})$ and $Q_y(\overline{\omega})$ depict as the stochastic transverse shear stress resultants. $R_x(\overline{\omega})$, $R_y(\overline{\omega})$ and $R_{xy}(\overline{\omega})$ denote the stochastic radii of curvature along the x and y directions and the radius of twist, respectively.

$$(P_1, P_2, P_3)(\overline{\omega}) = \sum_{i=1}^{n}\int_{z_{k-1}}^{z_k} \rho_k(\overline{\omega})[1, z, z^2]\,dz \qquad (8.13)$$

where n is the layer number of laminate and $\rho_k(\overline{\omega})$ is the random mass density of k-th layer; and z_k represents the k-th layer's distance from the midplane. The present study considers eight nodes in isoparametric quadratic element wherein five degrees of freedom (three translations and two rotations) is assumed at each nodal point. Considering Hamilton's principle (Meirovitch 1992) in conjunction to Lagrange's equation, the dynamic equilibrium equation of motion for free vibration can be expressed as

$$[M(\overline{\omega})][\ddot{\delta}] + [K(\overline{\omega})]\{\delta\} = 0 \qquad (8.14)$$

where $M(\bar{\omega})$, $[K(\bar{\omega})]$, $\{\delta\}$ are represented as mass matrix, elastic stiffness matrix and vector of generalized coordinates, respectively. The random natural frequencies $[\omega_m(\bar{\omega})]$ are derived from the standard eigenvalue problem (Bathe 1990) using QR algorithm and are obtained as

$$\omega_m^2(\bar{\omega}) = \frac{1}{\lambda_m(\bar{\omega})} \quad \text{where} \quad m = 1, 2, 3.............., n_m \tag{8.15}$$

where n_m denotes the mode number and $\lambda_m(\bar{\omega})$ indicates the m-th eigenvalue of matrix $A = K^{-1}(\bar{\omega}) \, M(\bar{\omega})$.

8.3 Support vector regression

The Support vector regression (SVR) model is derived from the Support Vector Machine (SVM) pertaining to regression analysis. Suppose the training data is given as $\{(x_1, y_1), (x_2, y_2)........(x_l, y_l)\} \subset \chi \times \Re$ where χ and $\Re$ denote the space of the input patterns and Euclidean space vector. In support vector regression, the primary objective is to find a function $\hat{f}(x)$ that has at most ε deviation from the actually obtained targets y_i for all these training data and at the same time, is as flat as possible. In the formulation in Fig. 8.2, only the points distributed outside the shaded zone are contributed to the cost linked with insensitive loss function (ζ) at points (m and n) wherein the deviations are penalized in a linear fashion. Thus the optimization problem is solved more easily in its dual formulation. Moreover, the dual formulation provides the key for extending SV machine to nonlinear functions. Hence it can be used as a standard dualization method utilizing Lagrange multipliers. The errors are neglected as long as they are less than the region of tolerance (say $\pm \varepsilon$) (refer to Fig. 8.2), but it will not accept any deviation larger than this limiting value. The SVR model is constructed by employing the subset sample data and support vectors wherein maximum deviation of ε from the function value of each training data exist. For a linear case, SVR model can be expressed as (Vapnik 1998)

$$\hat{f}(x) = \hat{Y}(x) = <W \cdot x> + b \tag{8.16}$$

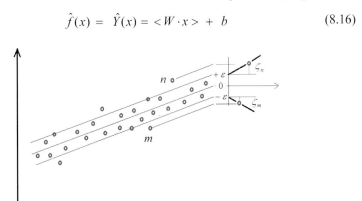

Fig. 8.2 Soft margin loss setting corresponding to a linear Support Vector machine (Fletcher 1989).

where $\hat{Y}(x)$, W and b indicate the predicted value of objective function, weight-vectors and bias, respectively while $<\cdot>$ denotes the inner product. The sample data points within the $\pm\varepsilon$ band (known as the ε-tube) are neglected, while the predictors are considered wherein the data points are found on or outside this region. The SVR prediction can be expressed as,

$$\hat{f}(x) = b + \sum_{i=1}^{k} W^{(i)} y^{(i)}(x, x^{(i)}) \qquad (8.17)$$

where $y^{(i)}$ and $W^{(i)}$ are the basis function and weights, respectively. For generalized prediction, it is therefore needed to develop the function with ε deviations from y as well as least complex. Despite reducing the risk of using training data for fitting, SVR reduces the upper bound on the calculated risk by employing ε-insensitive loss function, as constrained convex quadratic optimization problem proposed by Chandrashekhara (1989).

$$G(x) = \begin{cases} 0 & \left|Y(x) - \hat{Y}(x)\right| \le \varepsilon \\ \left|Y(x) - \hat{Y}(x)\right| - \varepsilon & Otherwise \end{cases} \qquad (8.18)$$

where ε and G parameters are selected based on the recommendation proposed by Cherkassky and Ma (2004). SVR model performs both linear as well as non-linear regression in conjunction to ε-insensitive loss function, simultaneously. It attempts to decrease the complexity by reducing the weighting vector as the objective function,

$$\text{Minimize } \frac{1}{2}\left|W\right|^2 \qquad (8.19)$$

$$\text{Subjected to } \begin{cases} Y_i - <W \cdot x^{(i)}> - b \le \varepsilon \\ <W \cdot x^{(i)}> + b - Y_i \le \varepsilon \end{cases}$$

A non-linear regression can be formed by replacing the $<\cdot>$ in Equation (8.16) with a kernel function, K as (Gunn 1997)

$$\hat{f}(x) = \sum_{i=1}^{k} (\alpha_i - \alpha_i^*) \, K(x_i, x) + b \qquad (8.20)$$

In the present study, Gaussian kernel function is used throughout the entire investigation.

8.4 Stochastic approach using SVR model

The dimension of cutout of laminated composite panel with respect to each layer can be defined as $C_o = c/a$ where c/a denotes the percentage of cutout with respect to overall panel-dimension. The effects of both single variable as well as multi-dimensional random variables are investigated in conjunction to different sizes of cutout in the present analysis as follows:

a) Variation of only ply-orientation angle:

$$\theta(\overline{\omega}) = \{\theta_1 \ \theta_2 \ \theta_3 ....\theta_i ......\theta_l\}$$

b) Variation of only twist angle:

$$\psi(\overline{\omega}) = \{\psi_1 \ \psi_2 \ \psi_3 ....\psi_i ......\psi_l\}$$

c) Variation of only thickness:

$$t(\overline{\omega}) = \{t_1 \ t_2 \ t_3 ....t_i ......t_l\}$$

d) Variation of only material properties:

$$p_m(\overline{\omega}) = \{p_{m(1)} \ p_{m(2)} \ p_{m(3)} ....p_{m(i)} ......p_{m(l)}\}$$

e) Combined variation of ply orientation angle, twist angle, thickness and materials properties:

$$g(\overline{\omega}) = \{\Phi_1(\theta_1...\theta_l), \Phi_2(\psi_1...\psi_l), \Phi_3(t_1...t_l), \Phi_4(p_{m(1)}...p_{m(l)})\}$$

where θ_i, ψ_i, t_i, $p_{m(i)}$ are the ply orientation angle, twist angle, thickness and material properties wherein material properties include $E_{1(i)}$, $E_{2(i)}$, $G_{12(i)}$, $G_{23(i)}$, $\mu_{12(i)}$ and ρ_i denoting the elastic modulus (longitudinal direction), elastic modulus (transverse direction), shear modulus (longitudinal direction), shear modulus (transverse direction), Poisson ratio and mass density, respectively. In conformity of the same, $\pm 5°$ variability is assumed for ply orientation and twist angle while $\pm 10\%$ variability from respective deterministic mean values of thickness and material properties are considered.

The flowchart of uncertainty quantification algorithm based on SVR with and without noise effect using SVR model is shown above in Fig. 8.3. Latin hypercube sampling is used for forming the sample dataset of the input space. The pronounced noise-effect on the proposed SVR based UQ algorithm is also accounted for by introducing different levels of noise as depicted in Fig. 8.3b. In the present investigation, Gaussian white noise is employed for SVR model formation

$$f_{ijN} = f_{ij} + p \times \xi_{ij} \tag{8.21}$$

where, p and f are the multiplication factor and natural frequency with the subscript i and j frequency number and sample number, respectively. A function generating random numbers (normal-distributed) with zero mean and unit standard deviation is represented as ξ_{ij}. Thus p (noise level) in the above expression basically represents the standard deviation of introduced noise level. Subscript N is used here to indicate the noisy frequency. Thus simulated noisy dataset (i.e., the sampling matrix for SVR model formation) is formed by introducing pseudo random noise in the responses, while the input design points are kept unaltered. Subsequently for each dataset, SVR based MCS is carried out to quantify uncertainty of composite laminates. The noise-effect has been investigated previously by some researchers (Friswell et al. 2015, Mukhopadhyay 2018a) mostly dealing with deterministic analysis. The assessment of SVR based uncertainty propagation with noise-effect is the first attempt of its kind to the best of the authors' knowledge. The root-causes of such inevitable noise-effect can be attributed to the fact of other unknown sources of uncertainty

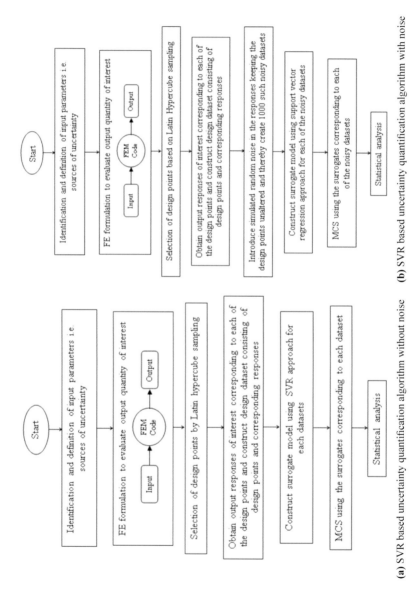

(a) SVR based uncertainty quantification algorithm without noise   **(b)** SVR based uncertainty quantification algorithm based on SVR including the effect of noise.

Fig. 8.3 Flowchart on uncertainty quantification algorithm based on SVR including the effect of noise.

such as measurement-errors, modelling-errors and computer simulation-errors and other system-specific epistemic uncertainties. Thus the present investigation is portrayed with a comprehensive idea about the robustness of the SVR based UQ algorithm including noise-effect.

8.5 Results and discussion

The present study deals with a three layered graphite–epoxy angle-ply composite cantilever spherical shallow panel with a central square shaped cutout. Four different types of panels are considered for detail analyses: plate ($R_x = R_y = \infty$), cylindrical ($a/R_x = \infty$, $b/R_y = 0.25$), spherical ($a/R_x = b/R_y = 0.25$) and hyperpolic paraboloid ($a/R_x = 0.25$, $b/R_y = 0.25$). The length, width and thickness of the composite laminate assumed in the present analyses are 1 m, 1 m and 5 mm, respectively and the dimension of the square shaped cutout size is considered as a percentage of its overall length and width [$C_l = c/a$ and $C_b = d/b$ wherein $c = d$ for square cutout] from 0.1 to 0.5 with a step of 0.1. Material properties of graphite–epoxy composite (Qatu and Leissa 1991) are considered with deterministic mean values as $E_1 = 138.0$ GPa, $E_2 = 8.96$ GPa, $G_{12} = 7.1$ GPa, $G_{13} = 7.1$ GPa, $G_{23} = 2.84$ GPa, $\mu = 0.3$, $\rho = 1600$ kg/m^3. A convergence study is carried out to validate the present formulation and to ascertain the optimal finite element mesh size as shown in Table 8.1 and Table 8.2. Table 8.1 presents the convergence study for non-dimensional fundamental natural frequencies of three layered graphite–epoxy untwisted angle-ply composite plates with finite element sizes (4×4), (6×6), (8×8) and (10×10), respectively in addition to comparison with the results obtained by Qatu and Leissa (1991a). In contrast, Table 8.2 presents the convergence of fundamental natural frequencies for a simply supported square plate with specific size of the cutout with finite element sizes (4×4), (8×8), (12×12), (16×16) and (20×20), respectively in addition to

Table 8.1 Convergence study for non-dimensional fundamental natural frequencies [$\omega = \omega_n L^2 \sqrt{(\rho/E_1 t^2)}$] of three layered ($\theta°/-\theta°/\theta°$) graphite-epoxy untwisted composite plates ($a/b = 1$ and $b/t = 100$).

θ	Present FEM							Qatu and Leissa (1991a)
	(4×4)	(6×6)	(8×8)	(10×10)	(12×12)	(16×16)	(20×20)	
0°	1.0112	1.0133	1.0107	1.0040	1.0031	1.0028	1.0022	1.0175
90°	0.2553	0.2567	0.2547	0.2542	0.2540	0.2533	0.2530	0.2590

Table 8.2 Convergence of fundamental natural frequencies ($\lambda = \omega a^2 \sqrt{\rho h / D_{22}}$) of simply supported square plate with cutout size of $c/a = 0.5$, $a/b = 1$, $b/h = 100$.

Shell type	Present FEM					Reddy (1982)
	(4×4)	(8×8)	(12×12)	(16×16)	(20×20)	
Isotropic	23.8432	23.570	23.4703	23.4364	23.4218	23.489
Orthotropic	51.8546	51.0597	50.7899	50.6944	50.6505	51.232
Composite	48.9546	48.2535	48.0650	48.0222	48.0064	48.414

comparison with the results obtained by Reddy (1982). Thus, Table 8.1 and Table 8.2 provide validation of the deterministic finite element model. A discretization of (8×8) mesh on plan area with 64 elements, 225 nodes with natural coordinates of an isoparametric quadratic plate bending element are considered for the present FEM analysis. In general, the number of expensive finite element analysis required for original Monte Carlo simulation based UQ approach is same as the sampling size. The present approach of SVR based uncertainty quantification develops a predictive and representative surrogate model relating each natural frequency to a number of stochastic input parameters.

SVR model for a particular mode represents the result encompassing each possible combination of all stochastic input parameters. A convergence study of sample size for SVR model formation with respect to original MCS is tabulated in Table 8.3 for the first three modes corresponding to individual (ply-orientation angle) and combined variation (ply-orientation angle, twist angle, thickness, elastic moduli, shear moduli, poission ratio and mass density). By analysing the statistical parameters presented in the Table 8.3 it is evident that sample size of 256 and 512 are adequate for the SVR model formation corresponding to individual and combined cases, respectively. Figure 8.4 and Fig. 8.5 present the scatter plot and probability density function plot, respectively considering the converged sample sizes for stochastic natural frequencies using SVR model and traditional MCS approach for angle-ply (45°/–45°/45°) composite curved panels with cutout corresponding to individual and for combined variation, respectively. It is evident from these figures that the results of the proposed SVR based approach are in good agreement with that of direct MCS simulations corroborating accuracy and validity of the proposed approach.

The probability density function plots for stochastic first three modes using SVR approach for individual variation of ply orientation angle considering angle-ply $(\theta°/–\theta°/\theta°)$ composite curved panels with cutout are presented in Fig. 8.6. The figure reveals that as the ply orientation angle (θ) of the angle-ply composite curved panels with a particular size of cutout increases, the stochastic first three natural frequencies are found to reduce.

In Fig. 8.4–8.12, the first three natural frequencies are referred as fundamental natural frequency (FNF), second natural frequency (SNF) and third natural frequency (TNF) which are stochastic in nature. Figure 8.7 presents the probability density function plot for stochastic fundamental natural frequency for only the individual variation of twist angle of angle-ply (45°/–45°/45°) composite curved panels with cutout. The twist angle in shell-panel structures causes reduction of stiffness which in turn decreases the values of the first three natural frequencies corresponding to a constant cutout size. The probability distributions of the first three natural frequencies for only variation of thickness corresponding to different sizes of cutout for composite shells are furnished in Fig. 8.8. The stochastic mean values of first three natural frequencies are found to reduce with increasing cutout size. Figure 8.9 presents probability distributions of first three natural frequencies for variation in all the material properties. It is observed that the combined effects

Table 8.3 Convergence study of natural frequencies corresponding to first three modes due to individual $[\theta(\bar{\omega})]$ and combined effect $[g(\bar{\omega})]$ of stochasticity for composite curved panels with cutout ($C_{o} = 0.1$).

Type	Mode	Method	Samples	Maximum	Minimum	Mean	SD
Individual variation (Only ply-orientation angle)	First	MCS	10,000	17.8750	13.9750	15.8501	0.7935
		Present method	64	17.8810	13.9782	15.8561	0.7948
			128	17.8604	13.9835	15.8464	0.7946
			256	17.8725	13.9905	15.8556	0.7929
			512	17.8770	13.9887	15.8570	0.7941
			1024	17.8758	13.9762	15.8512	0.7945
	Second	MCS	10,000	100.8807	79.8358	90.0001	4.2127
		Present method	64	100.7124	80.99413	90.1040	4.1351
			128	100.6888	80.95801	89.9539	4.1981
			256	100.8305	80.01091	90.0567	4.2017
			512	100.8747	79.91743	90.0316	4.2215
			1024	100.8813	79.84141	90.0016	4.2301
	Third	MCS	10,000	136.1571	122.0387	128.9192	2.7264
			64	136.1873	122.1453	129.9542	2.7171
		Present method	128	136.2222	121.9553	128.9793	2.7513
			256	136.1728	122.1545	129.9441	2.6950
			512	136.1651	122.0598	128.9398	2.7144
			1024	136.1578	122.0436	128.9291	2.7345
Combined variation	First	MCS	10,000	21.3550	13.1850	17.0607	1.3754
			128	21.2290	13.2270	17.08304	1.3595
		Present method	256	21.4815	13.1062	17.07911	1.3963
			512	21.4277	13.1694	17.0867	1.3848
			1024	21.3673	13.1416	17.07897	1.3822
			2048	21.3778	13.1873	17.07107	1.3787
	Second	MCS	10,000	110.3002	73.0035	90.6014	6.4184
			128	110.3709	72.9484	90.7125	6.3907
		Present method	256	110.9686	72.7500	90.6700	6.5456
			512	110.3437	73.1406	90.6806	6.4109
			1024	110.3767	73.0716	90.6662	6.4375
			2048	110.3034	73.0315	90.6418	6.4297
	Third	MCS	10,000	192.8565	123.9751	158.4033	14.0007
			128	192.6459	124.6592	158.5274	13.9243
		Present method	256	193.2740	124.7608	158.6299	13.6715
			512	192.5198	124.5556	158.4206	13.8215
			1024	193.0577	124.1068	158.3772	13.9229
			2048	192.8565	123.9751	158.4033	14.0007

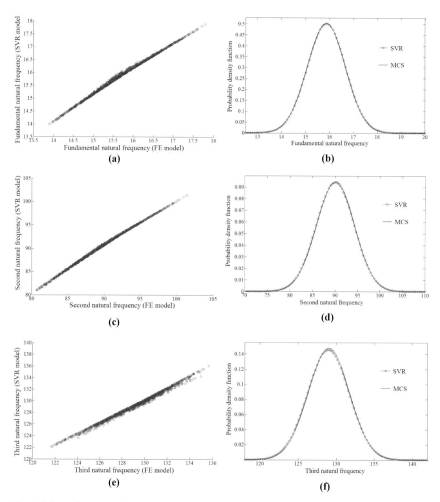

Fig. 8.4 (a, c, e) Scatter plot and (b, d, f) Probability density function plot for stochastic first three natural frequencies using SVR approach for individual variation of ply orientation angle [$\theta(\bar{\omega})$] of angle-ply (45°/–45°/45°) composite curved panel with cutout (C_o = 0.1).

of variation of all the material properties follow Gaussian distribution and the mean for all the three natural frequencies reduce with the increase in cutout size.

The probability density function plots of the first three modes with combined variation (ply-orientation angle, twist angle, thickness, elastic moduli, shear moduli, poission ratio and mass density) corresponding to different twist angles are shown in Fig. 8.10. The mean of stochastic natural frequencies are found to reduce with increase in twist angle. It is interesting to notice that the probabilistic distributions for combined variation (Fig. 8.10) follow Gaussian distribution, in contrast to the probabilistic characters depicted for individual variation of twist angle only (Fig. 8.7). Probability distributions for the first three modes with combined

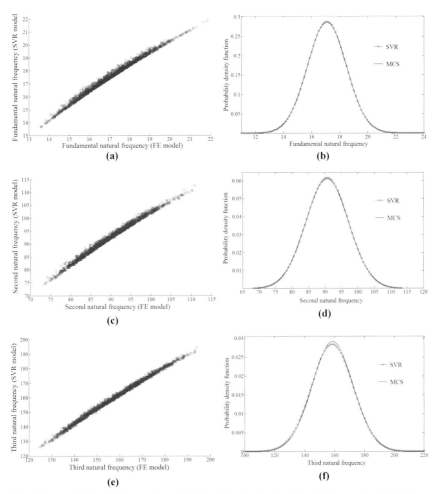

Fig. 8.5 (a, c, e) Scatter plot and (b, d, f) Probability density function plot for stochastic first three natural frequencies using SVR approach for combined variation of ply angle, elastic modulus, shear modulus, Poisson ratio and mass density [$g(\bar{\omega})$] of angle-ply (45°/–45°/45°) composite curved panel with cutout ($C_o = 0.1$).

variation corresponding to different sizes of cutout are shown in Fig. 8.11, wherein it is evident that the distributions are of Gaussian nature and mean of stochastic natural frequencies decrease with the increase in cutout size. The probability density function plots of the stochastic first three natural frequencies for combined variation corresponding to different geometry such as composite plate, cylindrical, spherical, hyperbolic paraboloid curved shells as shown in Fig. 8.12. Even though the probability distributions are found to follow Gaussian distributions, the mean and standard deviation of natural frequencies for different modes are highly dependent on the type of shell geometry. Figure 8.13 presents the coefficient of

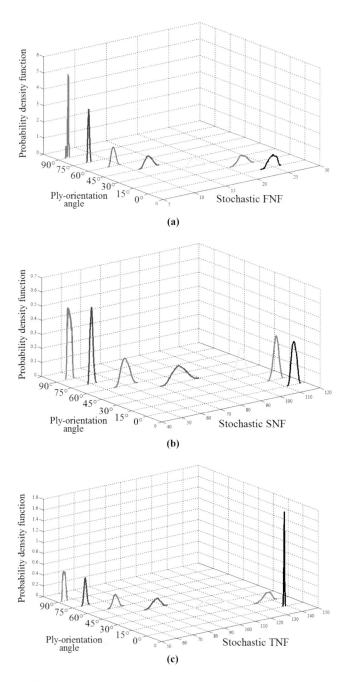

Fig. 8.6 Probability density function plot for first three stochastic natural frequencies using SVR approach for individual variation of ply orientation angle $[\theta(\bar{\omega})]$ of angle-ply $(\theta^\circ/-\theta^\circ/\theta^\circ)$ composite curved panel with cutout.

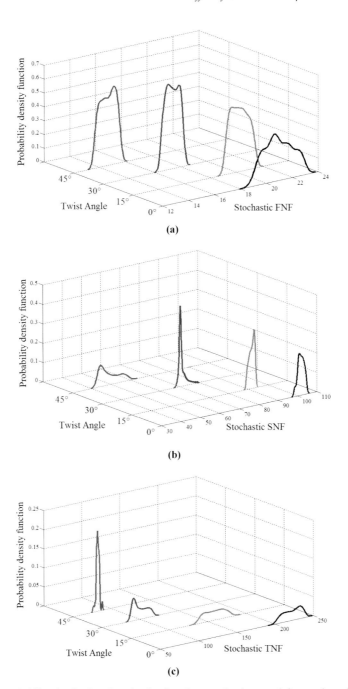

Fig. 8.7 Probability density function plot for first three stochastic natural frequencies using SVR approach for individual variation of twist angle $[\psi(\bar{\omega})]$ of angle-ply (45°/–45°/45°) composite curved shells with cutout.

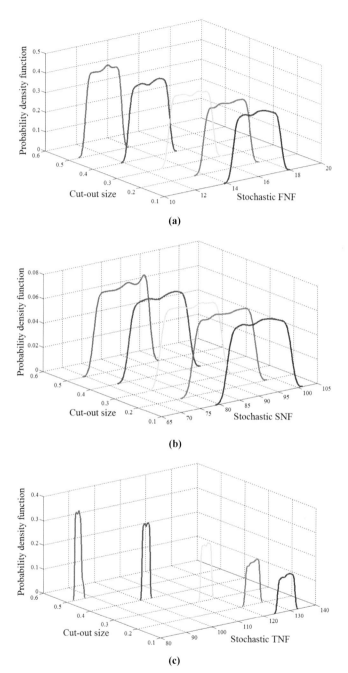

Fig. 8.8 Probability density function plots for variation of only thickness $[t(\bar{\omega})]$ corresponding to different cutout sizes for composite curved shells with cutout.

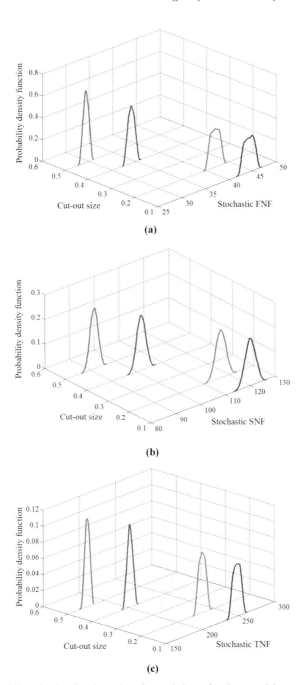

Fig. 8.9 Probability density function plots for variation of only material properties $[p_m(\bar{\omega})]$ corresponding to different cutout sizes for composite curved shells with cutout.

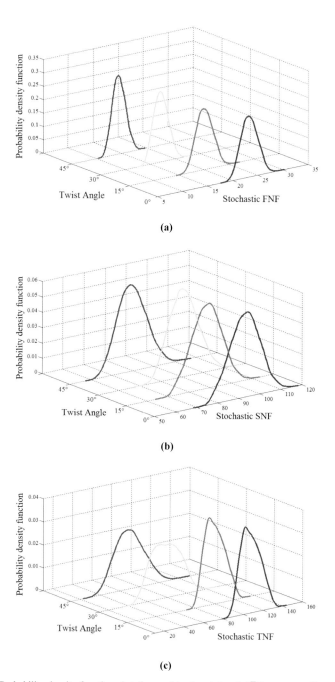

Fig. 8.10 Probability density function plots for combined variation $[g(\bar{\omega})]$ corresponding to different twist angle for composite curved shells with cutout.

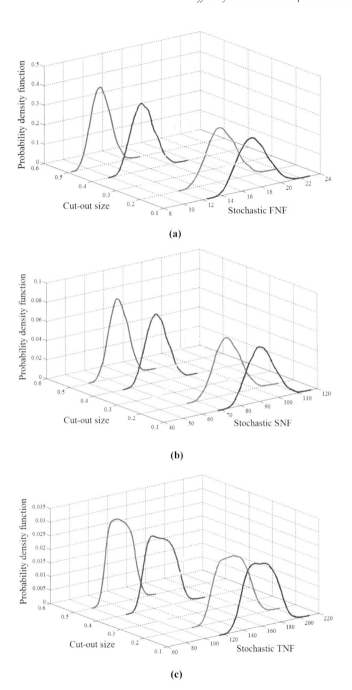

(a)

(b)

(c)

Fig. 8.11 Probability density function plots for combined variation of ply orientation angle, thickness, twist angle and material properties [$g(\bar{\omega})$] corresponding to different cutout sizes for composite curved shells with cutout.

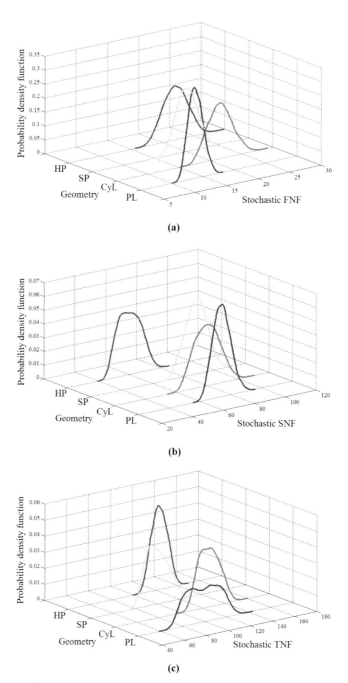

Fig. 8.12 Probability density function plots for combined variation $[g(\bar{\omega})]$ corresponding to different geometry for composite curved shells with cutout (PL-Plate, CyL-Cylindrical, SP-Spherical, HP-Hyperbolic Paraboloid).

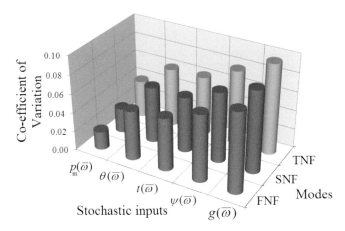

Fig. 8.13 Coefficient of variation for different of variations of input parameters.

variation (ratio of standard deviation and mean for a distribution) of the stochastic first three natural frequencies corresponding to individual and combined variation of stochastic input parameters. Coefficient of variation is highest for combined variation of all input parameters, as expected. The analysis presented in Fig. 8.13 provides a measure of relative sensitivity of different input parameters towards the natural frequencies. Among the stochastic geometric features, twist angle is found to be the most sensitive parameter followed by thickness, ply orientation angle and material properties, respectively. It is worthy to note here that all the probabilistic results furnished in this chapter are obtained on the basis of 10,000 simulations. Application of the support vector regression based approach allows us to obtain these results by means of efficient virtual simulations instead of actual expensive finite element simulation.

The number of samples to construct the SVR model are 512 and 256 for layer-wise combined variation and individual variation of the stochastic input parameters, respectively. Thus for the purpose of stochastic analysis, the same number of actual FE simulations are required in the present approach, in contrast with 10,000 FE simulations needed in the traditional MCS based approach. Therefore, the proposed SVR approach for uncertainty quantification is more computationally efficient than the traditional MCS approach in terms of FE simulations. Figure 8.14 shows the effect of noise on prediction using SVR considering combined variation of all input parameters. Representative results are presented in Fig. 8.15 showing the effect of noise on first three natural frequencies considering different levels of noise (p) ranging from 0 to 0.15. The results presented in this chapter are obtained on the basis of 1000 such noisy datasets, which involves construction of the SVR model and thereby performing MCS for each dataset using corresponding SVR models as explained in Fig. 8.3b. A comparative assessment of the effect of different levels of noise with noise free data ($p = 0$) provides a comprehensive idea about the influence of such noise in the probability distributions of first three natural frequencies.

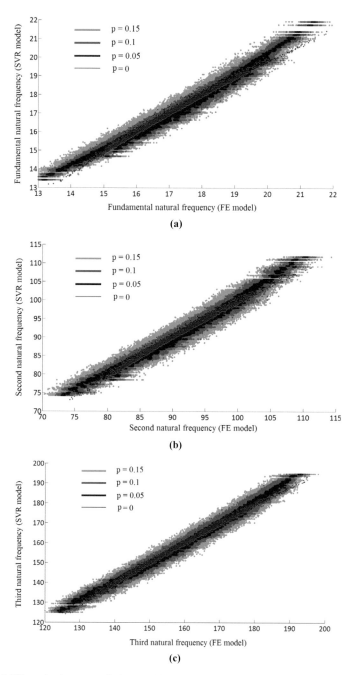

Fig. 8.14 Effect of noise on prediction capability of SVR model for first three natural frequencies considering combined stochasticity.

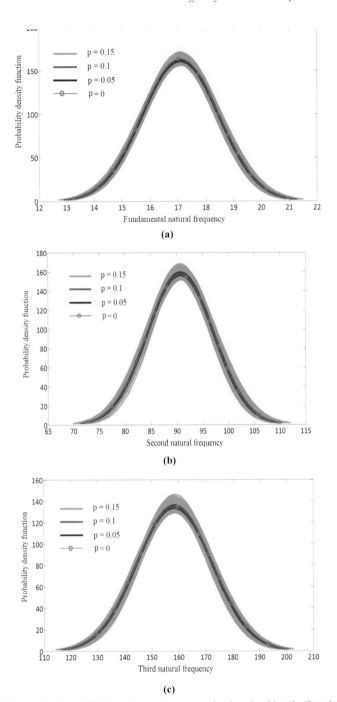

Fig. 8.15 Effect of noise on SVR based uncertainty quantification algorithm for first three natural frequencies considering combined stochasticity.

8.6 Summary

This chapter presents an efficient support vector regression based stochastic natural frequency analysis for laminated composite curved panels with cutout, including the effect of twist angle and variation in shell-panel geometry (such as cylindrical, spherical, hyperbolic paraboloid and plate). First three stochastic natural frequencies are analyzed considering the layer-wise variation of individual (low dimensional input parameter space) as well as the combined effect (relatively higher dimensional input parameter space) of random input parameters (such as ply orientation, twist angle, thickness and material properties). The computational time and cost are reduced significantly by using the present SVR based approach compared to the direct Monte Carlo simulation method. A sensitivity analysis among the stochastic material and geometric features is carried out to ascertain their relative importance. The effect of noise on the SVR based uncertainty quantification algorithm in characterizing the probabilistic distribution of natural frequencies is presented to account for the inevitable variability in practical applications of such structures. Even though this chapter focuses on stochastic natural frequency analysis of laminated composite curved panels, the SVR based approach for uncertainty quantification including the effect of noise can be extended to deal with other computationally intensive problems in different fields of science and engineering.

CHAPTER 9

Stochastic Dynamic Stability Analysis of Composite Laminates

9.1 Introduction

The previous chapters in this book address the effect of uncertainty on the dynamic responses of composite structures (such as stochastic natural frequency, mode shape and frequency response function). In this chapter, a stochastic dynamic stability analysis of composite panels is presented considering the effect of non-uniform partial edge loading. The system input parameters are randomized to ascertain the stochastic first buckling load and zone of resonance. Considering the effects of transverse shear deformation and rotary inertia, first order shear deformation theory is used to model the composite curved panels. Moving least square method is employed as a surrogate of the actual finite element model to reduce the computational cost. Statistical results are presented to show the effects of radius of curvatures, material properties, fibre parameters, and non-uniform load parameters on the stability boundaries.

Composite structures subjected to in-plane periodic forces may lead to parametric resonance because of certain random combinations in the values of uncertain system parameters. The instability may occur below the stochastic critical load of the structure under compressive loads over wide ranges of resonance frequencies. Specially the aerospace structures such as skin panels in wings, fuselage, submarine hulls and civil application has practical importance of stability analysis of doubly curved panels/open shells subjected to uncertain non-uniform loading condition. Traditionally, structural analysis is formulated with deterministic behavior of material properties, loads and other system parameters.

However, the real-life structures employed in aerospace, naval, civil, and mechanical applications are always subjected to intrusive uncertainties. The inherent sources of such uncertainties in real structural problems can be due to randomness in material properties, loading conditions, geometric properties and other random input parameters. As an inevitable consequence of the uncertainties in these system parameters, the response of structural system will always exhibit some degree of uncertainty. The traditional deterministic analysis based on an exact reliable model would not help in properly accounting the variation in the response and therefore, the analysis based on deterministic material properties may vary significantly from the real behavior. The incorporation of randomness of input parameters enables the prediction of the performance variation in the presence of uncertainties and more importantly their sensitivity for targeted testing and quality control. In order to provide useful and accurate information about the safe and reliable design of structures, it is essential to incorporate these uncertainties into account for modeling, design and analysis procedure. The steady development of powerful computational technologies in recent years has led to high-resolution numerical models of real-life engineering structural systems. It is also required to quantify uncertainties and robustness associated with a computational model. Hence, the quantification of uncertainties plays a key role in establishing the credibility of a numerical model. Therefore, the development of an efficient mathematical model possessing the capability to quantify the uncertainties present in the structures is extremely essential in order to accurately assess the laminated composite structures.

Structural elements under in-plane periodic forces may undergo unstable transverse vibrations, leading to parametric resonance, due to certain combinations of the values of in-plane load parameters and natural frequency of transverse vibration. Several means of combating parametric resonance such as damping and vibration isolation may be inadequate and sometimes dangerous with reverse results (Evan-Iwanowski 1965). A number of catastrophic incidents can be traced to parametric instability and is often studied in the spectrum of determination of natural frequency and critical load of structures. The stochasticity in the measurement of natural frequencies, critical load and ultimately the excitation frequencies during parametric resonance are of great technical importance in studying the instability behavior of dynamic systems. Many authors have addressed the parametric instability characteristics of laminated composite flat panel subjected to uniform loads (Iwatsubo et al. 1973, Moorthy and Reddy 1990, Chen and Yang 1990, Patel et al. 2009, Kochmann and Drugan 2009, Singha and Daripa 2009, Kim et al. 2013). In contrast, Bolotin (1964) and Yao (1965) studied the parametric resonance subjected to periodic loads. The Stochastic principal parametric resonance of composite laminated beam has been numerically investigated by Lan et al. (2014). Dey et al. (2018a) studied the stochastic dynamic stability of composite laminates. The influences of transverse shear (Andrzej et al. 2011) and rotary inertia (Ratko et al. 2012) on dynamic instability are studied for cross-ply laminated plates. The parametric dynamic stability analysis has been numerically investigated for composite beam (Meng-Kao and Yao 2004), plates (Dey and Singha 2006) or shells

(Bert and Birman, 1988) and stiffened panel (Sepe et al. 2016). Further studies have also been carried out for modelling mesoscopic volume fraction stochastic fluctuations in fiber reinforced composites (Guilleminot et al. 2008) and for parametric instability of graphite–epoxy composite beams under excitation (Yeh and Kuo 2004). Free vibration and dynamic stability analysis of rotating thin-walled composite beams (Saraviaa et al. 2011) and nonlinear thermal stability of eccentrically stiffened functionally graded truncated conical shells have been recently reported (Duc and Cong 2015). In contrast, many numerical investigations have been carried out using response surface methods such as moving least square (MLS) method and other methods for structural analysis (Choi et al. 2004, Wu et al. 2005, Park and Grandhi 2014, Shu et al. 2007, Kang et al. 2010). Some researchers specifically studied the moving least squares (MLS) approximation for the regression analysis (Lancaster and Salkauskas 1981, Breitkpf et al. 2005) instead of the conventional least squares (LS) approximation in conjunction to traditional response surface method (RSM) techniques (Mukhopadhyay et al. 2015b, Dey et al. 2015d).

The application of stochastic non-uniform loading on the structural component can significantly alter the global dynamic quantities of interests such as resonance frequency, buckling loads and dynamic stability region (DSR). Thus it is imperative to consider the effect of stochasticity for robust analysis, design and control of the system. Even though the perturbation method is an efficient way of stochastic analysis for relatively simpler structures (Kaminski 2013, Gadade et al. 2016a), this intrusive method can be mathematically quite cumbersome for complex problems like stochastic dynamic stability analysis of composite laminates. The main drawback of this method is that it can obtain only the statistical moments (not the entire probability distribution) of the stochastic output quantity of interest. If the nature of the output distribution is known to be Gaussian, the probability distribution can be obtained using the first two moments. However, the nature of distribution of the output parameter may not be known a priori in most engineering problems. Monte Carlo simulation, on the other hand, can obtain the entire probabilistic description of the stochastic output parameter. The main lacuna of traditional Monte Carlo simulation is its computational intensiveness. A surrogate based Monte Carlo simulation approach, as followed in this chapter, allows us to quantify the probabilistic descriptions in a computationally efficient manner. In the present study, a moving least square based approach is employed in conjunction with finite element formulation to figure out the random eigenvalue problem and quantify the probabilistic characteristics of the responses related to dynamic stability of composite laminates. The numerical results are shown for first random buckling load and stochastic fundamental resonance frequencies with individual and combined variation of the stochastic input parameters. This chapter is organized hereafter as, Section 9.2: importance of stochastic dynamic stability analysis in the context of composite laminates, Section 9.3: stochastic finite element formulation for the dynamic stability analysis of composite panels, Section 9.4: brief description of the moving least square based surrogate modelling approach,

Section 9.5: representation of stochastic input parameters and the description of the surrogate based uncertainty quantification algorithm, Section 9.6: results and discussion, Section 9.7: summary.

9.2 Importance of stochastic dynamic stability analysis in composite laminates

Engineering structures are often subjected to periodic loads. For examples, aerospace structures are subjected to wind load, rotating machine systems usually exert a periodic unbalanced inertia force, bridges are frequently subjected to the cyclic loads from the running vehicles, marine structures continuously suffer the periodic wave forces, etc. Structural components subjected to in-plane periodic forces undergo an unstable dynamic response known as dynamic instability or parametric instability or parametric resonance. Parametric resonance, may occur for certain combinations of natural frequency of transverse vibration, the frequency of the in-plane forcing functions and the magnitude of the in-plane load. A number of flight accidents can be traced due to parametric instability of structures. In comparison to the principal resonance, the parametric instability can take place not only at a single excitation frequency but even for small excitation amplitudes and combination of frequencies. The difference between good and bad vibration regimes of a structure under in-plane periodic loads can be found from the dynamic instability region (DIR) spectra. The computation of these spectra is usually studied in term of natural frequencies and static buckling loads. The parametric instability has a catastrophic effect on structures near critical regions of parametric instability. Hence, the parametric resonance characteristics of structures are of great technical importance for understanding the dynamic characteristics under periodic loads.

As discussed in the preceding paragraph, structures are subjected to dynamic loads more often than static loads. Dynamic load means the load varies with time. Periodic loading is one type of dynamic loading. This type of load occurs in repeated periods or cycles like sine and cosine functions. Structures subjected to in-plane periodic loads can be expressed in the form as suggested by Bolotin (1964): $P(t) = P_s + P_t\, cos\Omega t$, where P_s is the static portion of $P(t)$, P_t is the amplitude of the dynamic portion of $P(t)$ and Ω is the frequency of excitation. It can be noted here that the quantities P_s, P_t, Ω possess random values in practical systems. This, in turn, makes the time varying periodic load $P(t)$ random in nature. The present chapter aims to account such stochastic character of the time varying load along with other sources of stochasticity for a comprehensive probabilistic analysis of the system. Laminated composites being a complex structural form and susceptible to different forms of uncertainty, the compound effects of stochastic time varying loading and structural and material uncertainties associated with composites can be crucial in the intended performance for various engineering applications.

9.3 Governing equations for stochastic dynamic stability analysis

In the present study, a layered graphite–epoxy composite laminated simply supported shallow doubly curved shell is considered with thickness t, intensity of loading C, principal radii of curvature R_x, R_y along x- and y-direction, respectively and the radius of curvature R_{xy} in x-y plane, as furnished in Fig. 9.1. Using Hamilton's principle (Meirovitch 1992) for free vibration of composite shell structure subjected to in-plane loads, the equation of equilibrium can be expressed as

$$[M(\tilde{\omega})][\ddot{q}] + ([K_e(\tilde{\omega})] - F(\tilde{\omega})[K_g(\tilde{\omega})])\{q\} = 0 \qquad (9.1)$$

where $M(\tilde{\omega})$, $K_e(\tilde{\omega})$ and $K_g(\tilde{\omega})$ are mass, elastic stiffness and geometric stiffness matrices, respectively. Here $\tilde{\omega}$ is used to denote the element of probability space. Therefore, any quantity expressed as a function of $\tilde{\omega}$ is a random quantity (can be a scalar, vector or a matrix). The in-plane load $[F(\tilde{\omega})(t)]$ is periodic and can be expressed in the stochastic form (Patel et al. 2009)

$$F(\tilde{\omega})(t) = F_s(\tilde{\omega}) + F_t(\tilde{\omega})Cos\Omega t \qquad (9.2)$$

where $F_s(\tilde{\omega})$ and $F_s(\tilde{\omega})$ are the random static portion and the amplitude of the dynamic portion of stochastic in-plane load, respectively. The static buckling load of elastic shell $F_{cr}(\tilde{\omega})$ is the measure of the magnitude of $F_s(\tilde{\omega})$ and $F_t(\tilde{\omega})$

$$F_s(\tilde{\omega}) = \alpha(\tilde{\omega}) F_{cr}(\tilde{\omega}) \qquad F_t(\tilde{\omega}) = \beta(\tilde{\omega}) F_{cr}(\tilde{\omega}) \qquad (9.3)$$

where $\alpha(\tilde{\omega})$ and $\beta(\tilde{\omega})$ are known as static and dynamic load factors, respectively. The equation of motion can be expressed by employing equation (9.2) as

$$[M(\tilde{\omega})][\ddot{q}] + ([K_e(\tilde{\omega})] - \alpha(\tilde{\omega}) F_{cr}(\tilde{\omega})[K_g(\tilde{\omega})] - \beta(\tilde{\omega}) F_{cr}(\tilde{\omega})[K_g(\tilde{\omega})] cos\Omega t)\{q\} = 0 \qquad (9.4)$$

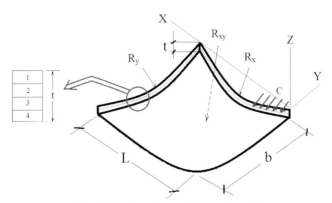

Fig. 9.1 Laminated composite curved panel.

It can be noted that the matrices involved in equation (9.4) are stochastic in nature. Depending on the degree of stochasticity, each element of the matrices is random in nature. The solution of equation (9.4) would obtain different results for each of the realizations of a Monte Carlo simulation depending on the respective set of input parameters. Thus probabilistic distributions can be obtained based on the results of different realizations following a non-intrusive method. This stochastic equation (9.4) indicates second order differential equations with periodic Mathieu-Hill type coefficients. The formation of zone of instability arises from Floquet's theory which establishes the existence of periodic solutions. The periodic solutions of period T and $2T$ derive the limiting bounds of the dynamic instability regions (where $T = 2\pi/\Omega$). The significant stochastic importance lies in the limiting bounds of primary instability regions with period $2T$ (Chen and Yang 1990) wherein the solution can be represented as the trigonometric series form

$$q(t) = \sum_{j=1,3,5}^{\infty} \left[\{a_j\} \sin\left(\frac{j\Omega t}{2}\right) + \{b_j\} \sin\left(\frac{j\Omega t}{2}\right) \right] \tag{9.5}$$

Considering this in equation (9.4) and first term of the above series, the equation (9.4) can be expressed by equating the coefficients of $Sin(\Omega t/2)$ and $Cos(\Omega t/2)$ as

$$\left[[K_e(\tilde{\omega})] - \alpha(\tilde{\omega}) \, F_{cr}(\tilde{\omega}) \, [K_g(\tilde{\omega})] \pm \beta(\tilde{\omega}) \, F_{cr}(\tilde{\omega}) \, [K_g(\tilde{\omega})] - \frac{\Omega^2}{4} [M] \right] \{q\} = 0 \tag{9.6}$$

The above equation represents an eigenvalue problem for known values of $\alpha(\tilde{\omega})$, $\beta(\tilde{\omega})$ and $F_{cr}(\tilde{\omega})$ as for $\Omega_j q_j = 0$ for j = 1,2,3.... Here the two conditions under a plus and minus sign represents the two limiting bounds of the dynamic instability region. The eigenvalues (Ω_j) provide the boundary frequencies of the instability regions for specific values of α and β and the reference stochastic static buckling load is computed accordingly (Ganapathi et al. 1999) and in contrast, the exact solution for doubly curved shells can also be carried out (Chaudhuri and Abuarja 1988). An eight-noded curved isoparametric element is employed with five degrees of freedom u, v, w, θ_x and θ_y per node. The present study employs the first order shear deformation theory and the shear correction coefficient for the nonlinear distribution of the thickness shear strains through the total thickness. The displacement field along the mid-plane is assumed to be straight before and after deformation, but it is not necessary to remain normal after deformation. The displacement components can be expressed as

$$\bar{u}\,(x, y, z) = u\,(x, y) + z\,\theta_x(x, y)$$

$$\bar{v}\,(x, y, z) = v\,(x, y) + z\,\theta_y(x, y) \tag{9.7}$$

$$\bar{w}(x, y, z) = w(x, y)$$

where the rotations of the mid-plane surface are represented by θ_x and θ_y. Here the displacement components in the x, y, z directions at any point and at the mid-plane surface are denoted as $\bar{u}$, $\bar{v}$, $\bar{w}$, and u, v and w, respectively. Thus the integrated relationship for the composite curved shell can be represented as

$$
\begin{Bmatrix} N_i(\tilde{\omega}) \\ M_i(\tilde{\omega}) \\ Q_i(\tilde{\omega}) \end{Bmatrix} = \begin{bmatrix} A_{ij}(\tilde{\omega}) & B_{ij}(\tilde{\omega}) & 0 \\ B_{ij}(\tilde{\omega}) & D_{ij}(\tilde{\omega}) & 0 \\ 0 & 0 & S_{ij}(\tilde{\omega}) \end{bmatrix} \begin{Bmatrix} \varepsilon_j \\ k_j \\ \gamma_j \end{Bmatrix}
\tag{9.8}
$$

where A_{ij}, B_{ij}, D_{ij} (where i, j = 1,2,6) and S_{ij} (where i, j = 4,5) are the extension-bending coupling, bending and transverse shear stiffness, respectively. The shear correction factor (= 5/6) is incorporated in S_{ij} in the numerical calculation. In the present analysis, shear deformable Sander's kinematic relation (Bathe 1990) is extended for doubly curved shells. The strain displacement equations of linear nature can be obtained as

$$
\varepsilon_{xl}(\tilde{\omega}) = \frac{\partial u}{\partial x} + \frac{w}{R_x(\tilde{\omega})} + z\,\kappa_x
$$

$$
\varepsilon_{yl}(\tilde{\omega}) = \frac{\partial v}{\partial y} + \frac{w}{R_y(\tilde{\omega})} + z\,\kappa_y
$$

$$
\gamma_{xyl}(\tilde{\omega}) = \frac{\partial u}{\partial y} + \frac{\partial v}{\partial x} + z\,\kappa_{xy}
\tag{9.9}
$$

$$
\gamma_{yz}(\tilde{\omega}) = \frac{\partial w}{\partial y} + \theta_y - C_1 \frac{v}{R_y(\tilde{\omega})}
$$

$$
\gamma_{xz}(\tilde{\omega}) = \frac{\partial w}{\partial x} + \theta_x - C_1 \frac{u}{R_x(\tilde{\omega})}
$$

where

$$
\kappa_x = \frac{\partial \theta_x}{\partial x} \text{ and } \kappa_y = \frac{\partial \theta_y}{\partial y}
\tag{9.10}
$$

$$
\kappa_{xy}(\tilde{\omega}) = \frac{\partial \theta_x}{\partial y} + \frac{\partial \theta_y}{\partial x} + \frac{1}{2} C_2 \left(\frac{1}{R_x(\tilde{\omega})} - \frac{1}{R_y(\tilde{\omega})} \right) \left(\frac{\partial v}{\partial x} - \frac{\partial u}{\partial y} \right)
$$

Here the formulation can be derived to shear deformable Love's first approximation and Donnell's theories from tracers (C_1 and C_2). Considering nonlinearity in strain, the element geometric stiffness matrix for doubly curved shells can be expressed as

$$\varepsilon_{xnl}(\tilde{\omega}) = \frac{1}{2}\left(\frac{\partial u}{\partial x} + \frac{w}{R_x(\tilde{\omega})}\right)^2 + \frac{1}{2}\left(\frac{\partial v}{\partial x}\right)^2 + \frac{1}{2}\left(\frac{\partial w}{\partial x} - \frac{u}{R_x(\tilde{\omega})}\right)^2 + \frac{1}{2}z^2\left[\left(\frac{\partial\theta_x}{\partial x}\right)^2 + \left(\frac{\partial\theta_y}{\partial x}\right)^2\right]$$

$$\varepsilon_{ynl}(\tilde{\omega}) = \frac{1}{2}\left(\frac{\partial u}{\partial y}\right)^2 + \frac{1}{2}\left(\frac{\partial v}{\partial y} + \frac{w}{R_y(\tilde{\omega})}\right)^2 + \frac{1}{2}\left(\frac{\partial w}{\partial y} - \frac{v}{R_y(\tilde{\omega})}\right)^2 + \frac{1}{2}z^2\left[\left(\frac{\partial\theta_x}{\partial y}\right)^2 + \left(\frac{\partial\theta_y}{\partial y}\right)^2\right]$$

$$(9.11)$$

$$\gamma_{xynl}(\tilde{\omega}) = \left[\left(\frac{\partial u}{\partial x} + \frac{w}{R_x(\tilde{\omega})}\right)\frac{\partial u}{\partial y} + \left(\frac{\partial v}{\partial y} + \frac{w}{R_y(\tilde{\omega})}\right)\frac{\partial v}{\partial x} + \left(\frac{\partial w}{\partial x} - \frac{u}{R_x(\tilde{\omega})}\right)\left(\frac{\partial w}{\partial y} - \frac{v}{R_y(\tilde{\omega})}\right)\right]$$
$$+ z^2\left[\left(\frac{\partial\theta_x}{\partial x}\right)\left(\frac{\partial\theta_x}{\partial y}\right) + \left(\frac{\partial\theta_y}{\partial x}\right)\left(\frac{\partial\theta_y}{\partial y}\right)\right]$$

The overall stochastic stiffness and mass matrices, i.e., $[K_e(\tilde{\omega})]$, $[K_g(\tilde{\omega})]$ and $[M(\tilde{\omega})]$ are obtained by assembling the corresponding element matrices by using skyline technique. The element mass and stiffness matrices of composite shells are computed wherein the geometric stiffness matrix is obtained as the function of in-plane stress distribution in the element due to applied edge loading. Due to non-uniformity in the stress field, plane stress analysis is carried out by using the finite element formulation. The possible shear locking is avoided by employing the reduced integration technique for the element matrices. The subspace iteration method (Bathe 1990) is utilized to solve the stochastic eigenvalue problems.

9.4 Moving least square method

In general, the polynomial regression models give the large errors in conjunction to non-linear responses while giving good approximations in small regions wherein the responses are less complex. Such features are found advantageous while implementing the method of moving least squares (MLS). Moreover, the least square method gives a good result to represent the original limit state but it creates a problem if anyone likes to fit a highly nonlinear limit function with this technique because this technique uses the same factor for approximation throughout the space of interest. To overcome this problem, the moving least square method is introduced. In this method, a weighted interpolation function or limit state function is employed to the response surface and some extra support points are also generated over least square method to represent perfectly the nonlinear limit surface. In stochastic analysis, uncertainties can be expressed as a vector of random variables, $x = [x_1, x_2, x_3, \ldots\ldots x_n]^T$, characterized by a probability density function (PDF) with a particular distribution such as normal or lognormal with limit state function of these random variables. To avoid the curse of dimensionality in dealing with random input variables, response surface methods (RSM) can be utilised to increase the computational efficiency. These methods approximate an implicit limit state function as a response surface function (RSF) in an explicit form, which is evaluated for a set of selected design points throughout a number of deterministic structural analyses. RSM approximates an implicit limit state function as a RSF in explicit form. It selects experimental points by an axial sampling scheme and fits these experimental points using a second order polynomial without cross terms expressed as

$$y(x) = \beta_o + \sum_{i=1}^{k}\beta_i \ x_i + \sum_{i=1}^{k}\sum_{j>i}^{k}\beta_{ij} \ x_i \ x_j + \sum_{i=1}^{k}\beta_{ii} \ x_i^2 \qquad (9.12)$$

where β_o, β_i, β_{ij} and β_{ii} are the unknown coefficients of the polynomial equation. The least squares approximation commonly used in the conventional RSM allots equal weight to the experimental points in evaluating the unknown coefficients of the RSF. The weights of these experimental points should consider the proximity to the actual limit state function so that MLS enables a higher weight to yield a more accurate output. The approximated RSF can be defined in terms of basis functions $b(x)$ and the coefficient vector $a(x)$ as

$$\tilde{L}(x) = b(x)^T \ a(x) \qquad (9.13)$$

The coefficient vector $a(x)$ is expressed as a function of the random variables x to consider the variation of the coefficient vector according to the change of the random variable at each iteration. The local MLS approximation at x is formulated as (Kang et al. 2010)

$$\tilde{L}(x,x_i) = b(x_i)^T \ a(x) \qquad (9.14)$$

where x_i denotes experimental points and the basis functions $B(x)$ are commonly chosen as

$$b(x) = \begin{bmatrix} 1 \ x_1 \ \ldots\ldots x_n \ x^2 \ldots\ldots x_n^2 \end{bmatrix}^T \qquad (9.15)$$

The vector of unknown coefficients $a(x)$ is determined by minimizing the error between the experimental and approximated values of the limit state function. This error is defined as

$$Err(x) = \sum_{i=1}^{n} w(x-x_i)\left[\tilde{L}(x,x_i) - L(x_i)\right]^2 = (Ba - L)^T \ W(x) \ (Ba - L) \quad (9.16)$$

where $L = \begin{bmatrix} L(x_1), L(x_2), \ldots\ldots\ldots L(x_n) \end{bmatrix}^T$, $B = \begin{bmatrix} b(x_1), b(x_2), \ldots\ldots\ldots b(x_n) \end{bmatrix}^T$ and $W(x) = diag.\begin{bmatrix} w_1(x_1 - x), w_2(x_2 - x), \ldots\ldots\ldots w_m(x_n - x) \end{bmatrix}$. Here $(n+1)$ is the number of sampling points and $(m+1)$ is the number of basis functions. Now for minimization of error with respect to $a(x)$, $\partial(Err)/\partial a = 0$ transforming the coefficient of vector $a(x)$ as

$$a(x) = (B^T \ W(x) B)^{-1} \ B^T \ W(x) L \qquad (9.17)$$

The approximated response surface function is obtained from equation (9.14) as

$$\tilde{L}(x) = b(x)^T \ (B^T \ W(x) B)^{-1} \ B^T \ W(x) L \qquad (9.18)$$

9.5 Random input representation and surrogate based uncertainty propagation

The random input parameters such as ply-orientation angle, radius of curvatures, material properties (both longitudinal and transverse elastic modulus, shear modulus, Poisson ratio, mass density), load, load factors (both static and dynamic) and combined variation of all these parameters are considered for composite doubly curved shells considering layer-wise stochasticity. It is assumed that the uniform random distribution of input parameters exists within a certain band of tolerance with their mean values. The following cases are considered in the present study:

a) Variation of ply-orientation angle only: $\theta(\tilde{\omega}) = \{\theta_1 \ \theta_2 \ \theta_3 ....\theta_i......\theta_l \}$

b) Variation of radius of curvatures only: $R(\tilde{\omega}) = \{R_x(\tilde{\omega}), R_y(\tilde{\omega})\}$

c) Variation of material properties only:

$P(\tilde{\omega}) = \{E_1(\tilde{\omega}), E_2(\tilde{\omega}), G_{12}(\tilde{\omega}), G_{23}(\tilde{\omega}), G_{13}(\tilde{\omega}), \mu(\tilde{\omega}), \rho(\tilde{\omega})\}$

d) Variation of intensity of load only: $\{F(\tilde{\omega})\}$

e) Variation of static load factor $\{\alpha(\tilde{\omega})\}$ and dynamic load factor $\{\beta(\tilde{\omega})\}$

f) Combined variation of ply orientation angle, radius of curvatures, material properties (namely, elastic moduli, shear moduli, Poisson's ratio and density), applied load and load factors (static and dynamic): $\{\theta, R, P, F, \alpha, \beta\}(\tilde{\omega})$.

In the present study, $\pm 5°$ variation for ply orientation angle, $\pm 10\%$ volatility in material properties (as per industry standard), applied load and load factors, respectively are considered from their respective deterministic values unless otherwise specified. Figure 9.2 presents a flowchart of the stochastic dynamic stability analysis using the MLS method (surrogate based Monte Carlo simulation) as followed in the present study. A brief description of the Monte Carlo simulation method is provided in the following paragraphs.

Uncertainty quantification is part of modern structural analysis problems. Practical structural systems are faced with uncertainty, ambiguity, and variability constantly. Even though one might have unprecedented access to information due to the recent improvement in various technologies, it is impossible to accurately predict future structural behaviour during its service life. Monte Carlo simulation, a computerized mathematical technique, lets us realize all the possible outcomes of a structural system leading to better and robust designs for the intended performances. The technique was first used by scientists working on the atom bomb; it was named after Monte Carlo, the Monaco resort town renowned for its casinos. Since its introduction in World War II, this technique has been used to model a variety of physical and conceptual systems across different fields such as engineering, finance, project management, energy, manufacturing, research and development, insurance, oil and gas, transportation and environment.

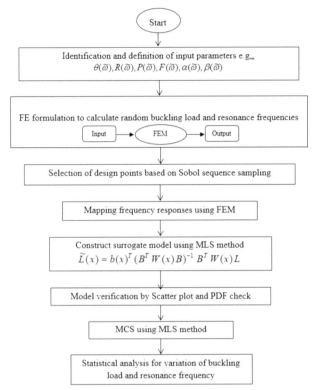

Fig. 9.2 Flowchart of stochastic dynamic stability analysis using MLS method.

Monte Carlo simulation furnishes a range of prospective outcomes along with their respective probability of occurrence. This technique performs uncertainty quantification by forming probabilistic models of all possible results accounting for a range of values from the probability distributions of any factor that has inherent uncertainty. It simulates the outputs over and over, each time using a different set of random values from the probability distribution of stochastic input parameters. Depending upon the nature of stochasticity, a Monte Carlo simulation could involve thousands or tens of thousands of recalculations before it can provide a converged result depicting the distributions of possible outcome values of the response quantities of interest. Each set of samples is called an iteration or realization, and the resulting outcome from that sample is recorded. In this way, Monte Carlo simulation provides not only a comprehensive view of what could happen, but how likely it is to happen, i.e., the probability of occurrence.

The mean or expected value of a function $f(x)$ of a n dimensional random variable vector, whose joint probability density function is given by (x), can be expressed as

$$\mu_f = E\big[f(x)\big] = \int_\Omega f(x)\phi(x)\,dx \tag{9.19}$$

Similarly the variance of the random function $f(x)$ is given by the integral below,

$$\sigma_f^2 = Var\left[f(x)\right] = \int_\Omega \left(f(x) - \mu_f\right)^2 \phi(x) dx \qquad (9.20)$$

The above multidimensional integrals, as shown in equation (9.19) and (9.20) are difficult to evaluate analytically for many types of joint density functions and the integrand function $f(x)$ may not be available in analytical form for the problem under consideration. Thus the only alternative way is to calculate it numerically. The above integral can be evaluated using MCS approach, wherein N sample points are generated using a suitable sampling scheme in the n-dimensional random variable space. The N samples drawn from a dataset must follow the distribution specified by $\varphi(x)$. Having the N samples for x, the function in the integrand $f(x)$ is evaluated at each of the N-sampling points x_i of the sample set $\chi = \{x_1, \ldots, x_N\}$. Thus, the integral for the expected value takes the form of averaging operator as shown below

$$\mu_f = E\left[f(x)\right] = \frac{1}{N}\sum_{i=1}^{N} f(x_i) \qquad (9.21)$$

Similarly, using sampled values of MCS, the equation (9.20) leads to

$$\sigma_f^2 = Var\left[f(x)\right] = \frac{1}{N-1}\sum_{i=1}^{N}\left(f(x_i) - \mu_f\right)^2 \qquad (9.22)$$

Thus the statistical moments can be obtained using a brute force Monte Carlo simulation based approach, which is often computationally very intensive due the evaluation of function $f(x_i)$ corresponding to the N-sampling points x_i, where $N \sim 10^3$. The noteworthy fact in this context is the adoption of surrogate based Monte Carlo simulation approach in the present study that reduces the computational burden of traditional (i.e., brute force) Monte Carlo simulation to a significant extent.

9.6 Results and discussion

The present study considers a simply supported four layered graphite–epoxy angle-ply (45°/–45°/45°/–45°) and cross-ply (0°/90°/0°/90°) composite doubly curved shallow shells. In finite element formulation, an eight noded isoparametric quadratic element is considered. For graphite–epoxy composite shells, the deterministic values of geometric properties are considered as $L = 1$ m, $b = 0.5$ m, $t = 0.005$ m, $C = 0.5$, $R_x = R_y = 10$, (for spherical shell), $\alpha = 0.5$, $\beta = 0.5$ and the material properties are assumed as $E_1 = 141$ GPa, $E_2 = 9.23$ GPa, $G_{12} = G_{13} = 5.95$ GPa, $G_{23} = 2.96$ GPa, $\rho = 1580$ Kg/m³, $v = 0.3$. Table 9.1 presents the non-dimensional buckling loads for the simply supported singly-curved cylindrical composite (0°/90°) panel for different b/R_y ratios (Baharlou and Leissa 1987). Table 9.2 presents the convergence study of non-dimensional fundamental natural frequencies of three layered graphite–epoxy untwisted composite plates (Qatu and Leissa 1991b). A close agreement with benchmarking results are obtained in conjunction to (4×4), (8×8) and (10×10) mesh size. Table 9.3 presents the non-dimensional natural frequencies for simply-supported symmetric cross-ply composite plates and spherical shells (Reddy 1984, Chandrashekhara 1989). It can be noted here that, the analysis of small constituent components is worthwhile and insightful to understand the structural behaviour of larger structures. For example, fuselage of aircraft consists of a cylindrical shell stiffened by circumferential frames and longitudinal stringers. Tests on full scale structure showed that adjacent panels across a frame vibrate independently of one another, with the frames acting as rigid boundaries (Clarson and Ford 1962). Hence, in compliance of the same, the present study considers a simple example

Table 9.1 Non-dimensional buckling loads for the simply supported singly-curved cylindrical composite (0°/90°) panel with $a = 0.25$ m, $b = 0.25$ m, $t = 0.0025$ m, $a/R_x = 0$, $E_1 = 2.07 \times 10^{11}$ N/m², $E_2 = 5.2 \times 10^9$ N/m², $G_{12} = 2.7 \times 10^9$ N/m², $v_{12} = 0.25$.

Structure	$b/R_y = 0.1$	$b/R_y = 0.2$	$b/R_y = 0.3$
Present method	17.612	32.5027	57.117
Baharlou and Leissa (1987)	17.49	32.17	56.62

Table 9.2 Convergence study for non-dimensional frequencies [$\omega = \omega_n L^2 \sqrt{(\rho/E_1 t^2)}$] without in-plane load of doubly curved (45°/–45°/45°) angle ply composite with $a/b = 1$, $b/t = 100$, $b/R_y = 0.5$, $E_1 = 138$ GPa, $E_2 = 8.96$ GPa, $G_{12} = 7.1$ GPa, $v_{12} = 0.3$.

Structure	Present FEM (4 × 4)	Present FEM (8 × 8)	Present FEM (10 × 10)	Qatu and Leissa (1991b)
Plate	0.4600	0.4581	0.4577	0.4607
Spherical Shell	1.3507	1.2977	1.2941	1.3063

Table 9.3 Non-dimensional fundamental frequencies [$\omega = \omega_n a^2 \sqrt{(\rho/E_2 t^2)}$] for the simply supported four layered cross-ply (0°/90°/90°/0°) composite with $E_{11}/E_{22} = 25$, $G_{23} = 0.2E_{22}$, $G_{12} = G_{13} = 0.5E_{22}$, $v_{12} = 0.25$.

Analysis	$a/t = 100$		$a/t = 10$	
	Plate	Spherical ($R/b = 1$)	Plate	Spherical ($R/b = 1$)
Present FEM	15.187	126.320	12.228	16.146
Reddy (1984)	15.184	126.330	12.226	16.172
Chandrashekhara (1989)	15.195	126.700	12.233	16.195

The problem is of a small component of laminated composite curved shells as a representative case to map the zone of dynamic instability due to stochastic variations on input parameters wherein the moving least square (MLS) model is employed to reduce the computational time and cost compared to Monte Carlo Simulation (MCS). However, in future, an extended work of the present study can be carried out to deal with the role of components in the overall stability of the whole large complex structural system.

The moving lease square based approach is validated with the original Monte Carlo simulation considering random variations of input parameters within upper and lower bounds (tolerance zone). Figure 9.3 presents the scatter plot which establishes the accuracy of present MLS model with respect to the original finite

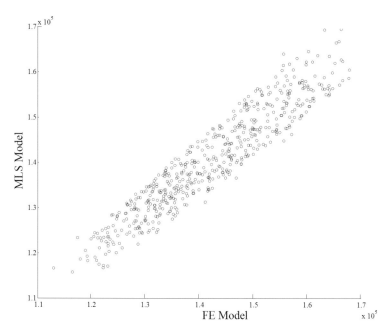

Fig. 9.3 Scatter plot for stochastic buckling loads corresponding to FE model and MLS model.

element model corresponding to stochastic first buckling load for combined variation of ply-orientation angle, radius of curvatures, material properties (both longitudinal and transverse elastic modulus, shear modulus, poisson ratio, mass density), load, and load factors (both static and dynamic). The present MLS surrogate model is used to determine the first stochastic buckling load and resonance frequencies corresponding to given values of input variables, instead of time-consuming deterministic finite element analysis. The probability density function is plotted as the benchmark of bottom line results. The variations of material properties, load intensity and factors are scaled in the range between the lower and the upper limit (tolerance limit) as ± 10% with respective mean values while for ply orientation angle as within ± 5° fluctuation (as per standard of composite manufacturing industry) with respective deterministic values. Due to paucity of space, only a few important representative results are furnished.

A sample size of 64 is considered in case of individual variation of stochastic input parameters while due to higher number of input variables for combined random variation, the subsequent sample size of 512 is found to meet the convergence criteria in the present MLS method. The sampling size of 10,000 is considered for direct MCS with 10,000 finite element (FE) iteration. In contrast, comparatively much lesser number of actual FE iteration (equal to number of design points required to construct the surrogate model) is carried out in case of MLS method. The surrogate model is formed employing MLS method, on which the full sample size of direct MCS is conducted. Hence, the computational time and effort expressed in terms of FE calculation is significantly reduced compared to full scale direct MCS. This provides an efficient and economic way to simulate the uncertainties in buckling load and resonance frequencies (both upper bound and lower bound) for dynamic stability analysis. The scatter plot is also presented for validation of the present MLS model with original FE model with respect to resonance frequencies (fundamental) of lower bound (Fig. 9.4a) and upper bound

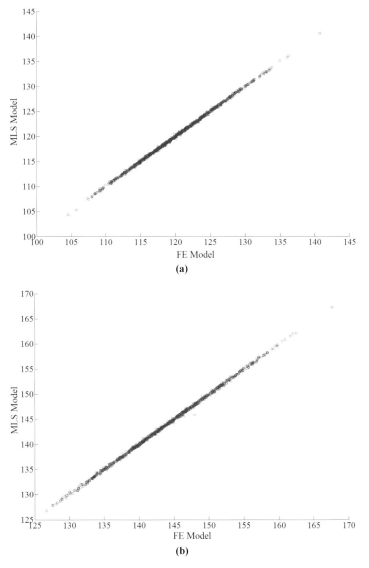

Fig. 9.4 Scatter plot for (a) lower bound and (b) upper bound of fundamental resonance frequencies corresponding to combined variation.

(Fig. 9.4b) corresponding to combined variation of ply-orientation angle, radius of curvatures, material properties, load, load factors (both static and dynamic). The probability density function (PDF) is plotted as the benchmark results due to individual and combined variation as depicted in Fig. 9.5 and Fig. 9.6, respectively. The confidence interval boundaries (95%, 97% and 99%) for mean and standard deviation of buckling load are shown in Table 9.4 for samples of direct MCS and MLS model.

   The MLS model is validated extensively for different laminate configurations as well as different forms of stochasticity (individual and combined) so that the computationally efficient surrogate is ensured to obtain accurate results in the uncertainty analysis. The combined variations of stochastic input parameters for both MCS as well as the present MLS method are carried out corresponding to both angle-ply (45°/–45°/45°/–45°) and cross-ply (0°/90°/0°/90°) composite spherical shells. Due to random variation of input parameters, the elastic stiffness of the laminated composite plate is found to be varied, which in turn influence the stochastic output irrespective of laminate configuration. Table 9.5 presents the comparative results of the Monte Carlo simulation (MCS) and present the MLS method for first buckling load and resonance frequencies (upper bound and lower bound) due to individual and combined variations of ply-orientation angle, radius of curvatures, material properties, intensity of load and load factors of a simply supported angle-ply (45°/–45°/45°/–45°) composite shallow spherical shells. The influence of static load factor and dynamic load factor on stochastic resonance frequencies due to combined variation of ply-orientation angle, radius of curvatures, material properties, loading for angle-ply (45°/–45°/45°/–45°) composite spherical shells are furnished in Fig. 9.7. It is observed that the width of the instability zone increases with the increase of static and dynamic load factors. Based on the rate of increment of the region of instability, it can be inferred that the dynamic load factor (β) is more sensitive to resonance frequencies than static load factor (α). Further, to explore the effect of degree of stochasticity on resonance frequency and the capability of the proposed MLS based approach for higher degree of variations in the stochastic input parameters, three different degree of stochasticities are considered: 5%, 10% and 15% variations in the stochastic input parameters with

Table 9.4 Confidence interval boundaries for mean and standard deviation of buckling load (KN/m) for samples of direct MCS and MLS model.

Confidence interval (%)		MLS		MCS	
		Mean	SD	Mean	SD
95	Min	1.35240×10^5	1.64351×10^4	1.35869×10^5	1.48764×10^4
	Max	1.36893×10^5	1.68972×10^4	1.36460×10^5	1.52946×10^4
97	Min	1.35205×10^5	1.64110×10^4	1.35837×10^5	1.48545×10^4
	Max	1.36928×10^5	1.69226×10^4	1.36492×10^5	1.53176×10^4
99	Min	1.35137×10^5	1.63645×10^4	1.35776×10^5	1.48124×10^4
	Max	1.36996×10^5	1.69718×10^4	1.36553×10^5	1.53621×10^4

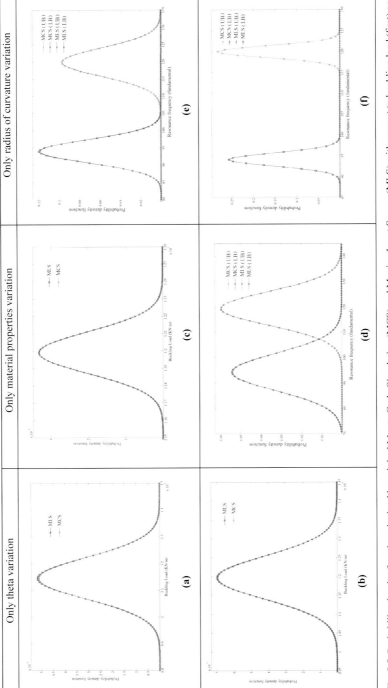

Fig. 9.5 Probability density function obtained by original Monte Carlo Simulation (MCS) and Moving Least Square (MLS) with respect to buckling load (first) and fundamental resonance frequencies [Upper Bound(UB), Lower Bound(LB)] due to individual variation of ply orientation angle, material properties and radius of curvatures for angle – ply (45°–45°/45°–45°) composite spherical shells.

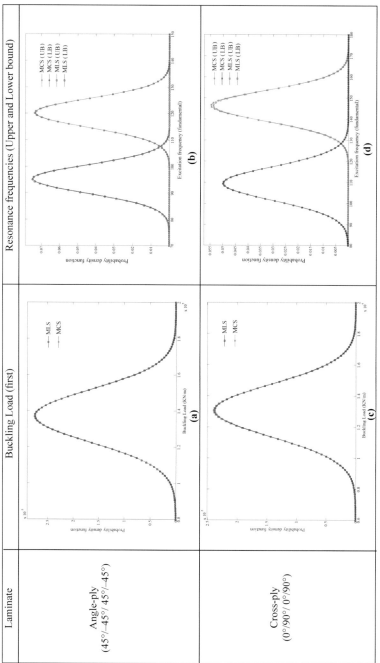

Fig. 9.6 Probability density function obtained by original Monte Carlo Simulation (MCS) and Moving Least Square (MLS) with respect to (a) buckling load (first) and (b) fundamental resonance frequencies [Upper Bound (UB), Lower Bound(LB)] due to combined variation for simply supported angle-ply and cross-ply composite spherical shells.

Table 9.5 Stochastic buckling load (first) and resonance frequencies (first and second) with error due to individual and combined variation of simply supported angle-ply (45°/–45°/45°/–45°) composite spherical shells considering $L = 1$ m, $b = 0.5$ m, $t = 0.005$ m, $c = 0.5$, $R_x = R_y = 10$ m, $E_1 = E_2 = 141$ GPa, $G_{12} = G_{13} = 5.95$ GPa, $G_{23} = 2.96$ GPa, $\rho = 1580$ Kg/m³, $v = 0.3$.

Parameter	Value	Buckling load (first)			Resonance frequency (first)						Resonance frequency (second)					
					Upper bound			Lower bound			Upper bound			Lower bound		
		MCS	MLS	Err %	MCS	MLS	Err %	MCS	MLS	Err %	MCS	MLS	Err %	MCS	MLS	Err %
$\theta(\widetilde{\omega})$	Max	128291.9	127596.5	0.54	137.13	136.76	0.27	114.56	115.37	–0.71	154.4566	154.68	–0.14	137.2272	137.7822	–0.40
	Min	88701.2	85767.6	3.31	102.14	101.13	0.99	77.68	76.36	1.70%	126.1999	124.61	1.26	103.3465	101.2563	2.02%
	Mean	114559.3	114586.8	–0.02	119.50	119.45	0.04	94.37	94.35	0.02%	143.71	143.68	0.02	124.66	124.73	–0.06
	SD	7812.6	7718.3	1.21	6.68	6.66	0.30	7.30	7.36	–0.82	4.83	4.86	–0.62	6.53	6.48	0.77%
$R(\widetilde{\omega})$	Max	122779.8	122735.2	0.04	123.17	123.12	0.04	97.61	97.59	0.02%	153.2791	153.15	0.08	129.4123	129.3734	0.03%
	Min	117567.8	117573.0	0.00	117.25	117.26	–0.01	91.12	91.12	0.00%	135.4901	135.42	0.05	121.0511	120.9995	0.04%
	Mean	119910.8	119919.8	–0.01	119.94	119.96	–0.02	94.08	94.09	–0.01	144.2207	144.25	–0.02	125.3795	125.3827	0.00
	SD	1202.1	1199.9	0.18	1.41	1.40	0.71	1.55	1.54	0.65%	3.74	3.71	0.80	1.78	1.78	0.00%
$P(\widetilde{\omega})$	Max	131650.3	131442.9	0.16	131.05	130.91	0.11	103.32	103.17	0.15%	157.5167	157.29	0.14	137.57	137.23	0.25%
	Min	108278.9	108146.1	0.12	109.54	110.03	–0.45	85.33	85.43	–0.12	131.5795	132.32	–0.56	114.05	114.24	–0.17
	Mean	119787.4	119748.1	0.03	119.92	119.89	0.03	94.01	93.98	0.03%	144.1147	144.07	0.03	125.41	125.36	0.04%
	SD	5678.4	5728.4	–0.88	4.11	4.10	0.24	3.35	3.31	1.19%	5.10	5.08	0.39	4.56	4.51	1.10%

Table 9.5.contd....

...*Table 9.5 contd.*

Para-meter	Value	Buckling Load (first)			Resonance frequency (First)						Resonance frequency (Second)					
					Upper bound			Lower bound			Upper bound			Lower bound		
		MCS	MLS	Err %	MCS	MLS	Err %	MCS	MLS	Err %	MCS	MLS	Err %	MCS	MLS	Err %
$F(\widetilde{\omega})$	Max	157639.4	157447.1	0.12	120.36	120.38	−0.02	96.25	96.28	−0.03	143.9863	144.01	−0.02	126.43	126.42	0.01%
	Min	119790.2	119692.4	0.08	119.81	119.79	0.01	93.94	93.92	0.02%	143.4501	143.43	0.01	125.31	125.34	−0.02
	Mean	137767.5	137644.7	0.09	120.08	120.08	−0.01	95.11	95.11	0.00%	143.7182	143.71	0.01	125.90	125.90	0.00%
	SD	12653.6	12272.1	3.01	0.27	0.27	0.00	0.89	0.87	2.25%	0.26	0.26	0.00	0.31	0.30	3.23%
$\{\theta, R, P, F, \alpha, \beta\}$	Max	177886.5	180808.1	−1.64	140.77	139.61	0.82	114.0825	117.94	3.38	169.3019	165.86	2.03	146.10	149.0155	−2.00
	Min	101954	101032.5	0.90	103.79	104.23	−0.42	77.42	78.28	−1.11	124.7183	124.71	0.01	106.07	107.45	−0.30
	Mean	137164.4	137238.2	−0.05	120.20	120.18	0.02	95.27	95.28	−0.01	144.0335	143.99	0.03	125.86	125.87	−0.01
	SD	14468.9	11919.0	17.62	5.40	5.44	−0.74	5.32	5.23	1.69	6.58	6.60	−0.30	6.09	6.02	1.15

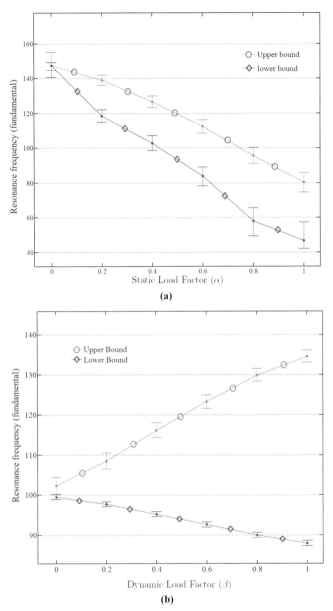

Fig. 9.7 Effect of static load factor and dynamic load factor on stochastic resonance frequencies (fundamental) due to combined variation of ply-orientation angle, radius of curvatures, material properties, loading for simply supported angle-ply (45°/–45°/45°/–45°) composite spherical shells.

respect to their respective deterministic values. Figure 9.8 presents the validation in resonance frequencies (fundamental) using the MLS model corresponding to different degree of stochasticities (5%, 10% and 15%) for combined variation of

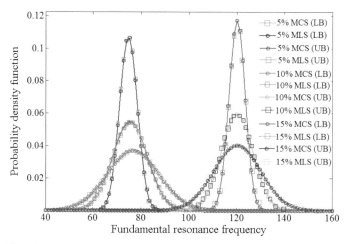

Fig. 9.8 Effect of percentage variation (5%, 10% and 15%) for combined variations of input parameters on resonance frequencies (fundamental) for simply supported angle-ply (45°/–45°/45°/–45°) composite spherical shells.

input parameters considering simply supported angle-ply spherical shells. The figure clearly depicts the increase in sparsity of resonance frequency (fundamental) due to increase in percentage of varitions of random input parameters. The figure also affirms that the proposed MLS based uncertainty quantification algorithm for composites produces quite satisfactory results for different degree of stochasticities in input parameters with respect to direct Monte Carlo simulations.

Depending on the geometry of doubly curved shells, a comparative study is carried out for cylindrical, hyperbolic paraboloid and spherical shells as furnished in Fig. 9.9 for both stochastic buckling load and random resonance frequencies (fundamental) due to combined variation of cross-ply (0°/90°/0°/90°) composite shells. The zone of resonance frequencies (fundamental) maps the different instability regions for different shell geometries. It is observed that the resonance frequency (fundamental) decreases with reduction of curvatures from a spherical shell to hyperbolic paraboloid shells while a single cylindrical shell shows the least stiffness compared to the other two. In order to address the influence of degree of shallowness ($R_x/a = R_y/b$ = 5, 10, 20) of the doubly curved shells, a spherical shell is considered to portray the instability regions as furnished in Fig. 9.10. It is identified that there is an increase of instability resonance frequencies with the decrease in radius of curvature along x and y directions (i.e., R_x and R_y values). The significant effects of degree of orthotropy on stochastic buckling load and resonance frequency (fundamental) due to the combined variation of ply-orientation angle, radius of curvatures, material properties, and loading for cross-ply composite spherical shells are furnished in Fig. 9.11. As the static parameter is increased, the dynamic instability zone tends to shift towards lower frequencies and become stipper. The effect of degree of orthotropy is studied for E_1/E_2 ratio = 15, 30, 45, by randomizing the other parameters.

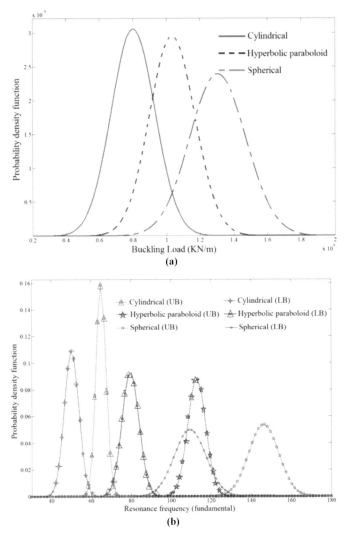

Fig. 9.9 Effect of shell geometry (Cylindrical, Hyperbolic paraboloid and Spherical) on stochastic (a) buckling load (first) and (b) resonance frequencies (fundamental) due to combined variation of for simply supported cross-ply ($0°/90°/0°/90°$) composite curved shells.

The study shows an increase of random resonance frequencies due to increase in degree or orthotropy. The boundary conditions of the composite shells are observed to have a significant influence on the dynamic instability regions. The influence of different boundaries (CCCC, SCSC, SSSS where C stands for clamped and S for simply supported) is investigated for stochastic buckling load and first resonance frequencies (lower and upper bounds) due to the combined variation of ply-orientation angle, radius of curvatures, material properties, loading for cross-ply

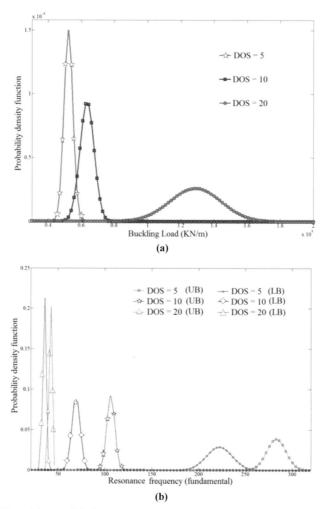

Fig. 9.10 Effect of degree of shallowness (DOS) ($R_x/a = R_y/b$) on stochastic (a) buckling load (first) and (b) resonance frequencies (fundamental) due to combined variation of for simply supported cross-ply (0°/90°/0°/90°) composite spherical shells

composite spherical shells by probability density function as furnished in Fig. 9.12. This study shows that the stochastic resonance frequencies are found minimum for simply supported and maximum for clamped edges due to the restraint at the edges while the SCSC boundary condition is found to be intermediate for both stochastic buckling load as well as zone of resonance frequencies.

The effect of individual variations and combined variation of different random parameters for angle-ply composite spherical shells on stochastic first buckling load are furnished in Fig. 9.13 wherein the maximum sparsity of buckling load is observed for only variation of load-intensity among all the individual parameters.

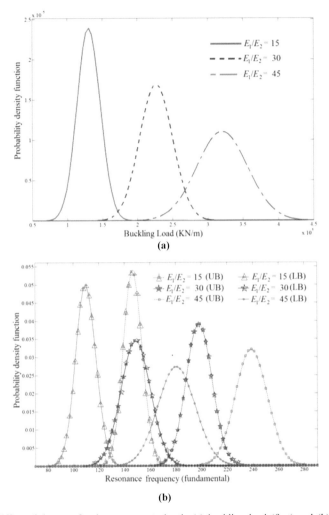

Fig. 9.11 Effect of degree of orthotropy on stochastic (a) buckling load (first) and (b) resonance frequencies (fundamental) due to combined variation for simply supported cross-ply (0°/90°/0°/90°) composite spherical shells.

Figure 9.14 represents the influence of aspect ratio (AR = a/b) on stochastic buckling load and resonance frequency (fundamental) due to the combined variation of ply-orientation angle, radius of curvatures, material properties, loading for cross-ply composite spherical shells. Because of the shear deformation, it is found that the width of instability region narrows down. It is also found that as the aspect ratio (a/b) increases, the resonance frequencies also increase and the width of the instability zone becomes wider. In the present study, the relative coefficient of variance (RCV) (normalized mean to standard deviation ratio) due to individual and

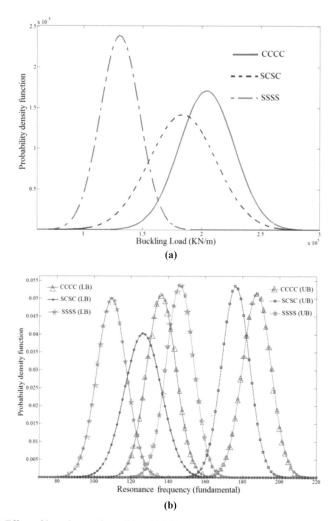

Fig. 9.12 Effect of boundary end condition (CCCC, SCSC, SSSS) on stochastic (a) buckling load and (b) resonance frequency (fundamental) due to combined variation for cross-ply (0°/90°/0°/90°) composite spherical shells.

combined variations is quantified for angle-ply laminate as furnished in Fig. 9.15. On the basis of individual variation of input parameters, ply orientation angle is found to be comparatively most sensitive, while loading parameter (for resonance frequencies) and radius of curvature (for buckling load) are found to have lesser sensitivity.

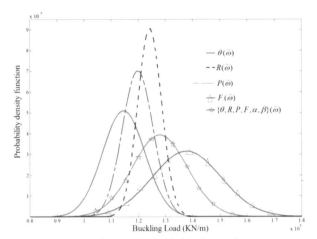

Fig. 9.13 Stochastic buckling load for combined variation for simply supported cross-ply spherical shells.

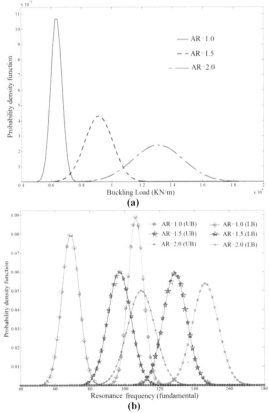

Fig. 9.14 Effect of aspect ratio (AR) on stochastic (a) buckling load and (b) resonance frequencies (fundamental) for combined variation for simply supported cross-ply spherical shells.

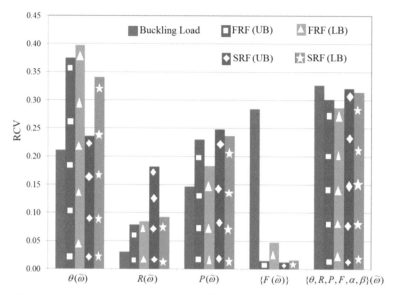

Fig. 9.15 Relative coefficient of variance (RCV) of buckling load and first resonance frequencies (FRF) and second resonance frequencies (SRF) due to individual variation of ply orientation angle, radius of curvatures, material properties, loading and combined variation for simply supported angle-ply (45°/–45°/45°/–45°) composite spherical shells.

9.7 Summary

This chapter presents an efficient stochastic dynamic stability analysis of laminated composite curved panels considering non-uniform partial edge loading. The ranges of variation in first stochastic buckling load and fundamental resonance frequencies are analyzed considering both individual and combined stochasticity of input parameters. Moving least square method is employed in conjunction with stochastic finite element analysis following a non-intrusive approach to achieve the computational efficiency. After utilizing the surrogate modelling approach, the number of finite element simulations is found to be significantly reduced compared to original Monte Carlo simulation without compromising the accuracy of results. The computational time is reduced to (1/157) times (for individual variation) and (1/20) times (for combined variation) of Monte Carlo simulation. The stochastic instability regions are found to shift to lower frequencies with the increase in static load factor showing wider random instability regions indicating a destabilization effect on the dynamic stability characteristics of composite spherical shells. It is observed that the zone of stochastic instability has significant influence due to the variation in degree of orthotropy, aspect ratio and boundary condition. The width of the stochastic instability region increases with the increase of degree of orthotropy and aspect ratio. The ply orientation angle is found to be most sensitive, while the least sensitive parameters are observed as loading parameter (for resonance frequencies) and radius of curvatures (for buckling load) compared to other parameters considered in this analysis. Laminated composites being a complex structural form and susceptible to different forms of uncertainty, the compound effects of stochastic time varying loading and structural and material uncertainties associated with composites are crucial for the intended performance in various applications. It is found that stochastic variations of input parameters has significant impact on dynamic instability of composite shell structures and thus the probabilistic effect of such sensitive parameters are required to be considered for an inclusive design. The numerical results obtained in this study provide a comprehensive idea for design and control of laminated composite curved panels. The efficient moving least square based approach of uncertainty quantification can be extended further to other computationally intensive analyses of composite structures.

CHAPTER 10

Uncertainty Quantification for Skewed Laminated Soft-core Sandwich Panels

10.1 Introduction

Having presented the stochastic dynamic analyses for composite laminates in the previous chapters, this chapter focuses on the stochastic analysis of laminated soft-core sandwich panels including the effect of skewness in the geometry. An efficient multivariate adaptive regression splines based approach for dynamics and stability analysis of sandwich plates is presented considering the random system parameters. The propagation of uncertainty in such structures has significant computational challenges due to inherent structural complexity and high dimensional space of input parameters. The theoretical formulation is developed based on a refined C^0 stochastic finite element model and higher-order zigzag theory in conjunction with multivariate adaptive regression splines. A cubical function is considered for the in-plane parameters as a combination of a linear zigzag function with different slopes at each layer over the entire thickness while a quadratic function is assumed for the out-of-plane parameters of the core and constant in the face sheets. Both the individual and combined stochastic effect of skew angle, layer-wise thickness, and material properties (both core and laminate) of sandwich plates are considered. Statistical analyses are carried out to illustrate the results of the first three stochastic natural frequencies and buckling load.

The application of sandwich structures has gained immense popularity in advanced engineering applications, especially in aerospace, marine, civil, and mechanical structures that require superior performances and outstanding properties like lightweight, high stiffness, high structural efficiency and durability. The construction of sandwich panels consisting of thin face sheets of high strength material separated by a relatively thick and low density material offers excellent mechanical properties such as high strength-to-weight ratio and high stiffness-to-weight ratios. The characteristic features of such structures are affected by their layered construction and variations in properties through their thickness and therefore, it is important to predict their overall response in a realistic manner considering all these features. The effect of shear deformation plays a vital role in the structural analysis of sandwich and composite constructions because of their low shear modulus compared to extensional rigidity with a large variation in material properties between the core and the face layers. Moreover, due to their special type of construction and behavior, sandwich structures possess high statistical variations in the material and geometric properties. These inherent uncertainties should be properly taken into account in the analysis in order to have more realistic and safe design. This cannot be mapped by the conventional deterministic analysis. In fact, accurate predictions of the vibration response of these structures become more challenging to the engineers in the presence of inherent scatter in stochastic input parameters consisting of both material and geometric properties. Stochastic natural frequencies of such sandwich structures consist of overall mode and localized ones or through the thickness that the classical deterministic theories lack to detect. Due to the dependency of a large number of parameters in complex production and fabrication processes, the system properties are inevitably random in nature resulting in uncertainty in the response of the sandwich plate. Therefore, there is a need for an efficient and accurate computational technique which takes into account the effects of parameter uncertainty on the structural response. In the deterministic analysis of structures, the variations in the system parameters are neglected and the mean values of the system parameters are used in the analysis. But the variations in the system parameters should not be ignored for accurate and realistic studies that require a probabilistic description in which the response statistics can be adequately achieved by considering the material and geometric properties to be stochastic in nature.

Many review articles (Noor et al. 1996, Bert 1991, Mallikarjuna and Kant 1993, Altenbach 1998) are published on deterministic analysis of sandwich composite plates. Several investigators (Alibeigloo and Alizadeh 2015, Carrera and Brischetto 2009, Mukhopadhyay et al. 2015c, Singh et al. 2015) studied deterministic bending, buckling and free vibration analysis of skew composite and sandwich plates and thereby optimizing such structures. Recently a study has been published on analytical development for free vibration analysis of sandwich panels with randomly irregular honeycomb cores investigated by Mukhopadhyay and Adhikari (2016e). Free vibration response of laminated skew sandwich plates has been investigated by Garg et al. (2006) using C_0 isoparametric finite element model based on HSDT. The vibration behaviour of imperfect sandwich plates with in-plane partial edge load is presented by Chakrabarti and Sheikh (2004) in a deterministic framework, while free vibration analyses of sandwich plates subjected to thermo-mechanical loads is studied by Shariyat (2010) using a generalized global-local higher order theory.

Many literatures (Elmalich and Rabinovitch 2012, Chalak et al. 2015, Singh and Chakrabarti 2013, Aguib et al. 2014, Zhen et al. 2010) are found which investigate dynamic and stability of soft core sandwich plates using the analytical or finite element method. Radial basis function is used by (Roque et al. 2006, Ferrera et al. 2005, 2008, 2011, Rodrigues et al. 2011–12) to analyse bending, buckling and free vibration characteristics of composite sandwich plates. The mesh-free moving Kriging interpolation method is presented by Bui et al. (2011) for analysis of natural frequencies of laminated composite plates while Yang et al. (2015) studied the vibration and damping analysis of thick sandwich cylindrical shells with a viscoelastic core. There is plenty of literature found which presents buckling and free vibration analyses of sandwich plates using the Rayleigh-Ritz method (Narita 2001, Carrera et al. 2011, Watkins and Barton 2009, Iurlaro et al. 2013). Probabilistic analyses of sandwich panels are presented by (Kumar et al. 2015, Ying et al. 2013, Dey et al. 2018b). Several recent reports investigate bending and buckling analysis of sandwich plates with functionally graded material (Taibi et al. 2015, Gulshan et al. 2014). Researchers reported their results using the Galerkin method (Liu et al. 2010), quadrature method (Zhang and Sainsbury 2000), State-space method (Makheda et al. 2002, Xiang and Wei 2004), Levy's method (Aydogdu and Ece 2006, Dehkordi et al. 2016), Navier's method (Xiang et al. 2011) or the exact solutions method (Douville and Le Grognec 2013) for buckling and free vibration analyses of laminated sandwich plates. Recently an analytical approach has been presented to obtain equivalent elastic properties of spatially irregular honeycomb panels (Mukhopadhyay and Adhikari 2016b, 2016c, 2017d, Mukhopadhyay et al. 2018b), which are often used as the core of sandwich panels. Such equivalent elastic properties of an irregular honeycomb core can be an attractive solution for stochastic analyses of honeycomb panels with a honeycomb core.

Most of the investigations carried out so far concerning the analysis of sandwich composite panels are deterministic in nature that lacks to cater the necessary insight on the behaviour of different structural responses generated from inherent statistical variations of stochastic material and geometric parameters. In general, Monte Carlo simulation (MCS) is commonly used for stochastic response analysis leading to complete probabilistic descriptions of the output parameters. But the traditional MCS based stochastic analysis approach is very expensive because it requires thousands of finite element simulations to be carried out to capture the random nature of parametric uncertainty. Hence, reduced order modelling (ROM) techniques are used to reduce the computational time and cost. In the past, reduced order computational models are found to be employed in stochastic structural analysis (Batou and Soize 2013, Mahata et al. 2016) and some of them are specifically applied in laminated composite plates and shells including the effect of noise (Mukhopadhyay 2018a, Chakraborty 2005, Dey et al. 2016c). In the present chapter, we have used a multivariate adaptive regression splines (MARS) based efficient uncertainty quantification algorithm for composite sandwich structures. In this approach the expensive finite element model for sandwich composite structures is effectively replaced by the computationally efficient MARS model making the overall process of uncertainty quantification much more cost-effective. Sudjianto et al. (1998) used the MARS model to emulate a conceptually intensive complex automotive shock tower model in fatigue life durability analysis while Wang et al. (1999) compared MARS to linear, second-order, and higher-order regression models for a five variable automobile structural analysis. Friedman (1991) integrated the MARS procedure to approximate the behaviour of performance variables in a simple alternating current series circuit. Literature suggests that the major advantages of using the MARS based reduced order modelling appears to be the accuracy and significant reduction in computational cost associated with constructing the surrogate model compared to other conventional emulators such as Kriging (Jin et al. 2001). MARS constructs the input/output relation from a set of coefficients and basis functions that are entirely driven from the regression data. The algorithm allows partitioning of the input space into regions, each with its own regression equation. This makes MARS particularly suitable for problems with high dimensional input parameter space. As finite element models of sandwich structures normally have large number of stochastic input parameters, MARS has the potential to be an efficient mapping route for the inputs and responses of such structures. In the present study, a stochastic analysis for free vibration and buckling of laminated sandwich skewed plates is carried out by solving the random eigenvalue problem through an improved higher-order zigzag theory in conjunction with MARS following a bottom up random variable framework. The effect of skewed geometry, which often turns out to be inevitable in many engineering applications, is investigated in the analysis. Subsequently the relative individual effect of different stochastic input parameters for natural frequencies and buckling load are discussed. The numerical results are presented for both individual and combined layer-wise variation of the stochastic input parameters. This chapter is organized hereafter as, Section 10.2: stochastic finite element formulation for laminated soft-core sandwich panels, Section 10.3: brief description of MARS based surrogate modelling algorithm, Section 10.4: MARS based uncertainty quantification algorithm, Section 10.5: results and discussion, Section 10.6: summary.

10.2 Theoretical formulation

Consider a laminated soft core sandwich plate (Fig. 10.1) with thickness 't' and skew angle 'ϕ' (as shown in Fig. 10.2), consisting of 'n' number of thin lamina, the stress-strain relationship considering plane strain condition of an orthotropic layer or lamina (say k-th layer) having any fiber orientation angle 'θ' with respect to structural axes system (X-Y-Z) can be expressed as:

$$\{\sigma(\tilde{\omega})\} = \{Q_k(\tilde{\omega})\}\{\varepsilon(\tilde{\omega})\}$$

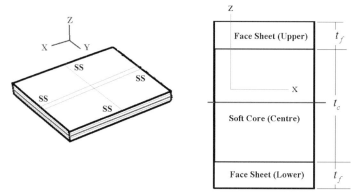

Fig. 10.1 Simply-supported (SS) soft core sandwich plate.

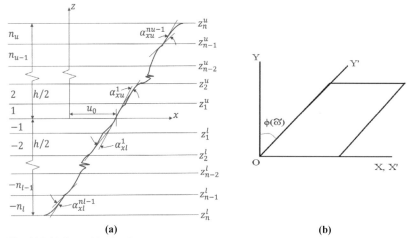

(a)   **(b)**

Fig. 10.2 (a) General lamination and displacement configuration (b) Skewed laminate geometry

$$
\begin{Bmatrix}
\sigma_{xx}(\tilde{\omega}) \\
\sigma_{yy}(\tilde{\omega}) \\
\sigma_{zz}(\tilde{\omega}) \\
\tau_{xy}(\tilde{\omega}) \\
\tau_{xz}(\tilde{\omega}) \\
\tau_{yz}(\tilde{\omega})
\end{Bmatrix}
=
\begin{bmatrix}
\bar{Q}_{11}(\tilde{\omega}) & \bar{Q}_{12}(\tilde{\omega}) & \bar{Q}_{13}(\tilde{\omega}) & \bar{Q}_{14}(\tilde{\omega}) & 0 & 0 \\
\bar{Q}_{21}(\tilde{\omega}) & \bar{Q}_{22}(\tilde{\omega}) & \bar{Q}_{23}(\tilde{\omega}) & \bar{Q}_{24}(\tilde{\omega}) & 0 & 0 \\
\bar{Q}_{31}(\tilde{\omega}) & \bar{Q}_{32}(\tilde{\omega}) & \bar{Q}_{33}(\tilde{\omega}) & \bar{Q}_{34}(\tilde{\omega}) & 0 & 0 \\
\bar{Q}_{41}(\tilde{\omega}) & \bar{Q}_{42}(\tilde{\omega}) & \bar{Q}_{43}(\tilde{\omega}) & \bar{Q}_{44}(\tilde{\omega}) & 0 & 0 \\
0 & 0 & 0 & 0 & \bar{Q}_{55}(\tilde{\omega}) & \bar{Q}_{56}(\tilde{\omega}) \\
0 & 0 & 0 & 0 & \bar{Q}_{65}(\tilde{\omega}) & \bar{Q}_{66}(\tilde{\omega})
\end{bmatrix}
\begin{Bmatrix}
\varepsilon_{x}(\tilde{\omega}) \\
\varepsilon_{y}(\tilde{\omega}) \\
\varepsilon_{z}(\tilde{\omega}) \\
\gamma_{xy}(\tilde{\omega}) \\
\gamma_{xz}(\tilde{\omega}) \\
\gamma_{yz}(\tilde{\omega})
\end{Bmatrix}
$$

$$(10.1)$$

where $\{\sigma(\tilde{\omega})\}$, $\{\varepsilon(\tilde{\omega})\}$ and $\{Q_k(\tilde{\omega})\}$ are random stress vector, random strain vector and random transformed rigidity matrix of k-th lamina, respectively. Here the symbol $\tilde{\omega}$ indicates the stochasticity of respective input parameters. Figure 10.2a represents the in-plane displacement field. The in-plane displacement parameters are expressed as

$$U_x(\tilde{\omega}) = u_o + z\theta_x + \sum_{i=1}^{n_u-1}(z - z_i^u)(\tilde{\omega})\,\Delta(z - z_i^u)\,\alpha_{xu}^i + \sum_{j=1}^{n_l-1}(z - z_j^l)(\tilde{\omega})\,\Delta(-z + z_{j_l}^l)\,\alpha_{xl}^j + \beta_x z^2 + \chi_x z^3 \tag{10.2}$$

$$V_x(\tilde{\omega}) = v_o + z\theta_y + \sum_{i=1}^{n_u-1}(z - z_i^u)(\tilde{\omega})\,\Delta(z - z_i^u)\,\alpha_{yu}^i + \sum_{j=1}^{n_l-1}(z - z_j^l)(\tilde{\omega})\,\Delta(-z + z_{j_l}^l)\,\alpha_{yl}^j + \beta_y z^2 + \chi_y z^3 \tag{10.3}$$

where, u_o and v_o are the in-plane displacements of any point in the X-axis and Y´-axis on the mid surface, θ_x and θ_y are the rotations of the normal to the middle plane about the Y-axis and X-axis respectively, n_u and n_l are the number of upper and lower layers, respectively while β_x, β_y, χ_x and χ_y are the higher order unknown co-efficient, α_{xu}^i, α_{yu}^i, α_{xl}^j and α_{yl}^j are the slopes of i-th and j-th layer corresponding to upper and lower layers respectively and $\Delta(z - z_i^u)$ and $\Delta(-z + z_{ji}^l)$ are the unit step functions. In the general lamination scheme, governing equations and displacement configuration are considered as per Pandit et al. (2008). The transverse displacements are assumed to vary quadratically through the core thickness and constant over the face sheets and it may be expressed as,

$$W(\tilde{\omega}) = \frac{z(z + t_l)}{t_u(t_u + t_l)}\,w_u(\tilde{\omega}) \; + \; \frac{(t_l + z)(t_u - z)}{t_u\,t_l}\,w_o(\tilde{\omega}) \; + \; \frac{z(t_u - z)}{-t_l(t_u + t_l)}\,w_l(\tilde{\omega}) \text{ (for core)} \tag{10.4}$$

$$W(\tilde{\omega}) = w_u(\tilde{\omega}) \text{ (for upper face layers)} \tag{10.5}$$

$$W(\tilde{\omega}) = w_l(\tilde{\omega}) \text{ (for lower face layers)} \tag{10.6}$$

where $w_u(\tilde{\omega})$, $w_o(\tilde{\omega})$ and $w_l(\tilde{\omega})$ are the values of the transverse displacement at the top layer, middle layer and bottom layer of the core, respectively. Utilizing the conditions of zero transverse shear stress at the top and bottom surfaces of the plate and imposing the conditions of the transverse shear stress continuity at the interfaces between the layers along with the conditions, $u = u_u$ and $v = v_u$ at the top and $u = u_l$ and $v = v_l$ at the bottom of the plate, the generalized displacement vector $\{\delta\}$ for the present plate model can be expressed as

$$\{\delta\} = \{u_o\ v_o\ w_o\ \theta_x\ \theta_y\ u_u\ v_u\ w_u\ u_l\ v_l\ w_l\}^T \text{ and } y_l = y_l'\cos[\phi(\tilde{\omega})] \tag{10.7}$$

where $\phi(\tilde{\omega})$ denotes the random skew angle (Fig. 10.2). For the skewed plates, the elements on the inclined edges may not be parallel to the global axes $(x_g - y_g - z_g)$. To determine the elemental stiffness matrix at skew edges, it becomes necessary to use edge displacements $(u_o, v_o, w_o, \theta_x, \theta_y, u_u, v_u, w_u, u_l, v_l$ and $w_l)$ in local coordinates $(x' - y' - z')$ (Fig. 10.2). It is thus required to transform the element matrices corresponding to global axes to local axes with respect to which the elemental stiffness matrix can be conveniently determined. The relation between the global and local degrees of freedom of a node on the skew edge can be obtained through the simple transformation rules and the same can be expressed as

$$\{\delta^L(\tilde{\omega})\}^T = [T_n(\tilde{\omega})]\,\{\delta\}^T \tag{10.8}$$

A nine noded isoparametric element is used for finite element formulation considering eleven degrees of freedom where, $\{\delta^{~l}(\tilde{\omega})\}$ and $[T_n(\tilde{\omega})]$ are the displacement vector in the localised coordinate system and node transformation matrix respectively. Using the node transformation matrix, the elemental transformation matrix $[T_{ele}(\tilde{\omega})]$ can be determined, which is used to transfer the elemental stiffness matrix of the skew edge elements from the global axes to local axis. The node transformation matrix $[T_n(\tilde{\omega})]$ and the elemental transformation matrix $[T_{ele}(\tilde{\omega})]$ are expressed as

$$[T_n(\tilde{\omega})] = \begin{bmatrix} m & -n & 0 & 0 & 0 & 0 & 0 & 0 & 0 & 0 & 0 \\ n & m & 0 & 0 & 0 & 0 & 0 & 0 & 0 & 0 & 0 \\ 0 & 0 & 1 & 0 & 0 & 0 & 0 & 0 & 0 & 0 & 0 \\ 0 & 0 & 0 & m & -n & 0 & 0 & 0 & 0 & 0 & 0 \\ 0 & 0 & 0 & n & m & 0 & 0 & 0 & 0 & 0 & 0 \\ 0 & 0 & 0 & 0 & 0 & m & -n & 0 & 0 & 0 & 0 \\ 0 & 0 & 0 & 0 & 0 & n & m & 0 & 0 & 0 & 0 \\ 0 & 0 & 0 & 0 & 0 & 0 & 0 & 1 & 0 & 0 & 0 \\ 0 & 0 & 0 & 0 & 0 & 0 & 0 & 0 & m & -n & 0 \\ 0 & 0 & 0 & 0 & 0 & 0 & 0 & 0 & n & m & 0 \\ 0 & 0 & 0 & 0 & 0 & 0 & 0 & 0 & 0 & 0 & 1 \end{bmatrix} \quad \text{and} \quad (10.9)$$

$$[T_{ele}(\tilde{\omega})] = \begin{bmatrix} [T_n(\tilde{\omega})] & 0 & 0 & 0 & 0 & 0 & 0 & 0 & 0 & 0 & 0 \\ & [T_n(\tilde{\omega})] & 0 & 0 & 0 & 0 & 0 & 0 & 0 & 0 & 0 \\ & & [T_n(\tilde{\omega})] & 0 & 0 & 0 & 0 & 0 & 0 & 0 & 0 \\ & & & [T_n(\tilde{\omega})] & 0 & 0 & 0 & 0 & 0 & 0 & 0 \\ & & & & [T_n(\tilde{\omega})] & 0 & 0 & 0 & 0 & 0 & 0 \\ & & & & & [T_n(\tilde{\omega})] & 0 & 0 & 0 & 0 & 0 \\ & \text{Sym.} & & & & & [T_n(\tilde{\omega})] & 0 & 0 & 0 & 0 \\ & & & & & & & [T_n(\tilde{\omega})] & 0 & 0 & 0 \\ & & & & & & & & [T_n(\tilde{\omega})] & 0 & 0 \\ & & & & & & & & & [T_n(\tilde{\omega})] & 0 \\ & & & & & & & & & & [T_n(\tilde{\omega})] \end{bmatrix}$$

$$(10.10)$$

where $m = \sin\theta(\tilde{\omega})$ and $n = \cos\theta(\tilde{\omega})$, wherein $\theta(\tilde{\omega})$ is the random fibre orientation angle.

Using linear strain-displacement relation, the strain field $\{\bar{\varepsilon}(\tilde{\omega})\}$ may be expressed in terms of unknowns (for the structural deformation) as

$$\{\overline{\varepsilon(\tilde{\omega})}\} = \left[\frac{\partial U(\tilde{\omega})}{\partial x} \frac{\partial V(\tilde{\omega})}{\partial y} \frac{\partial W(\tilde{\omega})}{\partial z} \frac{\partial U(\tilde{\omega})}{\partial x} + \frac{\partial V(\tilde{\omega})}{\partial y} \frac{\partial U(\tilde{\omega})}{\partial z} \frac{\partial W(\tilde{\omega})}{\partial x} \frac{\partial V(\tilde{\omega})}{\partial z} + \frac{\partial W(\tilde{\omega})}{\partial x} \right]$$

$$\text{i.e., } \{\overline{\varepsilon(\tilde{\omega})}\} = [H(\tilde{\omega})]\{\varepsilon(\tilde{\omega})\} \tag{10.11}$$

where,

$$\{\varepsilon\} = [u_0 v_0 w_0 \theta_x \theta_y u_u v_u w_u u_l v_l w_l (\partial u_0 / \partial x)(\partial u_0 / \partial y)(\partial v_0 / \partial x)(\partial v_0 / \partial y)$$

$$(\partial w_0 / \partial x)(\partial w_0 / \partial y)(\partial \theta_x / \partial x)(\partial \theta_x / \partial y)$$

$$(\partial \theta_y / \partial x)(\partial \theta_y / \partial y)(\partial u_u / \partial x)(\partial u_u / \partial y)(\partial v_u / \partial x)(\partial v_u / \partial y)(\partial w_u / \partial x)(\partial w_u / \partial y)$$

$$(\partial u_l / \partial x)(\partial u_l / \partial y)(\partial v_l / \partial x)(\partial v_l / \partial y)(\partial w_l / \partial x)(\partial w_l / \partial y)]$$

and the elements of [H] are functions of z and unit step functions. In the present problem, a nine-node quadratic element with eleven field variables (u_0, v_0, w_0, θ_x, θ_y, u_u, v_u, w_u, u_l, v_l and w_l) per node is employed. Using finite element method the generalized displacement vector $\{\delta(\tilde{\omega})\}$ at any point may be expressed as

$$\{\delta(\tilde{\omega})\} = \sum_{i=1}^{n} N_i(\tilde{\omega})\delta_i(\tilde{\omega}) \tag{10.12}$$

where, $\{\delta\} = \left\{ u_0 v_0 w_0 \theta_x \theta_y u_u v_u w_u u_l v_l w_l \right\}^{\mathrm{T}}$ as defined earlier, δ_i is the displacement vector corresponding to node i, N_i is the shape function associated with the node i and n is the number of nodes per element. With the help of equation (10.12), the strain vector $\{\varepsilon\}$ that appeared in equation (10.11) may be expressed in terms of unknowns (for the structural deformation) as

$$\{\varepsilon(\tilde{\omega})\} = [B(\tilde{\omega})]\{\delta(\tilde{\omega})\} \tag{10.13}$$

where [B] is the strain-displacement matrix in the Cartesian coordinate system.

From Hamilton's principle (Meirovitch 1992), the dynamic equilibrium equation for free vibration analysis can be expressed as

$$[K(\tilde{\omega})] \ \{\overline{\delta}\} = \lambda^2 [M(\tilde{\omega})]\{\overline{\delta}\} \tag{10.14}$$

where, $[\lambda(\tilde{\omega})]$ is the stochastic free vibration frequencies for different modes and the global mass matrix $[M(\tilde{\omega})]$ may be formed by assembling a typical element mass matrix as shown below,

$$[M(\tilde{\omega})] = \sum_{i=1}^{n_u + n_l} \iiint \rho_i(\tilde{\omega})[N]^T [P]^T [N][P] \, dxdydz \ = \ \iint [N]^T [R(\tilde{\omega})][N] dxdy \tag{10.15}$$

where $\rho_i(\tilde{\omega})$ is the random mass density of the i-th layer, , matrix $[P]$ is of order 3×11 contains z terms and some constant quantities, matrix $[N]$ is the shape function matrix and the matrix $[R(\tilde{\omega})]$ can be expressed as

$$[R(\tilde{\omega})] = \sum_{i=1}^{n_u + n_l} \rho_i(\tilde{\omega})\,[P]^T[P]\,dz \tag{10.16}$$

A numerical code is developed to implement the above mentioned operations involved in the proposed finite element model to determine the vibration response of laminated skew composite sandwich plates. The skyline technique is used to store the global stiffness matrix in a single array. Simultaneous iteration technique (Corr and Jennings 1976) is used in free vibration analysis. In the present study, a nine noded isoparametric element with eleven degrees of freedom at each node is considered for finite element formulation. The elemental potential energy can be expressed by (Chalak et al. 2015)

$$\Pi_e = U_s - U_{ext} \tag{10.17}$$

where U_s and U_{ext} are the strain energy and the energy due to external in-plane load, respectively.

$$\Pi_e = \frac{1}{2}\iint\{\delta\}^T\left[B(\tilde{\omega})\right]^T\left[D(\tilde{\omega})\right]\left[B(\tilde{\omega})\right]\{\delta\}\,dx\,dy$$

$$- \frac{1}{2}\iint\{\delta\}^T\left[B(\tilde{\omega})\right]^T\left[G(\tilde{\omega})\right]\left[B(\tilde{\omega})\right]\{\delta\}\,dx\,dy$$

$$= \frac{1}{2}\{\delta\}^T\left[K_e(\tilde{\omega})\right]\{\delta\} - \frac{1}{2}\lambda\{\delta\}^T\left[K_G(\tilde{\omega})\right]\{\delta\} \tag{10.18}$$

where, $\left[K_e(\tilde{\omega})\right] = \int\left[B(\tilde{\omega})\right]^T\left[D(\tilde{\omega})\right]\left[B(\tilde{\omega})\right]dx$ and

$$\left[K_G(\tilde{\omega})\right] = \int\left[B(\tilde{\omega})\right]^T\left[G(\tilde{\omega})\right]\left[B(\tilde{\omega})\right]dx$$

Here $[B(\tilde{\omega})]$ is the random strain-displacement matrix while $[K_e(\tilde{\omega})]$ and $[K_G(\tilde{\omega})]$ are the stochastic elastic stiffness matrix and geometric stiffness matrix, respectively. The equilibrium equation can be obtained by minimizing Π_e as given in equation (10.18) with respect to $\{\delta\}$ as

$$\left[K_e(\tilde{\omega})\right]\{\delta\} = \lambda(\tilde{\omega})\left[K_G(\tilde{\omega})\right]\{\delta\} \tag{10.19}$$

where $\lambda(\tilde{\omega})$ is a stochastic buckling load factor. The skyline technique has been used to store the global stiffness matrix in a single array and the simultaneous iteration technique is used for solving the stochastic buckling equation (10.19).

10.3 Formulation of Multivariate Adaptive Regression Splines (MARS)

MARS (Friedman 1991) provides an efficient mathematical relationship between input parameters and output feature of interest for a system under investigation based on few algorithmically chosen samples. MARS is a nonparametric regression procedure that makes no assumption about the underlying functional relationship between the dependent and independent variables. MARS algorithm adaptively selects a set of basis functions for approximating the response function through a forward and backward iterative approach. The MARS model can be expressed as

$$Y = \sum_{k=1}^{n} \alpha_k \, H_k^f(x_i)$$

(10.20)

$$\text{with } H_k^f(x_1, x_2, x_3 \ldots \ldots x_n) = 1, \text{ for } k = 1$$

where α_k and $H_k^f(x_i)$ are the coefficient of the expansion and the basis functions, respectively. Thus the first term of equation (10.20) becomes α_1, which is basically an intercept parameter. The basis function can be represented as

$$H_k^f(x_i) = \prod_{i=1}^{i_k} \left[z_{i,k} \left(x_{j(i,k)} - t_{i,k} \right) \right]_{tr}^q$$

(10.21)

where i_k is the number of factors (interaction order) in the k-th basis function, $z_{i,k} = \pm 1$, $x_{j(i,k)}$ is the j-th variable, $1 \leq j(i,k) \leq n$, and $i_{i,k}$ is a knot location on each of the corresponding variables. q is the order of splines. The approximation function Y is composed of basis functions associated with k sub-regions. Each multivariate spline basis function $H_k^f(x_i)$ is the product of univariate spline basis functions $z_{i,k}$, which is either order one or cubic, depending on the degree of continuity of the approximation. The notation "tr" means the function is a truncated power function.

$$\left[z_{i,k} \left(x_{j(i,k)} - t_{i,k} \right) \right]_{tr}^q = \left[z_{i,k} \left(x_{j(i,k)} - t_{i,k} \right) \right]^q \quad \textit{for } \left[z_{i,k} \left(x_{j(i,k)} - t_{i,k} \right) \right] < 0 \quad (10.22)$$

$$\left[z_{i,k} \left(x_{j(i,k)} - t_{i,k} \right) \right]_{tr}^q = 0, \qquad \textit{Otherwise} \quad (10.23)$$

Here each function is considered as piecewise linear with a trained knot 'tr' at each $x_{(i,k)}$. By allowing the basis function to bend at the knots, MARS can model functions that differ in behaviour over the domain of each variable. This is applied to interaction terms as well. The interactions are no longer treated as global across the entire range of predictors but between the sub-regions of every basis function generated. Depending on fitment, the maximum number of knots to be considered, the minimum number of observations between knots, and the highest order of interaction terms are determined. The screening of automated variables occur as a result of using a modification of the generalized cross-validation (GCV) model fit criterion, developed by Craven and Wahba (1979). MARS finds the location and number of the needed spline basis functions in a forward or backward stepwise fashion. It starts by over-fitting a spline function through each knot, and then by removing the knots that least contribute to the overall fit of the model as determined by the modified GCV criterion, often completely removing the most insignificant variables. The equation depicting the lack-of-fit (L_f) criterion used by MARS is

$$L_f(Y_{\tilde{k}}) = G_{cv}(\tilde{k}) = \frac{\frac{1}{n}\sum_{i=1}^{n}[Y_i - Y_{\tilde{k}}(x_i)]^2}{\left[1 - \frac{\tilde{c}(\tilde{k})}{n}\right]^2} \tag{10.24}$$

$$\text{where } \tilde{c}(\tilde{k}) = c(\tilde{k}) + M.\tilde{k}$$

where 'n' denotes the number of sample observations, $\tilde{c}(\tilde{k})$ is the number of linearly independent basis functions, $\tilde{k}$ is the number of knots selected in the forward process, and 'M' is a cost for basis-function optimization as well as a smoothing parameter for the procedure. Larger values of 'M' result in fewer knots and smoother function estimates. The best MARS approximation is the one with the highest GCV value. Thus MARS is also compared with parametric and nonparametric approximation routines in terms of its accuracy, efficiency, robustness, model transparency, and simplicity and it is found suitable methodologies because it is more interpretable than most recursive partitioning, neural and adaptive strategies wherein it distinguishes well between actual and noise variables. Moreover, the MARS are reported by Crino and Brown (2007) to work satisfactorily in terms of computational cost irrespective of dimension (low-medium-high) and noise.

10.4 Random input representation

The layer-wise random input parameters such as ply-orientation angle, skew angle, thickness and material properties (e.g., mass density, elastic modulus, Poisson's ratio) of both core and face sheet are considered for sandwich plates. It is assumed that the random uniform distribution of input parameters exists within a certain band of tolerance with their deterministic values. The individual and combined cases wherein the input variables considered in both soft core and each layer of face sheet of sandwich are as follows:

(Case-a) Variation of ply-orientation angle only: $\theta(\tilde{\omega}) = [\{\theta_1 \ \theta_2 \ \theta_3.......\theta_i......\theta_l \}](\tilde{\omega})$

(Case-b) Variation of thickness only: $t_{tot}(\tilde{\omega}) = [\{t_c \}, \{t_{fs(1)} \ t_{fs(2)}..t_{s(l)} \}](\tilde{\omega})$

(Case-c) Variation of mass density only: $\rho(\tilde{\omega}) = [\{\rho_{fs(1)} \ \rho_{c(2)} \}](\tilde{\omega})$

(Case-d) Variation of skew angle only: $\varphi(\tilde{\omega})$

(Case-e) Variation of material properties:

$$P(\tilde{\omega}) = [E_{x(fs,c)}, E_{y(fs,c)}, E_{z(fs,c)}, G_{12(fs,c)}, G_{13(fs,c)}, G_{23(fs,c)}, \mu_{12(fs,c)}, \mu_{21(fs,c)},$$
$$\cdots \mu_{13(fs,c)}, \mu_{23(fs,c)}, \mu_{32(fs,c)}, P_{(fs,c)}](\tilde{\omega})$$

(Case-f) Combined variation of ply orientation angle, thickness, mass density, skew angle, elastic moduli, shear moduli, Poisson ratios and mass density for both core and face sheet (total 63 numbers random input variables):

$$C(\tilde{\omega}) = [\theta(\tilde{\omega}), t_{tot}(\tilde{\omega}), \rho(\tilde{\omega}), \varphi(\tilde{\omega}), P(\tilde{\omega})]$$

where θ, t, ρ and φ are the ply orientation angle, thickness, mass density and skew angle respectively. The subscripts c and fs are used to indicate core and face sheet respectively. 'l' denotes the number of layer in the laminate, where $i = 1, 2, \ldots, l$. Six different cases are considered for the analysis: layer-wise stochasticity in ply orientation angle ($\theta(\tilde{\omega})$), combined effect for thickness of core and face sheet ($t_{tot}(\tilde{\omega})$), combined effect for mass density of core and face sheet ($\rho(\tilde{\omega})$), skew angle ($\varphi(\tilde{\omega})$), combined effect for material properties of core and face sheet ($P(\tilde{\omega})$) and combined variation of all parameters ($C(\tilde{\omega})$). In the present study, $\pm5°$ variation for ply orientation angle and skew angle and $\pm10\%$ variation in material properties are considered from their respective deterministic values unless mentioned otherwise for some analyses. Figure 10.3 presents the flowchart of proposed stochastic frequency analysis using MARS model for the laminated soft core sandwich structure.

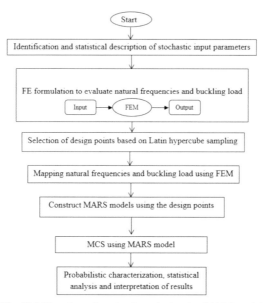

Fig. 10.3 Flowchart of stochastic analysis using MARS model.

10.5 Results and discussion

The present study considers a sandwich composite plate with soft core (both upper and lower as 0°) and two facesheets with four layered cross-ply (90°/0°/90°/0°) laminate covering the core in both top and bottom side. A nine noded isoparametric plate bending element is considered for finite element formulation. For the analysis, dimensions and boundary conditions considered for the sandwich composite plate are as follows: length (L) = 1 m, width (b) = 0.5 m, thickness (t) = L /10; with simply supported boundary conditions (unless otherwise mentioned). The considered material properties of the sandwich plate are provided in Table 10.1. The present MARS model is employed to find a predictive and representative surrogate model relating each natural frequency to a number of stochastic input variables. The MARS based surrogate models are used to determine the first three natural frequencies corresponding to a set of input variables, instead of time-consuming and expensive finite element analysis. The probability density function is plotted as the benchmark of bottom line results. The variation of geometric and material properties are considered to fluctuate within the range of lower and upper limit (tolerance limit) as ± 10% with respective mean values while for ply orientation angles and skew angles as within ±5° fluctuation (as per industry standard) with respect to their deterministic mean values. A layer-wise random variable approach is employed for generating the set of random input variables which are considered for surrogate based numerical finite element iteration to obtain the respective set of random output parameters accordingly. The transverse shear stresses vanish only at top and bottom surfaces of the laminate irrespective of the considered boundary conditions, e.g., for clamped boundary condition, all the kinematic variables vanish at clamped edges. Results are presented for stochastic natural frequencies and stochastic buckling load for the sandwich plate.

Table 10.1 Material properties for core and face sheet of sandwich plate.

Material properties	Core	Face sheet
E_1	0.5776 GPa	276 GPa
E_2 and E_3	0.5776 GPa	6.9 GPa
G_{12} and G_{13}	0.1079 GPa	6.9 GPa
G_{23}	0.2221 GPa	6.9 GPa
v_{12} and v_{13}	0.25	0.25
v_{21} and v_{31}	0.00625	0.00625
v_{23} and v_{32}	0.25	0.25
ρ	1000 kg/m³	681.8 kg/m³

10.5.1 Stochastic natural frequency analysis

Mesh convergence and validation of the finite element model for the sandwich plate is conducted first considering a deterministic analysis. The optimum mesh size is finalized on the basis of a mesh convergence study as presented in Fig. 10.4, wherein a mesh size of (14 × 14) is found to be adequate. The non-dimensional natural frequencies ($\varpi = 100 \times \omega L \sqrt{\dfrac{\rho_c}{E_{2f}}}$, where ρ_c is the density of the core layer) for first two modes based on the present model are obtained for various skew angles and are tabulated in Table 10.2 along with the previous results obtained by Wang et al. (2000) and Kulkarni and Kapuria (2008). Table 10.3 presents the results for non-dimensional natural frequencies of a four layered clamped symmetric (0°/90°/90°/0°) laminated composite plate obtained from the present analysis for various aspect ratios with respect to the previous analyses reported by Wang et al. (2000) and Khandelwal et al. (2013). The results corroborate good agreement of the deterministic natural frequencies obtained using the present finite element model with respect to previous works. The validation of the MARS model as a surrogate of the actual finite element model is presented using scatter plots and probability

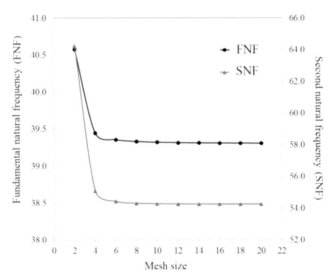

Fig. 10.4 Mesh convergence study of finite element analysis with different mesh sizes with respect to fundamental and second natural frequencies of sandwich skewed plates (FNF – First natural frequency, SNF – Second natural frequency).

Table 10.2 Non-dimensional natural frequencies of a four layered (0°/90°/0°/90°) anti-symmetric composite plate.

Skew angle	Mode	Present analysis	Wang et al. (2000)	Kulkarni and Kapuria (2008)
30°	1	1.8889	1.9410	1.9209
	2	3.4827	2.9063	3.5353
45°	1	2.5806	2.6652	2.6391
	2	3.7516	3.2716	4.1810

Table 10.3 Non-dimensional natural frequencies of four layered clamped symmetric $(0°/90°/90°/0°)$ laminated composite plate.

Aspect ratio	Mode	Present analysis	Kulkarni and Kapuria (2008)	Khandelwal et al. (2013)
10	1	18.0843	18.2744	17.9550
	2	28.9441	28.9047	28.9674
20	1	23.4534	24.1130	23.9339
	2	37.0587	36.7473	37.0614

density function plots (refer to Fig. 10.5 and Fig. 10.6). The low deviation of points from the diagonal line in the scatter plot (Fig. 10.5) corroborates the high accuracy of prediction capability of the MARS model with respect to the finite element model for all the random input parameter sets (combined effect of 63 numbers of random input parameters). The probability density function plots presented in Fig. 10.6 shows a negligible deviation between the MARS model and the original MCS model indicating validity and high level of precision for the present surrogate based approach further. It is noteworthy that the proposed MARS based approach requires 256 numbers of original finite element simulations for the layerwise individual variation of stochastic input parameters, while due to increment in number of input variables, 512 FE simulations are found to be adequate for combined random variation of input parameters. Here, although the same sample size as in direct MCS (10,000 samples) is considered for characterizing the probability distributions of natural frequencies, the number of actual FE simulations is much less compared to the direct MCS approach. Hence, the computational time and effort expressed in terms of expensive finite element simulations is reduced significantly compared to full scale direct MCS. This provides an efficient affordable way for simulating the uncertainties in natural frequency. The optimum number of FE simulations (i.e., the number of design points in Latin hypercube sampling) required to construct the MARS models are decided based on a convergence study as presented in Table 10.4.

In the present analysis, all the layer-wise individual cases of stochasticity are studied as described in Section 10.4. It is however noticed that skew angle, mass density and transverse shear modulus are the three most sensitive factors for first three stochastic natural frequencies (refer to Fig. 10.7) by analyzing the relative coefficient of variations. The relative combined effect of the other parameters are (longitudinal and transverse elastic modulus, ply orientation angle, thickness, longitudinal shear modulus and Poisson ratio) also shown in Fig. 10.7 for the first three natural frequencies. As the effect of other parameters has negligible sensitivity on stochastic natural frequencies, representative results are furnished for stochastic effect of two most effective parameters (skew angle and mass density) for analysis of individual cases. Probability distributions for first three stochastic natural frequencies of a simply supported composite sandwich plate due to only variation in skew angles are furnished in Fig. 10.8. As the skew angle increases the

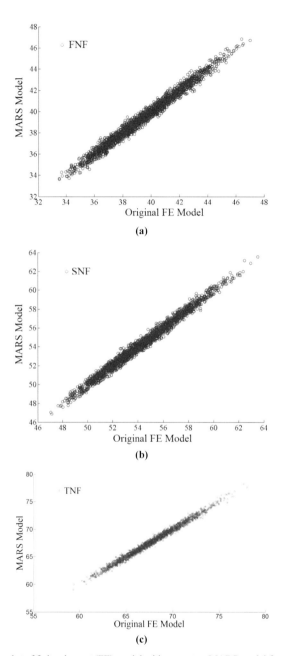

Fig. 10.5 Scatter plot of finite element (FE) model with respect to MARS model for (a) Fundamental natural frequency (FNF), (b) Second natural frequency (SNF) and (c) Third natural frequency (TNF) of simply supported sandwich skewed plates considering combined variation (total 63 numbers of random input variables) for $\phi(\widetilde{\omega}) = 45°$.

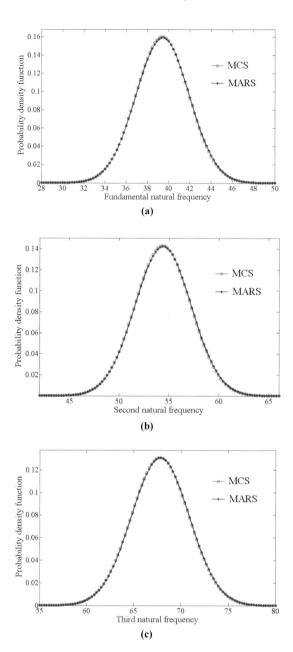

Fig. 10.6 Probability density function for MCS as well as MARS model for first three natural frequencies of simply supported compsoite sandwich skewed plates considering combined variation (total 63 numbers random input variables) for $\phi(\widetilde{\omega}) = 45°$.

Table 10.4 Convergence study of first three modes due to individual and combined variation of inputs for simply supported sandwich plates.

Individual	Value	f_1				f_2				f_3			
		MCS (10,000)	MARS (Sample run)			MCS (10,000)	MARS (Sample run)			MCS (10,000)	MARS (Sample run)		
			64	128	256		64	128	256		64	128	256
$\theta(\widetilde{\omega})$	Max	39.6407	39.6614	39.6527	39.6493	54.6218	54.7123	54.6531	54.6338	67.8206	67.9845	67.9214	67.8584
	Min	38.9549	38.9420	38.9456	38.9485	53.8200	53.8064	53.8119	53.8168	67.4350	67.4220	67.4307	67.4321
	Mean	39.3039	39.3115	39.3101	39.3092	54.2480	54.2311	54.2342	54.2381	67.6402	67.6998	67.6831	67.6587
	SD	0.1186	0.1194	0.1195	0.1196	0.1458	0.1473	0.1468	0.1461	0.0668	0.0735	0.0698	0.0681
$t_s(\widetilde{\omega})$	Max	39.6858	39.7564	39.7313	39.7257	54.7669	54.8214	54.7917	54.7764	68.4852	68.6154	68.5978	68.5432
	Min	38.8943	38.8021	38.8264	38.8409	53.7134	53.3116	53.3982	53.4918	66.8631	66.8167	66.8227	66.8497
	Mean	39.3044	39.3164	39.3114	39.3081	54.2644	54.4951	54.3718	54.3083	67.6574	67.6831	67.7952	67.7098
	SD	0.1159	0.1173	0.1176	0.1179	0.1527	0.1584	0.1562	0.1552	0.2391	0.2487	0.2423	0.2416
$\rho(\widetilde{\omega})$	Max	40.3206	41.1121	40.9821	40.5127	55.6577	56.1064	55.9561	55.7473	69.3890	69.9983	69.8134	69.4835
	Min	38.4011	38.3942	38.3964	38.9873	53.0081	52.6876	52.7942	52.9264	66.0856	65.8421	65.9226	65.9928
	Mean	39.3299	39.6734	39.5154	39.4212	54.2901	54.5876	54.5083	54.4221	67.6840	67.8674	67.8050	67.7213
	SD	0.4905	0.6154	0.5584	0.5129	0.6772	0.7954	0.7054	0.6997	0.8444	0.9533	0.8624	0.8517
$\phi(\widetilde{\omega})$	Max	41.7226	42.2134	42.0219	41.8687	56.7749	57.1259	56.9641	56.7963	70.1767	70.9897	70.6245	70.3516
	Min	37.2399	36.8276	36.9893	37.0867	52.1238	51.7383	51.9767	52.1013	65.5452	64.9984	65.1137	65.4194
	Mean	39.3663	39.6124	39.5483	39.4468	54.3251	54.6682	54.5437	54.4198	67.7213	67.9457	67.8438	67.7438
	SD	1.2740	1.3130	1.3030	1.2991	1.3218	1.3356	1.3286	1.3264	1.3164	1.3552	1.3487	1.3258

Table 10.4 contd....

...*Table 10.4 contd.*

Combined		MCS (10,000)	MARS (Sample run)			MCS (10,000)	MARS (Sample run)			MCS (10,000)	MARS (Sample run)		
			128	256	512		128	256	512		128	256	512
$C(\tilde{\omega})$	Max	46.59067	47.00219	46.9832	46.9265	62.80511	63.51472	63.2164	63.05806	77.35152	78.02273	77.9516	77.8462
	Min	33.0163	33.50135	33.4134	33.3372	46.86064	47.0558	46.9671	46.9493	59.25218	59.32715	59.2971	59.2832
	Mean	39.43564	39.39828	39.4002	39.4138	54.4066	54.36105	54.3883	54.3921	67.8260	67.77762	67.7884	67.7935
	SD	2.5081	2.4872	2.4889	2.4992	2.8085	2.7907	2.7921	2.7944	3.0515	3.0439	3.0476	3.0497

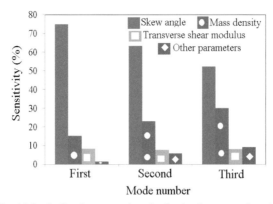

Fig. 10.7 Sensitivity for first three natural modes for simply supported sandwich plates.

mean of stochastic natural frequencies are also found to increase, while probability distributions corresponding to different skew angles vary considerably. Figure 10.9 represents the stochastic first three natural frequencies of a simply supported sandwich composite skewed plate (for skew angle $\phi(\widetilde{\omega}) = 45°$) due to only variation of mass density (layer-wise) with different degree of stochasticity. As the percentage of stochasticity in mass density increases the response bounds are found to increase accordingly, while the mean does not change for different percentage of variation in mass density. The effect of combined stochasticity in all input parameters (referred as $C(\widetilde{\omega})$ in Section 10.4) is also analyzed for different skew angles in addition to individual effect of the input parameters for stochastic natural frequencies of sandwich plates. In Fig. 10.10, the stochastic first three natural frequencies are presented for simply supported sandwich composite plates with different skew angles considering combined variation of input parameters $C(\widetilde{\omega})$ (total 63 numbers random input variables), wherein a general trend is noticed that the mean and response bounds increase with the increase in skew angle. Response bounds of the first three natural frequencies due to combined variation are noticed to be higher than individual variation of input parameters in all (total 63 numbers random input variables) cases. The stochastic first three natural frequencies of sandwich composite skewed plates with different boundary conditions (C-Clamped, S-Simply supported, F- Free) are shown in Fig. 10.11 considering combined variation of input parameters to investigate the effect of boundary conditions. The probability distributions are found to vary significantly depending on the considered boundary condition. Both mean and standard deviation of CCCC boundary condition are found to be highest for combined variation of all input parameters.

10.5.2 Stochastic buckling load analysis

Mesh convergence and validation of the finite element model for deterministic buckling load is presented above in Fig. 10.12. The convergence study on finite element mesh size is conducted to obtain the optimum mesh size considering

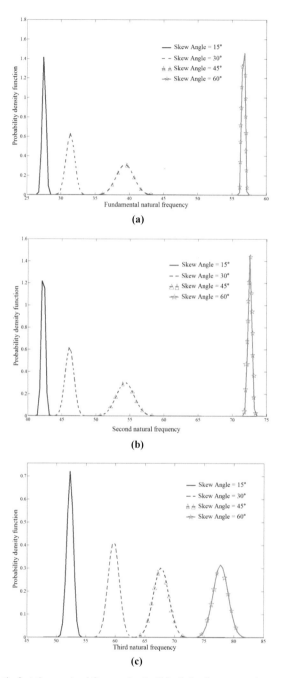

Fig. 10.8 Stochastic first three natural frequencies (rad/s) of simply supported composite sandwich plates due to only variation of skew angles.

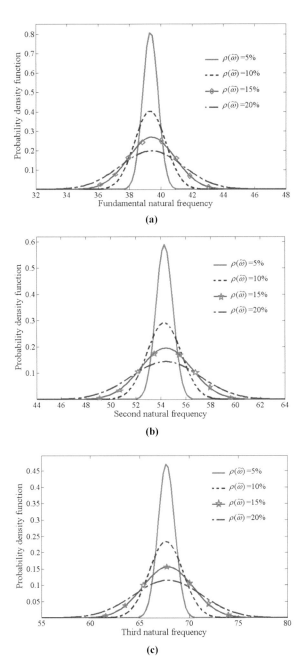

Fig. 10.9 Stochastic first three natural frequencies (rad/s) of simply supported sandwich composite skewed plates for $\phi(\widetilde{\omega}) = 45°$ due to only variation of mass density with different degree of stochasticity.

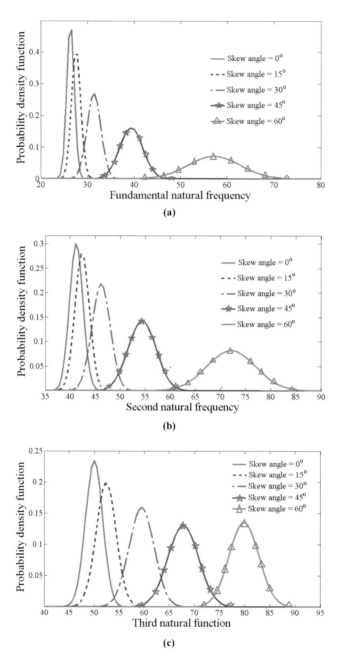

Fig. 10.10 Stochastic first three natural frequencies (rad/s) of simply supported sandwich composite plates for different skew angles considering combined variation of input parameters (total 63 numbers random input variables).

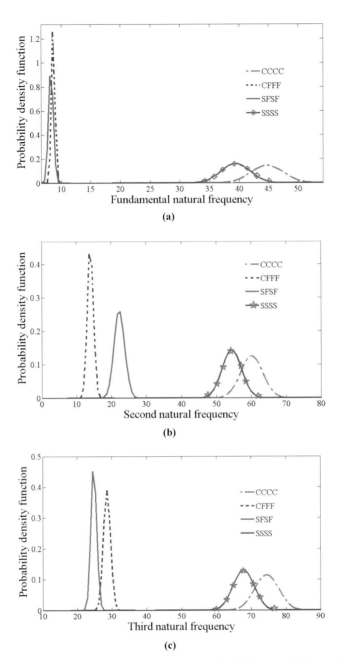

Fig. 10.11 Stochastic first three natural frequencies (rad/s) of sandwich composite skewed plates for $\phi(\widetilde{\omega}) = 45°$ with different boundary conditions considering combined variation.

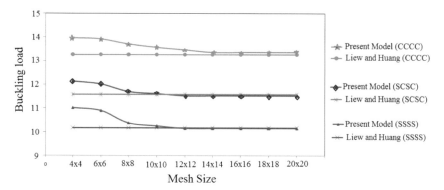

Fig. 10.12 Mesh convergence study and validation for comparison of non-dimensionalised critical bi-axial buckling load $[\bar{\lambda} = (\lambda l^2)/(h^2 E_{Tf})$ where λ, l, h and E_{Tf} are the buckling load factor, depth of the plate and transverse modulus of elasticity of face layer, respectively] for laminated sandwich plates with different boundary conditions.

different boundary conditions such as CCCC, SCSC and SSSS (where S-simply supported, C-clamped, indicating boundary condition of four sides). As the computational iteration time increases with the increase of mesh size, a (14×14) optimal mesh size is considered in the present study. The present buckling load are also validated with the results obtained by Liew and Huang (2003). The results corroborate good agreement of the buckling load obtained using the present finite element model with respect to previous works of Liew and Huang irrespective of imposed bounary conditions.

Further, the MARS model that is employed to achieve computational efficiency, is validated with traditional Monte Carlo simulation (MCS). Representative results are furnished for combined variation of all input parameters (512 samples) using probability density function plots and scatter plot as shown in Fig. 10.13. The figures indicate high degree of precision and accuracy of the present MARS model with respect to the original finite element model. The results for buckling load are presented hereafter (Figs. 10.13–10.19) as a ratio of stochastic buckling load and deterministic buckling load to provide a clear and direct interpretation for stochasticity in different input parameters.

The effects of variation of core thickness and face sheet thickness on stochastic normalized buckling load of sandwich plates are shown in Fig. 10.14 and Fig. 10.15, respectively. It is found that as the percentage of variation of both core and face sheet thickness increases, the response bound of stochastic buckling load also increases, while the mean does not vary. The sparsity of stochastic normalized buckling load due to variation of core thickness is observed to be significantly higher than that of the same due to variation of face sheet thickness. The effect of variation of all core material properties on stochastic buckling load of sandwich plates is furnished in Fig. 10.16, while Fig. 10.17 presents the effect of ply orientation angle of face sheet on stochastic normalized buckling load of sandwich plates. Likewise with variation of core and face sheet thickness (Figs. 10.14–10.15), mean value for

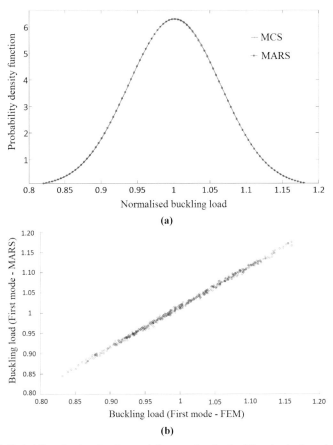

Fig. 10.13 Probability density function and Scatter plot for buckling load of sandwich plates considering combined variation of all input parameters $(C(\widetilde{\omega}))$.

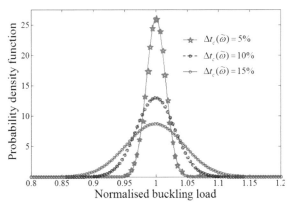

Fig. 10.14 Effect of variation of core thickness on normalized buckling load of sandwich plates with SCSC.

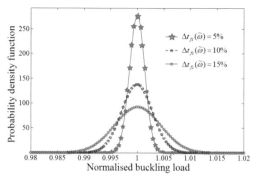

Fig. 10.15 Effect of variation of face sheet thickness on normalized buckling load of sandwich plates with SCSC.

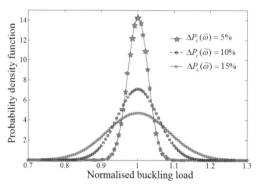

Fig. 10.16 Effect of variation in material properties of core on normalized buckling load of sandwich plates with SCSC.

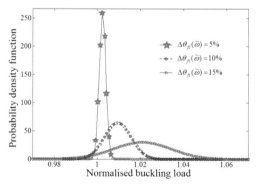

Fig. 10.17 Effect of variation of ply orientation angle of face sheet on normalized buckling load of sandwich plates with SCSC.

stochastic buckling load remain unaltered with different degree of stochasticity in core material properties, while the standard deviation increases with increase in degree of stochasticity. In contrast, both mean and standard deviation of stochastic buckling load are found to increase with increasing degree of stochasticity in ply orientation angle. The variation in buckling load due to stochasticity of core material properties (Fig. 10.16) is found to be higher than the other three individual cases (Figs. 10.14–10.15, 10.17). However, the maximum variation in normalized buckling load is observed in case of combined stochasticity of core and face sheet thickness, ply-orientation angle of face sheet and material properties (Fig. 10.18). The effect of different boundary conditions (CCCC, CFCF, SCSC and SSSS; where S-simply supported, C-clamped and F-fixed end condition) on normalized stochastic buckling load of sandwich plates is presented in Fig. 10.19. Even though the response bounds for different boundary conditions for normalized buckling load does not vary, the probability distributions for buckling loads in actual values will vary significantly depending on their deterministic values. Coefficient of variation

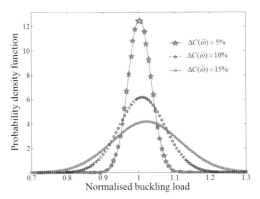

Fig. 10.18 Effect of combined variation of stochastic input parameters (core and face sheet thickness, ply-orientation angle of face sheet and material properties) on normalized buckling load of sandwich plates.

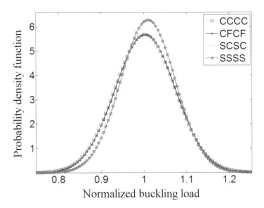

Fig. 10.19 Effect of boundary condition on normalized buckling load of sandwich plates.

corresponding to different degree of stochasticity for different cases considered in this study is plotted in Fig. 10.20. From the figure it is evident that the effect on buckling load due to stochastic variation of different input parameters in a decreasing order is: combined variation of all stochastic input parameters, core material properties, core thickness, ply orientation angle and face sheet thickness. Slope of the curves for different parameters corresponding to different degree of stochasticity provide a clear interpretation about their relative sensitivity towards buckling load.

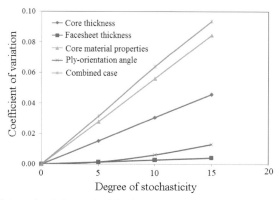

Fig. 10.20 Coefficient of variation on buckling load with respect to degree of stochasticity of input parameters for simply supported sandwich plates.

10.6 Summary

This chapter illustrates the layer-wise random variable based uncertainty quantification in sandwich skewed plates in an efficient surrogate based bottom-up framework. The probability distributions of first three natural frequencies and buckling load are analyzed considering both individual and combined stochasticity in input parameters. A multivariate adaptive regression splines (MARS) based approach is employed in conjunction with finite element modelling to map the variation of first three natural frequencies and buckling load caused due to uncertain input parameters, wherein it is found that the number of finite element simulations is exorbitently reduced compared to direct Monte Carlo simulation without compromising the accuracy of results. The computational expense is reduced by (1/78) times (individual effect of stochasticity) and (1/39) times (combined effect of stochasticity) than direct Monte Carlo simulation. The skew angle is found to be most sensitive to the frequencies corresponding to the first three modes. The mass density and transverse shear modulus are other two effective factors for the first three natural frequencies among the considered stochastic input parameters respectively. Combined effect of the material properties of the soft-core has the most sensitivity for buckling load, followed by core thickness, ply orientation angle and face sheet thickness respectively. The numerical results presented in this chapter shows that stochasticity in different material and geometric properties of laminated sandwich plates has considerable influence on the dynamics and stability of the structure. Thus it is of prime importance to incorporate the effect of stochasticity in subsequent analyses, design and control of such structures. The proposed MARS based uncertainty quantification algorithm can be extended further to explore other stochastic systems in the future course of research.

CHAPTER 11

Fuzzy Uncertainty Quantification in Composite Laminates

11.1 Introduction

We have focused on the probabilistic approaches of uncertainty quantification in composite laminates so far in this book. A non-probabilistic uncertainty propagation approach (fuzzy) for composites is presented in this chapter. Probabilistic descriptions of uncertain model parameters are not always available due to lack of data. This chapter investigates the uncertainty propagation in dynamic characteristics (such as natural frequencies, frequency response function and mode shapes) of laminated composite plates by using the fuzzy approach. A non-intrusive Gram–Schmidt polynomial chaos expansion (GPCE) method is adopted in the uncertainty propagation, wherein the parameter uncertainties are represented by fuzzy membership functions. A domain in the space of input data at zero-level of membership functions is mapped to a zone of output data with the parameters determined by D-optimal design. The obtained meta-model (GPCE) can also be used for higher α-levels of fuzzy membership function. The most significant input parameters such as ply orientation angle, elastic modulus, mass density and shear modulus are identified and then fuzzified. Fuzzy analysis of the first three natural frequencies is presented to illustrate the results and its performance. The proposed fuzzy approach is applied to the problem of fuzzy modal analysis for frequency response function of a simplified composite cantilever plates. The fuzzy mode shapes are also depicted for a typical laminate configuration. The GPCE based approach is found more efficient compared to the conventional global optimization approach in terms of computational time and cost.

Composite materials have gained immense popularity in application of aerospace, marine, automobile and construction industries due to its weight sensitivity and cost-effectiveness. Such structures are prone to considerable uncertainty in their fibre and material parameters. Therefore, it is important to investigate the structural behavior of composites due to the variability of parameters in each constituent laminate level. Uncertainty can be modelled either by probabilistic or non-probabilistic approach. Due to availability of limited sample experimental or testing data (crisp inputs), it will be more realistic to follow a non-probabilistic approach rather than a probabilistic approach in certain situations. As discussed in the previous chapters, significant advances have been made in representing uncertainty in composite material properties by probabilistic models. However, these methods usually need large volumes of data, which are expensive and computational costs are high. An alternative approach, assuming that large quantities of test data are not available, would be to use non-probabilistic methods such as interval and fuzzy. Fuzzy finite element analysis presented by Moens and Hans (2011), and Zadeh (1975) aims to combine the power of finite element method and uncertainty modelling capability of fuzzy variables. A little attention is drawn in the use of non-probabilistic models such as fuzzy in the context of composite structures (Dey et al. 2016d). One way to view a fuzzy input-output variable is the universality of an interval variable. It should be noted that the intervals do not represent the values of the variable, but the knowledge about the range of possible values that a variable can take. For an uncertain variable represented by interval, the values of the parameters can be observed within the two bounds (lower and upper). A membership function is introduced in fuzzy approach (Zadeh 1975). In real-life problems, the original Monte Carlo simulation is expensive due to high computational time. Therefore, the aim of the majority of current research is to reduce the computational cost. Under the possibilistic interpretation of fuzzy sets (Moller and Beer 2004) and uncertainty environment (Denga et al. 2012), fuzzy variables would become generalized interval variables. Consequently techniques employed in interval analysis such as classical interval arithmetic (Moore 1966), affine analysis (Dengrauwe et al. 2010) or vertex theorems (Qiu et al. 2005) can be used. The Neumann expansion (Lallemand et al. 1999), the transformation method (Hanss 2002), and response surface based methods (Demunck et al. 2008) are proposed for fuzzy analysis. In this context, fuzzy analysis has been employed recently to deal with uncertainties in engineering problems using only available data by Babuska et al. (2014). Earlier, fuzzy approach has been applied to safety analysis by Rao and Annandas (2008), random system properties by Lal and Singh (2011) and optimal design Diaz-Madronero et al. (2014). The PCE approach (Umesh and Ganguli 2013) and High Dimensional Model Representation (HDMR) approach (Chowdhury and Adhikari 2012) are proposed for the propagation of fuzzy uncertain variables. The coupling of fuzzy concepts (Zadeh 1965, Hanss and Willner 2000) and the modeling of arbitrary uncertainties using Gram-Schmidt polynomial chaos (Witteveen and Bijl 2006) provides an efficient means for fuzzy uncertainty propagation.

In practice, the fuzzy models can be used when there is lack of data to estimate an accurate PDF of the uncertain parameters. The present study employs the Gram-Schmidt algorithm based polynomial chaos expansion in propagation of structural uncertainties of composite structures, when parameter uncertainties are represented by fuzzy membership functions. The polynomial chaos expansion acts as a surrogate model (meta-model) for the full finite element model of composite structure. The regression coefficients of the PCE are then determined by first sampling in the space of input parameters (Montgomery 1991) and then a least square technique. Since the widest range of input parameters are considered to be at zero membership function, the regression coefficients can be obtained once at this level and used for all highest α-cuts. The finite element code of laminated composites is combined non-intrusively with the proposed method to treat uncertainty associated with complex systems like laminated composite structures. In the present study, a four layered graphite-epoxy composite laminated cantilever plate is considered as shown in Fig. 11.1. This chapter is organized hereafter as, Section 11.2: theoretical formulation for fuzzy uncertainty propagation in composites based on GPCE, Section 11.3: representation of fuzzy input parameters, Section 11.4: results and discussion, Section 11.5: summary.

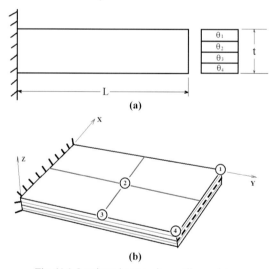

Fig. 11.1 Laminated composite cantilever plate.

11.2 Theoretical formulation for fuzzy uncertainty propagation in composites

In the fuzzy concept (Witteveen and Bijl 2006), a set of transitional states between the members and non-members are defined via a membership function $[\mu_{pi}]$ that indicates the degree to which each element in the domain belongs to the fuzzy set. The fuzzy number $[\widetilde{p}_i(\widetilde{\omega})]$ considering triangular membership function can be expressed as,

$$\tilde{p}_i(\tilde{\omega}_\alpha) = [\,p_i^U, p_i^M, p_i^L\,] \tag{11.1}$$

where p_i^M, p_i^U and p_i^L denote the mean value, the upper bound and lower bounds, respectively. $(\tilde{\omega})_\alpha$ indicates the fuzziness corresponding to α-cut where α is known as membership grade or degree of fuzziness ranging from 0 to 1.

The Gaussian probability function may be approximated by a triangle by equating the area under the normalised Gaussian distribution function (Hanss and Willner 2000) with the area under triangular membership function as shown in Fig. 11.2. This is just an example to show how a fuzzy membership function can be constructed from a given PDF. In practice, when there is limited measured data, the fuzzy membership function of uncertain data can be constructed using histogram. As a result of this approximation the triangular fuzzy membership function can be defined as

$$\mu_{p(i)} = \max\left[\,0,1 - \frac{\left|X_i^{(j)} - \hat{X}_i\right|}{\zeta}\right] \tag{11.2}$$

where $\zeta = \sqrt{2\pi}\sigma_X$, $\hat{X}_i$ and σ_X are the mean and standard deviation of the equivalent Gaussian distribution. In this chapter, the triangular shaped membership function is employed. Now the membership function $[\mu_{p(i)}]$ can be expressed as

$$\mu_{p(i)} = 1 - (p_i^M - p_i)/(p_i^M - p_i^L), \qquad for \ \ p_i^L \le p_i \le p_i^M$$
$$\mu_{p(i)} = 1 - (p_i - p_i^M)/(p_i^U - p_i^M), \qquad for \ \ p_i^M \le p_i \le p_i^U \tag{11.3}$$
$$\mu_{p(i)} = 0 \qquad\qquad\qquad\qquad\qquad Otherwise$$

The fuzzy input number p_i can be expressed into the set P_i of $(m+1)$ intervals $P_i^{(j)}$ using the α-cut method

$$P_i(\tilde{\omega}_\alpha) = [\,p_i^{(0)}, p_i^{(1)}, p_i^{(2)} \ldots\ldots p_i^{(j)} \ldots\ldots p_i^{(m)}\,] \tag{11.4}$$

where m denotes the number of α-cut levels. The interval of the j-th level of the i-th fuzzy number is given by

$$p_i^{(j)} = [\,p_i^{(j,L)}, p_i^{(j,U)}\,] \tag{11.5}$$

where $p_i^{(j,L)}$ and $p_i^{(j,U)}$ denote the lower and upper bounds of the interval at the j-th level, respectively. At $j=m$, $p_i^{(m,L)}=p_i^{(m,U)}=p_i^M$. The superscripts L and U denote the lower and upper bounds, respectively.

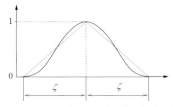

Fig. 11.2 Linear approximation of a Gaussian distribution by triangular fuzzy.

　　In order to propagate uncertainty in a system where uncertain model parameters are represented by fuzzy input numbers, one may apply a numerical procedure of interval analysis at a number of α-levels (Moens and Hanss 2011). In this case, the range of the response components on a specific level of membership function is searched within the same α-level on the input domain, which means that the analysis at each α-cut corresponds to an interval analysis for the system. In the present analysis, the orthogonal polynomial chaos basis functions, derived from Gram-Schmidt algorithm (Witteveen and Bijl 2006) is employed for uncertainty propagation, as depicted in Fig. 11.3, which Fshows the scheme for a particular case of two input parameters and one output but the idea can be readily generalised for the case of multi-inputs multi-outputs. The solution to fuzzy generalised equation at each α-level may be expanded into a polynomial chaos expansion as follows:

$$\mathbf{y}^{(\alpha)} = \mathbf{B}^{(\alpha)}\ \mathbf{\psi}(\mathbf{p}^{(\alpha)}) \qquad \text{for } \alpha = \alpha_k, \quad k = 1,2,...,r \qquad (11.6)$$

where $\mathbf{y}^{(\alpha)} = \left[y_1^{(\alpha)}\ y_2^{(\alpha)} ... y_n^{(\alpha)} \right]^T \in \mathfrak{R}^{n\times 1}$ denotes the assembled vector of output data at $\alpha = \alpha_k$ (the subscript k is removed from α_k for reason of simplicity), $\mathbf{\psi}(\mathbf{p}^{(\alpha)}) = \text{vec}\left(\mathbf{\Psi}(\mathbf{p}^{(\alpha)})\right) \in \mathfrak{R}^{p\times 1}$ denotes the assembled vector of Gram-Schmidt polynomial chaos basis functions, to be explained in the sequel and $\hat{\mathbf{A}}$ is expressed as

$$\mathbf{B} = \begin{bmatrix} \beta_0^{(1)} & \beta_1^{(1)} & \beta_2^{(1)} & .... & \beta_p^{(1)} \\ \beta_0^{(2)} & \beta_1^{(2)} & \beta_2^{(2)} & .... & \beta_p^{(2)} \\ \beta_0^{(3)} & \beta_1^{(3)} & \beta_2^{(3)} & .... & \beta_p^{(3)} \\ . & . & . & & . \\ . & . & . & & . \\ . & . & . & & . \\ \beta_0^{(n)} & \beta_1^{(n)} & \beta_2^{(n)} & .... & \beta_p^{(n)} \end{bmatrix} \qquad (11.7)$$

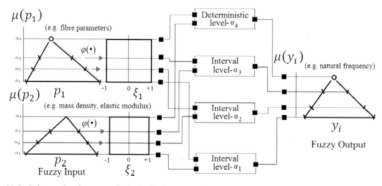

Fig. 11.3 Scheme for fuzzy analysis including transformation $\varphi(\cdot)$ of a fuzzy variable p_i to $\xi \in [-1,1]$ for different α-cuts.

where $\beta_k^{(i)}$ are the coefficients of polynomial expansion with $k=1,2,3....p$ (p is the number of terms retained in the expansion), n is the number of output parameters, $\mathbf{p}^{(\alpha)} = \begin{bmatrix} p_1^{(\alpha)} & p_2^{(\alpha)} & ... & p_m^{(\alpha)} \end{bmatrix}^T \in \mathfrak{R}^{m\times 1}$ is an m-dimensional vector of interval variables at $\alpha = \alpha_k$. As it is already mentioned in this chapter, we represent a fuzzy variable with a set of interval variables via the membership function. The lower and upper bounds of interval variables at different α-levels (i.e., $p_i^{(\alpha,L)}$, $p_i^{(\alpha,U)}$) can be transformed into the normalized values of -1 and 1, respectively. As explained by Adhikari and Haddad (2014), this is because the solution to the Legendre's differential equation is convergent when the random variables are between -1 and 1. This mapping can be done for any other type of probability distribution function, e.g., Hermite polynomials. The mapping process is shown in Fig. 11.3. The transformation function $\varphi(\bullet)$ and its inverse function can be obtained as

$$\xi_i = \varphi\left(p_i\right) = 2\left(\frac{p_i^{\alpha} - p_i^{(\alpha,L)}}{p_i^{(\alpha,U)} - p_i^{(\alpha,L)}} - \frac{1}{2}\right) \tag{11.8}$$

$$p_i^{\alpha} = \varphi^{-1}\left(\xi_i\right) = \frac{\left(p_i^{(\alpha,U)} - p_i^{(\alpha,L)}\right)}{2}\xi_i + \frac{\left(p_i^{(\alpha,U)} + p_i^{(\alpha,L)}\right)}{2} \tag{11.9}$$

The fuzzy propagation starts with a deterministic solution at $\alpha = 1$ and will continue by interval analysis at lower α-level cuts by substituting p_i^{α} in Equation (11.6) for all $i = 1,2,...,m$ and all alpha-cuts ($\alpha = \alpha_k$, $k = 1,2,...,r$). Consequently the fuzzy propagation can be expressed in terms of the vector valued variable ($\xi = \begin{bmatrix} \xi_1,\xi_2,...,\xi_m \end{bmatrix}^T \in \mathfrak{R}^{m\times 1}$) as indicated in equation (11.10).

The Gram-Schmidt algorithm used in this analysis to derive polynomial expressions $\psi(\xi)$ that map a domain in the space of input data 'ξ' to a zone of output data. In this case, Equation (11.6) can be written as

where $\psi(\xi)$ $\psi(\xi) = \text{vec}\left(\Psi(\xi)\right)\in \mathfrak{R}^{p\times 1}$ and $\psi(\xi)$ is a matrix that can be obtained by tensor product of the 'm' one dimensional orthogonal polynomials $\{\psi_0(\xi_i),\psi_1(\xi_i),...,\psi_h(\xi_i)\}$. The one dimensional polynomials are computed from classical Gram–Schmidt algorithm (Witteveen and Bijl 2006). In this method, the polynomial terms are represented as $\psi_j(\xi_i)= \xi_i^j +O\left(\xi_i^{j-1}\right)$ where $j = 0,1,...,h$. This results in $\psi_0(\xi_i)= 1$ and the remaining terms are computed using the following recursive equations:

$$\psi_j(\xi_i)= e_j(\xi_i)- \sum_{k=0}^{j-1} c_{jk}\psi_k(\xi_i) \tag{11.11}$$

where

$$c_{jk} = \frac{\int\limits_{-1}^{1} e_j(\xi_i)\psi_k(\xi_i)d\xi_i}{\int\limits_{-1}^{1} \psi_k^2(\xi_i)d\xi_i}$$

The D-optimal design approach (Mukhopadhyay 2015b) is employed to evaluate the input design points for the respective samples at each α-cut and subsequently those values are called for finite element iteration. The goodness of the present model obtained by least squares regression analysis is the minimum generalized variance of the estimates of the model coefficients. D-optimality is achieved if the determinant of $(Z^T Z)^{-1}$ is minimal where Z denotes the design matrix as a set of value combinations of coded parameters and Z^T is the transpose of Z.

In this chapter, the application of the above proposed method is demonstrated in a laminated composite cantilever plate considering uncertain input parameters as ply orientation angle, elastic modulus, mass density and shear modulus while output parameters are considered as natural frequency, modal frequency response function and mode shapes $[\mathbf{Y}^{(\alpha)}$ in Equation (11.10)]. The dynamic equation of motion of the system shown in Fig. 11.1 can be expressed as:

$$\mathbf{M}(\tilde{\omega}_\alpha)\ddot{\delta} + \mathbf{C}(\tilde{\omega}_\alpha)\dot{\delta} + \mathbf{K}(\tilde{\omega}_\alpha)\delta = 0 \tag{11.12}$$

where $\mathbf{M}(\tilde{\omega}_\alpha)$ is the mass matrix, $\mathbf{C}(\tilde{\omega}_\alpha)$ is damping matrix, $\mathbf{K}(\tilde{\omega}_\alpha)$ is the stiffness matrix and $\ddot{\delta}$, $\dot{\delta}$ and $\ddot{\delta}$ are displacement, velocity and acceleration vectors. The governing equations are derived based on Mindlin's theory incorporating rotary inertia and transverse shear deformation (Dey and Karmakar 2012c, Dey et al. 2015f) using an eight noded isoparametric plate bending element (Bathe 1990). The composite plate is assumed to be lightly damped and the natural frequencies of the system are obtained as:

$$\omega_j^2(\tilde{\omega}_\alpha) = \frac{1}{\lambda_j(\tilde{\omega}_\alpha)} \quad j = 1,...,n_r \tag{11.13}$$

where $\lambda_j(\tilde{\omega}_\alpha)$ is the j-th eigenvalue of matrix $\mathbf{A} = \mathbf{K}^{-1}(\tilde{\omega}_\alpha)\mathbf{M}(\tilde{\omega}_\alpha)$ and n_r indicates the number of modes retained in this analysis. Using the transformation $\delta(t) = [\mathbf{\Phi}]\mathbf{q}(t)$, Equation (11.12) can be decoupled in the modal coordinates as:

$$\ddot{q}_j(t) + 2\bar{\zeta}_j \omega_j \dot{q}_j(t) + \omega_j^2 q_j(t) = 0 \tag{11.14}$$

where ζ_j is the damping factor, $[\mathbf{\Phi}] \in \Re^{n \times n_r}$ is a matrix whose columns are the eigenvectors of the system and $q_j(t)$ is the j-th component of vector $\mathbf{q}(t)$. The generalized proportional damping model expresses the damping matrix as a linear combination of the mass and stiffness matrices, that is

$$\mathbf{C}(\tilde{\omega}_\alpha) = \alpha_1 \quad \mathbf{M}(\tilde{\omega}_\alpha) \tag{11.15}$$

where $\alpha_1 = 0.005$ is constant damping factor. In this case, the damping is said to be proportional damping. The components of transfer function matrix of the system with proportional damping can be obtained as

$$H_{ik}(j\omega)(\tilde{\omega}_\alpha) = \sum_{l=1}^{n} \frac{\Phi_{l_i}(\tilde{\omega}_\alpha)\,\Phi_{l_k}(\tilde{\omega}_\alpha)}{-\omega^2 + 2\,i\,\omega\,\zeta_l\omega_l + \omega_l^2} \tag{11.16}$$

where $\Phi_{l_i}(\tilde{\omega}_\alpha)$ and $\Phi_{l_k}(\tilde{\omega}_\alpha)$ are the i-th and the k-th components of the l-th fuzzy mode shape $\Phi_l(\tilde{\omega}_\alpha)$. Therefore, the dynamic response of a proportionally damped system can be expressed as a linear combination of the undamped mode shapes. As mentioned earlier, the Fuzzy outputs in Equation (11.10) are the natural frequencies in Equation (11.13), the component of mode shapes, e.g., $\Phi_{l_i}(\tilde{\omega}_\alpha)$ and the components of frequency response function given by Equation (11.16).

11.3 Fuzzy input representation

The fuzziness of input parameters such as ply-orientation angle, elastic modulus, mass density and shear modulus at each layer of laminate are considered for composite cantilever plates. It is assumed that the distribution of fuzzy input parameters exists within a certain tolerance zone with their crisp values. The cases wherein the fuzzy input variables considered in each layer of laminate are as follows:

a) Variation of ply-orientation angle only: $\theta(\tilde{\omega}_\alpha) = \{\theta_1 \; \theta_2 \; \theta_3 ....\theta_i ......\theta_l\}$

b) Variation of longitudinal elastic modulus only: $E_1(\tilde{\omega}_\alpha) = \{E_{1(1)} \; E_{1(2)} \; E_{1(3)} ....E_{1(i)} ......E_{1(l)}\}$

c) Variation of mass density only: $\rho(\tilde{\omega}_\alpha) = \{\rho_1 \; \rho_2 \; \rho_3 ....\rho_i ......\rho_l\}$

d) Variation of longitudinal shear modulus only: $G_{12}(\tilde{\omega}_\alpha) = \{G_{12(1)} \; G_{12(2)} \; G_{12(3)} ....G_{12(i)} ...G_{12(l)}\}$

e) Combined variation of ply orientation angle, longitudinal elastic modulus, mass density and shear modulus (longitudinal):

$$g[\theta, E_1, \rho, G_{12}(\tilde{\omega}_\alpha)] = [\Phi_1(\theta_1 ...\theta_l), \Phi_2(E_{1(1)} ...E_{1(l)}), \Phi_3(\rho_1 ..\rho_l), \Phi_4(G_{12(1)} ..G_{12(l)})]$$

where $\theta_{(i)}$, $E_{1(i)}$, $\rho_{(i)}$ and $G_{12(i)}$ are the ply orientation angle, elastic modulus (longitudinal), mass density and shear modulus (longitudinal), respectively and 'l' denotes the number of layer in the laminate. For individual and combined cases, the number of variables (n_v) are considered as 4 and 16, respectively. In present study, $\pm 5°$ for ply orientation angle and $\pm 10\%$ tolerance for material properties respectively from fuzzy crisp values are considered. The membership grades are considered as 0 to 1 in step of 0.1. Figure 11.4 presents the flowchart of the present fuzzy approach.

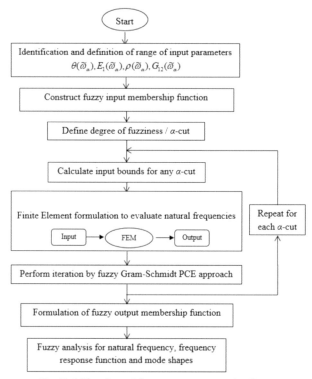

Fig. 11.4 Flowchart of frequency responses using fuzzy.

11.4 Results and discussion

The present study considers four layered graphite-epoxy angle-ply [$(\theta°/-\theta°/\theta°/-\theta°)$] and cross-ply ($0°/90°/0°/90°$) composite cantilever plates. Material properties of graphite–epoxy composite considered with deterministic value as E_1=138 GPa, E_2=8.96 GPa, v=0.3, G_{12}=G_{13}=7.1 GPa, G_{23}=2.84 GPa, ρ=1600 kg/m³. In general, the performance function is not available as an explicit function of the random design variables for complex composite structures. The fuzzy response in terms of natural frequencies of the composite structure can only be evaluated numerically at the end of a structural analysis procedure such as the finite element method which is often time-consuming. The present fuzzy model is employed to find non-probabilistic responses by predefined range of variations in input parameters. The fuzzy membership functions are used to determine the first three natural frequencies corresponding to given values of input variables with different degree of fuzziness. The uncertainty propagation of fuzzy variables can be carried out by the interval approach and the global optimisation approach. In the first approach, a fuzzy variable is considered as an interval variable for each α-cut by employing the classical interval arithmetic (Hanss 2005). Moreover, for multiple occurace of interval valued variables, the fuzzy expression maximizes the deviation of results from true values and thus makes the overestimation effect. Such overestimation will not be applicable to design decisions. In contrast, for the global optimisation based approach, two optimisation problems are solved to find fuzzy output quantities for each α-cut. If there are multiple local optima for the objective function in the optimisation problem, the global optimization is used to find the globally best solution. In the present study, the fuzzy polynomial chaos expansion (PCE) approach is adopted for uncertainty propagation in composite structures wherein the large number of fuzzy input variables are considered to optimize the upper and lower bound for output quantity of interest (natural frequency). The first three fuzzy natural frequencies are approximated by using the proposed fuzzy PCE method described in Section 11.2. The present computer code for fuzzy model is validated with the results available in the open literature. Table 11.1 presents the convergence study of non-dimensional fundamental natural frequencies of three layered ($\theta°/-\theta°/\theta°$) graphite-epoxy untwisted composite plates (Qatu and Leissa 1991a). Based on convergence study, a typical discretization of (6×6) mesh on plan area with 36 elements 133 nodes with natural coordinates of an isoparametric quadratic plate bending element are considered for the present finite element method. The present study investigates on a reliable representation for uncertainty quantification of frequency responses of laminated composite plates using fuzzy membership function approach. The propagation of uncertainties is also demonstrated in the estimation of structural responses of composite cantilever plates. The variations of fuzzy input variables of composite plate namely, ply-orientation angle, elastic modulus, mass density and shear modulus are considered as furnished in Fig. 11.5. Another convergence study is carried out in conjunction to PCE model, as depicted in Table 11.2. There is a trade off between accuracy and computational cost found with the increase of order of polynomial, i.e., a higher order polynomial yields a slightly higher level of accuracy but it costs comparatively much higher computational time and vice-versa. Hence, to balance between accuracy and computational cost, percentages of errors in results (i.e., maximum and minimum stochastic first three natural frequencies) are kept well below 1%. In order to maintain adequate level of accuracy and to optimize the computational time, the second order of polynomial is selected in the present study. Due to paucity of space, only a few important representative results are furnished.

Table 11.1 Convergence study for non-dimensional fundamental natural frequencies [$\omega=\omega_n L^2 \sqrt{(\rho/E_1 t^2)}$] of three layered ($\theta°/-\theta°/\theta°$) graphite-epoxy untwisted composite plates, a/b=1, b/t=100, considering $E_1 = 138$ GPa, $E_2 = 8.96$ GPa, $G_{12} = 7.1$ GPa, $v_{12} = 0.3$.

Ply orientation angle, θ	Present FEM (4 x 4)	Present FEM (6 x 6)	Present FEM (8 x 8)	Present FEM (10 x 10)	Qatu and Leissa (1995)
0°	1.0112	1.0133	1.0107	1.004	1.0175
45°	0.4556	0.4577	0.4553	0.4549	0.4613
90°	0.2553	0.2567	0.2547	0.2542	0.2590

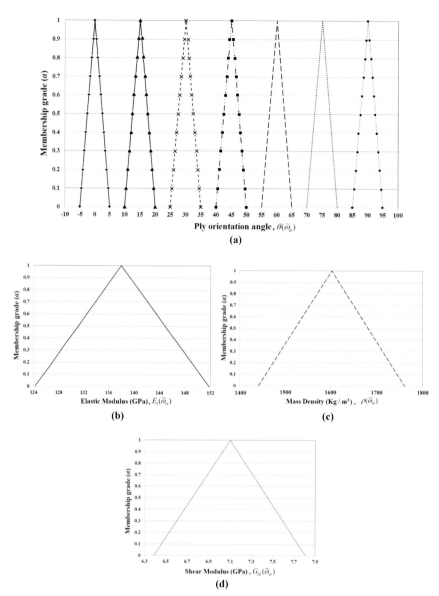

Fig. 11.5 Variation of four fuzzy input variables namely, ply-orientation angle, elastic modulus, mass density and shear modulus for four layered graphite-epoxy angle-ply ($\theta°/-\theta°/\theta°/-\theta°$) composite cantilever plate considering E_1=138 GPa, E_2=8.9 GPa, G_{12}=G_{13}=7.1 GPa, G_{23}=2.84 GPa, ρ=1600 Kg/ m³, t=0.006 m, v=0.3.

The variation of elastic modulus, mass density and shear modulus are scaled in the range with the lower and the upper limit (tolerance limit) as ± 10% variability (as per standard of composite manufacturing industry) with respective mean values

Table 11.2 Convergence study for order of polynomial in PCE model for maximum and minimum values of first three natural frequencies considering combined variation of ply-orientation angle, elastic modulus, shear modulus and mass density for four layered graphite-epoxy angle-ply (45°/–45°/45°/–45°) composite cantilever plate (FNF – Fundamental natural frequency, SNF – Second natural frequency, TNF – Third natural frequency).

Mode	Value	Order of polynomial				
		1	2	3	4	5
FNF	Max	6.3868	6.1996	6.1969	6.1719	6.1706
	Min	5.5876	5.3000	5.3231	5.2881	5.2891
SNF	Max	17.5101	17.9752	17.8513	17.8873	17.9041
	Min	16.9653	16.1847	16.4643	16.4278	16.4072
TNF	Max	38.9485	38.0000	38.3018	38.2333	38.2382
	Min	33.9329	33.0011	33.1779	33.0291	33.0316

while for ply orientation angle $[\theta(\tilde{\omega}_\alpha)]$ ranging from 0° to 90° in step of 15° in each layer of the composite laminate at the lower bound is considered as within ±5° fluctuation (as per standard of composite manufacturing industry) with respective mean deterministic values. The fuzzy models are formed to generate the maximum and minimum bounds for each α-cut of the first three natural frequencies for graphite-epoxy composite cantilever plates. Both angle-ply and cross-ply laminated composite cantilever plates are considered for the present analysis. For variation in only ply orientation angle, it is found that the fundamental natural frequency decreases as the ply orientation angle increases from 0° to 90° in the step of 15° irrespective of α-cut as furnished in Fig. 11.6. This is expected as the fundamental mode is bending and the bending stffiness of the composite plate falls when ply orientation angle increases. In contrast, for variation in only ply orientation angle, the second and third natural frequencies increases as the ply orientation angle increases from 0° to 30° in the step of 15° and subsequenyly deceases as the ply orientation angle increases from 45° to 90° in the step of 15° irrespective of α-cut. This is owing to the fact that the second and third modes are torsion and the maximum torsion stiffness is expected to be at about 45° while the minimum values are around 0° and 90°.It is also observed that the least variation in natural freuqency at α=0 is identified for $\theta(\tilde{\omega}_\alpha) = 90°$ for the first three natural frequencies. This can be attributed to the fact that the sensitivity of elastic stiffness is minimum for the fuzziness of ply orientation angle at $\theta(\tilde{\omega}_\alpha) = 90°$. For only variation in ply orientation angle, the maximum bound width (difference between maximum and minimum natural frequencies) is observed at α=0 and the minimum bound width is identified at α=1 for the first three natural frequencies as furnished in Fig. 11.7. The maximum bound width for fundamental natural frequency is found for $\theta(\tilde{\omega}_\alpha) = 15°$ while the minimum bound width is identified for $\theta(\tilde{\omega}_\alpha) = 90°$ irrespective of α-cut. In contrast, the maximum bound width for second and third natural frequencies are observed at $\theta(\tilde{\omega}_\alpha) = 60°$ and $\theta(\tilde{\omega}_\alpha) = 45°$, respectively while the minimum bound width is consistantly identified for $\theta(\tilde{\omega}_\alpha) = 90°$ irrespective of α-cut. This

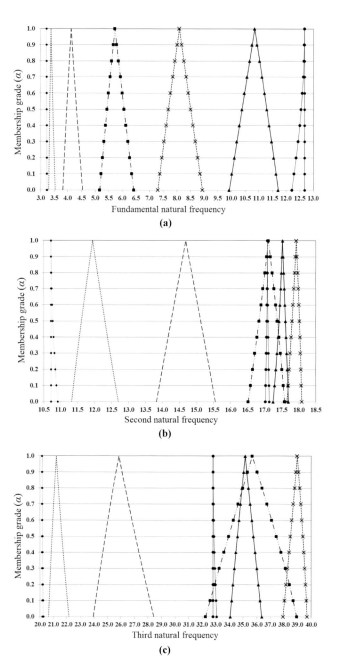

Fig. 11.6 Variation of first three natural frequencies due to only variation in ply-orientation angle for four layered graphite-epoxy angle-ply ($\theta°/-\theta°/\theta°/-\theta°$) composite cantilever plate considering E_1=138 GPa, E_2=8.9 GPa, G_{12}=G_{13}=7.1 GPa, G_{23}=2.84 GPa, ρ=1600 Kg/m³, t=0.006 m, v=0.3, n_v=4.

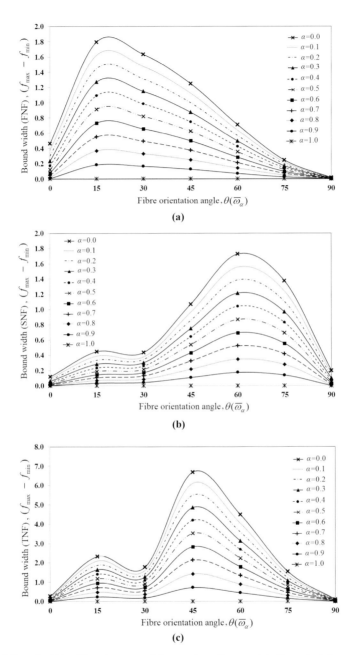

Fig. 11.7 Bound width of first three natural frequencies with respect to fibre orientation angle due to only variation of ply-orientation angle for four layered graphite-epoxy angle-ply ($\theta°/-\theta°/\theta°/-\theta°$) composite cantilever plate considering E_1=138 GPa, E_2=8.9 GPa, G_{12}=G_{13}=7.1 GPa, G_{23}=2.84 GPa, ρ=1600 Kg/m³, t=0.006 m, v=0.3, n_v=4. (FNF-Fundamental natural frequency, SNF-Second natural frequency, TNF-Third natural frequency).

could be attributed to the fact that the elastic stiffness for each α-cut of the fuzzy variation leads to such variation in natural frequencies and consequently the range of frequency response corresponding to ply orientation of the composite laminate. The maximum bound width or ranges of first three natural frequencies for both angle-ply (45°/–45°/45°/–45°) and cross-ply (0°/90°/0°/90°) composite cantilever plate are consistently observed (Fig. 11.8) for combined variation of ply-orientation angle, elastic modulus, mass density and shear modulus [θ, E_1, ρ, G_{12} ($\tilde{\omega}_\alpha$)] irrespective of fuzzy α-cut except for α=1 which indicates the respective deterministic value of natural frequencies. For the cases of only variation of input parameters [i.e., $\theta(\tilde{\omega}_\alpha)$, $E_1(\tilde{\omega}_\alpha)$, $\rho(\tilde{\omega}_\alpha)$ and G_{12} ($\tilde{\omega}_\alpha$)], the maximum range of frequency is identified for first and third modes at different α-cuts of the only variation of ply-orientation angle [$\theta(\tilde{\omega}_\alpha)$] for angle-ply (45°/–45°/45°/–45°) composite plate. For cross-ply (0°/90°/0°/90°) composite plate, the ranges of fundamental natural frequencies are in the order as $E_1(\tilde{\omega}_\alpha) > \rho(\tilde{\omega}_\alpha) > \theta(\tilde{\omega}_\alpha) > G_{12}$ ($\tilde{\omega}_\alpha$) (in case of only variation of any single input parameter) invariant to fuzzy α-cut. It is also noted that G_{12} ($\tilde{\omega}_\alpha$) has negligible effect on variation of fundamental natural frequency for cross-ply composite plate. In contrast, the ranges of second and third natural frequencies of cross-ply composite plates are in the order as $\rho(\tilde{\omega}_\alpha) > E_1(\tilde{\omega}_\alpha) > G_{12}$ ($\tilde{\omega}_\alpha$) > $\theta(\tilde{\omega}_\alpha)$ (in case of only variation of any single input parameter) irrespective of α-cut. Interestingly, the least influence of $\theta(\tilde{\omega}_\alpha)$ on range of natural frequencies in cross-ply is observed in case of cross-ply composite laminate while G_{12} ($\tilde{\omega}_\alpha$) is found to be least effective on range of first three natural frequencies for angle-ply composite plates.

The comparative studies for angle-ply (45°/–45°/45°/–45°) and cross-ply (0°/90°/0°/90°) composite cantilever plates with respect to maximum and minimum values of first three natural frequencies at α=0 and 0.5 are carried out using gobal optimization (GO) approach and present fuzzy Gram-Schmidt polynomial chaos expansion (GSPCE) approach as furnished in Table 11.3 (individual case) and Table 11.4 (combined case), respectively. The computational time required in the proposed approach is observed to be around (1/156) times (for individual variation of inputs) and (1/78) times (for combined variation of inputs) of global optimization approach. It should be noted that, standard Genetic Algorithm (GA) is used for global optimization. GA has been found as one of the powerful and robust global optimization method that search for global solutions to problems that contain multiple maxima or minima. Once the PCE model is formed that is capable of representing the entire design domain, global optimization algorithms like GA can be efficiently applied. For the purpose of comparison, results obtained using classical polynomial chaos expansion (CPCE) are also furnished in Table 11.3 and Table 11.4 that indicate that there is not much difference between the maximum and minimum values of responses following fuzzy GSPCE and CPCE approach. The deviation or difference between these two results corroborates the fact that the present PCE approach is accurate and computationally efficient. The frequency response function (FRF) plot is furnished in Fig. 11.9 indicating simulation bound,

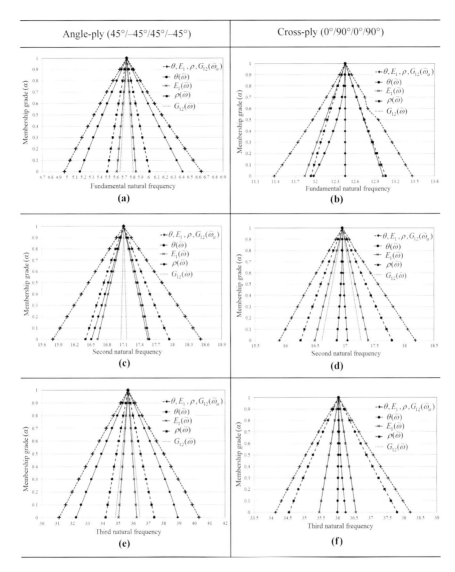

Fig. 11.8 Variation of first three natural frequencies due to only variation of ply-orientation angle, elastic modulus, shear modulus and mass density and combined variation for graphite-epoxy angle-ply (45°/–45°/45°/–45°) and cross-ply (0°/90°/0°/90°) composite cantilever plate considering E_1=138 GPa, E_2=8.9 GPa, G_{12}=G_{13}=7.1 GPa, G_{23}=2.84 GPa, ρ=1600 Kg/m³, t=0.006 m, v=0.3, n_v=4 (for individual cases) and n_v=16 (for combined case).

simulation mean and deterministic values corresponding to different α-cut for combined stochaticity in ply-orientation angle, elastic modulus, shear modulus and mass density of angle-ply (45°/–45°/45°/–45°) composite cantilever plate. The maximum simulation bound of frequency responses due to combined variation of

Table 11.3 Comparative study between global optimization (GO) approach and present fuzzy GSPCE and CPCE approach for α-cut=0 and 0.5 with respect to variation of first three natural frequencies due to only variation of ply-orientation angle [$\theta(\widetilde{\omega}_\alpha)$] for four layered graphite-epoxy angle-ply (45°/−45°/45°/−45°) and cross-ply (0°/90°/0°/90°) composite cantilever plate considering E_1=138 GPa, E_2=8.9 GPa, G_{12}=G_{13}=7.1 GPa, G_{23}=2.84 GPa, ρ=1600 Kg/m³, t=0.006 m, v=0.3, n_v=4 (in each cases) (FNF – Fundamental natural frequency, SNF – Second natural frequency, TNF – Third natural frequency, GSPCE – Gram-Schmidt polynomial chaos expansion, CPCE – Classical polynomial chaos expansion).

α-cut	Method	Angle-ply						Cross-ply					
		FNF		SNF		TNF		FNF		SNF		TNF	
		Max	Min	Max	Min	Max	Min	Max	Min	Max	Min	Max	Min
0	GO Approach	6.4111	5.1612	17.5827	16.5128	38.9342	32.2457	12.4517	11.9790	17.0457	16.8653	36.2476	36.0004
	Fuzzy GSPCE Approach	6.4252	5.1047	17.5887	16.4475	39.0405	31.8866	12.4495	11.9711	17.0625	16.8604	36.2891	35.9981
	% deviation	−0.22%	1.09%	−0.03%	0.40%	−0.27%	1.11%	0.02%	0.07%	−0.10%	0.03%	−0.11%	0.01%
	Fuzzy CPCE Approach	6.3212	5.2456	17.6324	16.3212	38.7645	32.4423	12.2314	11.8974	17.2301	16.6329	36.112	36.1432
	% deviation	1.40%	−1.64%	−0.28%	1.16%	0.44%	−0.61%	1.77%	0.68%	−1.08%	1.38%	0.37%	−0.40%
0.5	GO Approach	6.0607	5.4366	17.3675	16.8286	37.4502	33.9303	12.4517	12.3274	16.9816	16.9345	36.0859	36.0230
	Fuzzy GSPCE Approach	6.0719	5.4184	17.3710	16.8120	37.4868	33.8297	12.4515	12.3258	16.9850	16.9336	36.0929	36.0224
	% deviation	−0.18%	0.33%	−0.02%	0.10%	−0.10%	0.30%	0.00%	0.01%	−0.02%	0.01%	−0.02%	0.00%
	Fuzzy CPCE Approach	6.1265	5.3254	17.4235	16.6923	37.4321	33.7988	12.5978	12.2312	16.8765	16.8865	36.1238	36.1867
	% deviation	−1.09%	2.05%	−0.32%	0.81%	0.05%	0.39%	−1.17%	0.78%	0.62%	0.28%	−0.11%	−0.45%

Table 11.4 Comparative study between global optimization (GO) approach and fuzzy GSPCE and CPCE approach for α-cut=0 and 0.5 with respect to variation of first three natural frequencies due to combined variation of ply-orientation angle, elastic modulus, shear modulus and mass density for four layered graphite-epoxy angle-ply (45°/-45°/45°/-45°) and cross-ply (0°/90°/0°/90°) composite cantilever plate considering E_1=138 GPa, E_2=8.9 GPa, $G_{12}=G_{13}$=7.1 GPa, G_{23}=2.84 GPa, ρ=1600 Kg/m³, t=0.006 m, v=0.3, n_v=16 (FNF – Fundamental natural frequency, SNF – Second natural frequency, TNF – Third natural frequency, GSPCE – Gram-Schmidt polynomial chaos expansion, CPCE– Classical polynomial chaos expansion).

α-cut	Method	Angle-ply						Cross-ply					
		FNF		SNF		TNF		FNF		SNF		TNF	
		Max	Min	Max	Min	Max	Min	Max	Min	Max	Min	Max	Min
0	GO Approach	6.6376	4.9701	18.5214	15.8036	40.3142	31.1363	13.4631	11.3848	18.1851	15.9033	38.1998	34.1395
	Fuzzy GSPCE Approach	6.6504	4.8989	18.6893	15.7116	40.6208	30.5514	13.4791	11.1632	18.2047	15.6801	38.4355	33.9291
	% deviation	-0.19%	1.43%	-0.91%	0.58%	-0.76%	1.88%	-0.12%	1.95%	-0.11%	1.40%	-0.62%	0.62%
	Fuzzy CPCE Approach	6.6120	4.8932	18.8432	15.6543	40.4532	31.0876	13.3343	11.4532	18.0876	15.7643	38.3452	34.0054
	% deviation	0.38%	-1.71%	-1.71%	0.95%	-0.34%	0.15%	0.96%	-0.59%	0.54%	0.88%	-0.38%	0.39%
0.5	GO Approach	6.1616	5.3326	17.8166	16.4582	38.0229	33.3266	12.9504	11.9618	17.5578	16.4166	37.0791	35.0451
	Fuzzy GSPCE Approach	6.1996	5.3000	17.9752	16.1847	38.0000	33.0011	12.9907	11.9000	17.5889	16.2929	37.2481	34.9710
	% deviation	-0.62%	0.61%	-0.89%	1.66%	0.06%	0.98%	-0.31%	0.52%	-0.18%	0.75%	-0.46%	0.21%
	Fuzzy CPCE Approach	6.1494	5.2656	18.2107	16.7285	38.3598	33.1743	12.7521	11.8201	17.77572	16.3433	37.674	35.564
	% deviation	0.19%	1.25%	-2.21%	-1.64%	-0.88%	0.45%	1.55%	1.19%	-1.12%	0.44%	-1.57%	-1.46%

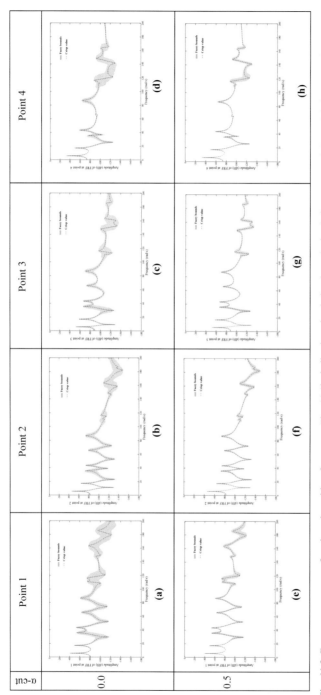

Fig. 11.9 Frequency response function plot with reference to Fig. 11.1b, indicating simulation bound, simulation mean and deterministic values for combined stochaticity in ply-orientation angle, elastic modulus, shear modulus and mass density of angle-ply ($45°/-45°/45°/-45°$) composite cantilever plate considering E_1=138 GPa, E_2=8.9 GPa, G_{12}=G_{13}=7.1 GPa, G_{23}=2.84 GPa, ρ=1600 Kg/m³, t=0.006 m, ν=0.3, n_v=16.

input parameters is found at $\alpha=0$. As the value of α-cut increases the simulation bound of frequency response function also decreases and finally at $\alpha=1$, it shows the deterministic value without any simulation bound as expected. For a given amount of fuzziness in the input parameters, more changes in the fuzzy output quantities are observed in the higher frequency ranges. Figure 11.10 presents the representative

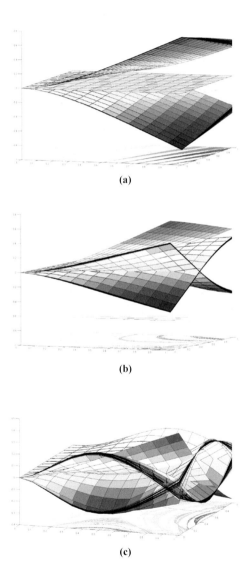

(a)

(b)

(c)

Fig 11.10 Stochastic modeshapes of first three modes due to combined stochaticity in ply-orientation angle, elastic modulus, shear modulus and mass density for four layered graphite-epoxy angle-ply (45°/–45°/45°/–45°) composite cantilever plate at α-cut=0 considering E_1=138 GPa, E_2=8.9 GPa, G_{12}=G_{13}=7.1 GPa, G_{23}=2.84 GPa, ρ=1600 Kg/m³, t=0.006 m, v=0.3, n_v=16.

modeshapes of the first three natural frequencies considering combined fuzzy-variation in ply orientation angle, elastic modulus (longitudinal), mass density and shear modulus. The fundamental natural frequency corresponds to first spanwise bending and as the mode increases the combined effect of torsion and spanwise

bending is predominently observed for the second and third modes. The normalised component of mode at point 3 (as indicated in Fig. 11.1) of the first three modes due to combined variation of ply-orientation angle, elastic modulus,

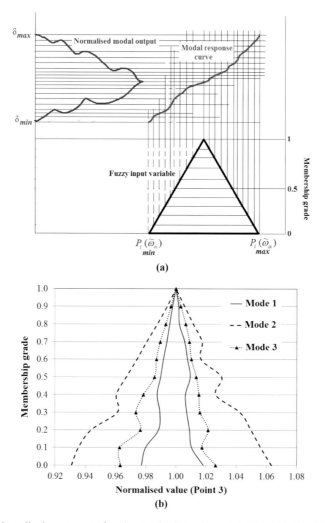

Fig 11.11 Normalised component of modes at point 3 (as shown in in Fig.11.1) of first three modes due to combined stochaticity in ply-orientation angle, elastic modulus, shear modulus and mass density for four layered graphite-epoxy angle-ply (45°/–45°/45°/–45°) composite cantilever plate at α-cut=0 considering E_1=138 GPa, E_2=8.9 GPa, $G_{12}=G_{13}$=7.1 GPa, G_{23}=2.84 GPa, ρ=1600 Kg/m³, t=0.006 m, v=0.3, n$_v$=16.

shear modulus and mass density for four layered graphite-epoxy angle-ply composite cantilever plate for different α-cut are furnished in Fig. 11.11. The normal component of the first three modes portrays the representative variation of fuzzy mode shapes corresponding to different membership grades. As it is illustrated in Fig. 11.11b, even though the input fuzzy numbers are defined by a triangular membership function, the resulting membership functions of output fuzzy numbers may not have a triangular shape.

11.5 Summary

This chapter presents a GPCE based approach for fuzzy uncertainty propagation in composite laminates. The uncertainty quantification of natural frequency and the frequency response functions with fuzzy variables are derived implicitly using the finite element method. The computational time and cost is reduced by using the present fuzzy PCE approach compared to the global optimization method. The maximum ranges of the first three natural frequencies are consistantly found for combined variation of ply-orientation angle, elastic modulus, mass density and shear modulus compared to individual variation of any input parameter irrespective of fuzzy α-cut. Due to combined variation of input parameters, the maximum simulation bound of frequency response function is obtained at $\alpha=0$ which decreases with increase of fuzzy α-cut and it shows the deterministic value without any simulation bound finally at $\alpha=1$. The fundamental natural frequency corresponds to first spanwise bending and as the mode increases the combined effect of torsion and bending is predominently observed for the second and third modes. The normal component of first three modes are portrayed as the representative variation of fuzzy mode shapes corresponding to different membership grades. The concepts presented in this chapter can be extended for future research to deal with a more complex system considering a large number of fuzzy variables.

CHAPTER 12

Comparative Assessment of Kriging Variants for Stochastic Analysis of Composites

12.1 Introduction

The previous chapters in this book deal with various aspects of uncertainties associated with composite structures (both probabilistic and non-probabilistic) following different surrogate based approaches. However, one important aspect in surrogate based uncertainty quantification of composites is the comparative assessment of different surrogate models and their variants. Kriging based surrogate modelling is widely used in different computationally intensive problems of science and engineering. This chapter presents a critical comparative assessment of Kriging model variants for surrogate based uncertainty propagation considering stochastic natural frequencies of laminated composite shells. The five Kriging model variants studied here are: Ordinary Kriging, Universal Kriging based on pseudo-likelihood estimator, Blind Kriging, Co-Kriging and Universal Kriging based on marginal likelihood estimator. First three stochastic natural frequencies of the composite shell are analysed by using a finite element model that includes the effects of transverse shear deformation based on Mindlin's theory in conjunction with a layer-wise random variable approach. The comparative assessment is carried out to address the accuracy and computational efficiency of five Kriging model variants. Comparative performance of different covariance functions is also studied. Subsequently the effect of noise in uncertainty propagation is addressed by using the Stochastic Kriging. Representative results are presented for both individual and combined stochasticity in layer-wise input parameters to address performance of various Kriging variants for low dimensional and relatively higher dimensional input parameter spaces. The error estimation and convergence studies are conducted with respect to original Monte Carlo Simulation to justify merit of the present investigation. The study reveals that Universal Kriging coupled with marginal likelihood estimate yields the most accurate results, followed by Co-Kriging and Blind Kriging. As far as computational efficiency of the Kriging models is concerned, it is observed that for high-dimensional problems, CPU time required for building the Co-Kriging model is significantly less as compared to other Kriging variants.

The inherent problem of computational modelling dealing with empirical data is germane to many engineering applications particularly when both input and output quantities are uncertain in nature. In empirical data modeling, a process of induction is employed to build up a model of the system from which it is assumed to deduce the responses of the system that have yet to be observed. Ultimately both quantity and quality of the observations govern the performance of the empirical model. By its observational nature, data obtained is finite and sampled; typically when this sampling is non-uniform and due to the high dimensional nature of the problem, the data will form only a sparse distribution in the input space. Due to the inherent complexity of composite materials, such variability of output with respect to randomness in input parameters is difficult to map computationally by means of deterministic finite element models. The well-known Monte Carlo simulation (MCS) technique (Arregui–Mena et al. 2016, Venkatram 1988) is employed universally to characterize the stochastic output parameter considering large sample size and thus, thousands of model evaluations (finite element simulations) are generally needed. Therefore, the traditional MCS based uncertainty quantification approach is inefficient for practical purposes and incurs high computational cost. To avoid such lacuna, the present investigation includes the Kriging model as surrogate for reducing the computational time and cost in mapping the uncertain natural frequencies. In this approach of uncertainty quantification, the computationally expensive finite element model is effectively replaced by an efficient mathematical model and thereby, thousands of virtual simulations can be carried out in a cost-effective manner.

Significant volume of scientific studies is found to be reported on Kriging model and its applications in various engineering problems (Venkatram 1988, Fedorov 1989, Diamond 1989, Carr 1990, Deutsch 1996, Cressie 1990, Mantheron 1963, Cressie 1993). Further studies on the basis of design and analysis of experiments have also been carried out (Montgomery 1991, Michael and Norman 1974, Martin and Simpson 2005). Several literatures are also found to address different problems related to optimization (Lee and Kang 2006, Sakata et al. 2004) and geo-statistics (Ryu et al. 2002) using Kriging model. Many researchers proposed a lot of improved methods to solve the problems with computational efficiency based on sampling. Importance Sampling (Bayer and Bucher 1999, Yuan et al. 2013, Au and Beck 1999) is the generic method to achieve comparatively better sampling efficiency and ease of implementation wherein equal attention is provided in both low and high density zone of sample points. In contrast, the important sampling increases the simulation efficiency to a certain extent, but the number of effective samples required to map the actual performance is still excessive. Some studies have concentrated on different variants of the Kriging model (Au and Beck 1999, Kaminiski 2015, Angelikopoulos et al. 2015, Peter and Marcelet 2008, Dixit et al. 2016, Huang et al. 2016, Tonkin et al. 2016) such as Universal Kriging, Co-Kriging, and Stochastic Kriging dealing with various engineering problems. Kersaudy et al. (2015) have studied a new surrogate modelling technique combining Kriging and polynomial chaos expansions for uncertainty analysis. Khodaparast et al. (2011) have used interval model updating with irreducible uncertainty using the Kriging predictor. Sensitivity analysis based on Kriging models for robust stability of brake systems has been conducted by Nechak et al. (2015). Pigoli et al. (2016) have studied manifold-valued random fields by Kriging model. The uncertainty quantification in DIC with Kriging regression is found to be studied by Wang et al. (2016). A range of Kriging based investigations have been reported for different field of applications with probabilistic approach (Shinkyu et al. 2005, Bayraktar and Turalioglu 2005, Hertog et al. 2006, Emery 2005, Martin and Simpson 2004). The Co-Kriging has recently been employed for dealing with optimization of the cyclone separator geometry for minimum pressure drop (Elsayed 2015). An improved moving Kriging-based mesh-free method is employed for static, dynamic and buckling analyses of functionally graded isotropic and sandwich plates by Thai et al. (2016). Kriging is found to be employed for reliability analysis by Yang et al. (2015b) and Gaspar et al. (2014). Huang et al. (2016) have investigated the assessment of small failure probabilities by combining Kriging and Subset Simulation. Kwon et al. (2015) have studied to find the trended Kriging model with R^2 indicator and application to design optimization.

Apart from the application of Kriging models encompassing different domains of engineering in general as discussed above, few recent applications of the Kriging model in the field of composites can also be found. Sakata et al. (2008) have studied the Kriging-based approximate stochastic homogenization analysis for composite materials. Kriging based surrogate modeling has been used for stochastic dynamic analysis of composite structures (Dey et al. 2015b, Mukhopadhyay et al. 2017c). Luersen et al. (2015) have carried out Kriging based optimization for the path of curve fiber in composite laminates.

In this chapter, the finite-element method is employed to study the stochastic free vibration characteristics of graphite–epoxy composite cantilever shallow curved shells by using five different Kriging surrogate model variants namely, Ordinary Kriging, Universal Kriging based on pseudo-likelihood estimator, Blind Kriging, Co-Kriging and Universal Kriging based on marginal likelihood estimator. The finite element formulation considering layer-wise random system parameters in a non-intrusive manner is based on consideration of eight noded isoparametric quadratic element with five degrees of freedom at each node. The selective representative samples for construction Kriging models are drawn using the Latin hypercube sampling algorithm (Mckay et al. 2000) over the entire domain ensuring good prediction capability of the constructed surrogate model. The distinctive comparative studies on efficiencies of aforesaid five Kriging model variants have been carried out in this chapter to assess their individual merits on the basis of accuracy and computational efficiency. Both individual and combined stochasticity in input parameters such as ply orientation angle, elastic modulus, mass density, shear modulus and Poisson's ratio have been considered for this computational investigation. The precision and accuracy of results obtained from the constructed surrogate models have been verified with respect to original finite element simulation. Detail description of the stochastic finite element model for composite doubly curved shells can be found in Chapter 5. Hereafter this chapter is organized as, Section 12.2: formulation of the five Kriging model variants; Section 12.3: stochastic approach for uncertain natural frequency characterization using finite element analysis in conjunction with the Kriging model variants; Section 12.4: results on comparative assessments considering different crucial aspects and discussion; and Section 12.5: summary.

12.2 Kriging model variants

In general, a surrogate is an approximation of the Input/Output (I/O) function that is implied by the underlying simulation model. Surrogate models are fitted to the I/O data produced by the experiment with the simulation model. This simulation model may be either deterministic or random (stochastic). The Kriging model was initially developed in spatial statistics by Danie Gerhardus Krige (1951) and subsequently extended by Matheron and Cressie. Kriging is a Gaussian process based modelling method, which is compact and cost effective for computation. The basic idea of this method is to incorporate interpolation, governed by prior covariances, to obtain the responses at the unknown points. However as pointed out by various researchers (Hengl et al. 2007, Matias and Gonzalez 2005, Omre and Halvorsen 1989, Tonkin and Larson 2002), results obtained using ordinary Kriging are often erroneous, specifically for problems that are highly nonlinear in nature. In order to address this issue, the universal Kriging has been proposed in (Warnes 1986, Stein and Corsten 1991, Olea 2011, Ghiasi and Nafisi 2015, Li et al. 2015). In this method, the unknown response is represented as:

$$y(x) = y_0(x) + Z(x) \qquad (12.1)$$

where $y(x)$ is the unknown function of interest, x is an m dimensional vector (m design variables), $y_0(x)$ is the known approximation (usually polynomial) function and $Z(x)$ represents the realization of a stochastic process with mean zero, variance, and nonzero covariance. In the model, the local deviation at an unknown point ($\mathbf{x}$) is expressed using stochastic processes. The sample points are interpolated with the Gaussian random function as the correlation function to estimate the trend of the stochastic processes. The $y_0(x)$ term is similar to a polynomial response surface, providing a global model of the design space (Sakata et al. 2008). However, the maximum order of the polynomial $y_0(x)$ is often randomly chosen in universal Kriging. This often renders the universal Kriging inefficient and erroneous. In order to address this issue associated with universal Kriging, blind Kriging has been proposed (Joseph et al. 2008, Hung 2011, Couckuyt et al. 2012). In this method, the polynomial $y_0(x)$ is selected in an adaptive manner. As a consequence, blind Kriging is highly robust. Other variants of Kriging surrogate includes co-Kriging (Koziel et al. 2014, Elsayed 2015, Clemens and Seifert 2015, Perdikaris et al. 2015) and stochastic Kriging (Kaminiski 2015, Qu and Fu 2014, Chen and Kim 2014, Wang et al. 2014, Wang et al. 2013, Chen et al. 2012, 2013). While co-Kriging yields multi-fidelity results, stochastic Kriging incorporates the noise present in experimental data. Brief description of various Kriging variants has been provided next.

12.2.1 Ordinary Kriging

In ordinary Kriging, one seeks to predict the value of a function at a given point by computing a weighted average of the known values of the function in the neighbourhood of the point. Kriging is formulated by (i) assuming some suitable covariances, (ii) utilizing the Gauss-Markov theorem to prove the independence of the estimate and the error. Provided suitable covariance function is selected, kriging yields the best unbiased linear estimate.

Remark 1: Simple Kriging is a variant of the ordinary Kriging. While in simple Kriging, $y_0(x) = 0$ is assumed, the ordinary Kriging assumes $y_0(x) = a_0$, where α_0 is unknown constant.

12.2.2 Universal Kriging

A more generalised version of the ordinary Kriging is the universal Kriging. Here, $y_0(x)$ is represented by using a multivariate polynomial as:

$$y_0(x) = \sum_{i=1}^{p} a_i b_i(\mathbf{x}) \tag{12.2}$$

where $b_i(\mathbf{x})$ represents the ith basis function and a_i denotes the coefficient associated with the ith basis function. The primary idea behind such a representation is that the regression function captures the Largent variance in the data (the overall trend) and the Gaussian process interpolates the residuals. Suppose $X = \{x^1, x^2, ..., x^n\}$ represents a set of n samples. Also assume $Y = \{y_1, y_2, ..., y_n\}$ to be the responses at sample points. Therefore, the regression part can be written as a $n \times p$ model matrix F,

$$F = \begin{pmatrix} b_1(x^1) & \cdots & b_p(x^1) \\ \vdots & \ddots & \vdots \\ b_1(x^n) & \cdots & b_p(x^n) \end{pmatrix} \tag{12.3}$$

whereas the stochastic process is defined using a $n \times n$ correlation matrix Ψ

$$\Psi = \begin{pmatrix} \psi\left(x^1, x^1\right) & \cdots & \psi\left(x^1, x^n\right) \\ \vdots & \ddots & \vdots \\ \psi\left(x^n, x^1\right) & \cdots & \psi\left(x^n, x^n\right) \end{pmatrix} \tag{12.4}$$

where ψ is a correlation function, parameterised by a set of hyperparameters θ. The hyperparameters are again identified by maximum likelihood estimation (MLE). Brief description of MLE is provided towards the end of this section. The prediction mean and prediction variance of are given as:

$$\mu(x) = M\alpha + r(x)\Psi^{-1}(y - F\alpha) \tag{12.5}$$

and

$$s^2(x) = \sigma^2 \left[1 - r(x)\Psi^{-1}r(x)^T + \frac{\left(1 - F^T\Psi^{-1}r(x)^T\right)}{F^T\Psi^{-1}F}\right] \tag{12.6}$$

where $M = \left(b_1\left(x_p\right) \quad \cdots \quad b_p\left(x_p\right)\right)$ is the modal matrix of the predicting point x_p,

$$\alpha = \left(F^T\Psi F\right)^{-1} F^T\Psi^{-1}Y \tag{12.7}$$

is a $p \times 1$ vector consisting the unknown coefficients determined by generalised least squares regression and

$$r(x) = \left(\psi\left(x_p, x^1\right) \quad \cdots \quad \psi\left(x_p, x^p\right)\right) \tag{12.8}$$

is an $1 \times n$ vector denoting the correlation between the prediction point and the sample points. The process variance σ^2 is given by

$$\sigma^2 = \frac{1}{n}(Y - F\alpha)^T \Psi^{-1}(Y - F\alpha) \tag{12.9}$$

Remark 2: Note that the universal Kriging, as formulated above, is an interpolation technique. This can be easily validated by substituting the ith sample point in Equation (12.5) and considering that $r(x^i)$ is the ith column of Ψ:

$$\mu\left(x^i\right) = M\alpha + y^i - M\alpha = y^i \tag{12.10}$$

Remark 3: One issue associated with the universal Kriging is selection of the optimal polynomial order. Conventionally, the order of the polynomial is selected empirically. As the consequence, the Kriging surrogate formulated may not be optimal.

12.2.3 Blind Kriging

This variant of Kriging incorporates the Bayesian feature selection method into the framework of universal Kriging. The basic goal is to efficiently determine the basis function b_i that captures the maximum variance in the sample data. To this end, a set of candidate functions is considered from which to choose from. In the ideal scenario, the trend function completely represents the sample data and the stochastic process has no influence. However, this is an extremely rare event. Suppose, an existing kriging model $Y(x)$ with a constant regression function is already available; the basic idea of blind Kriging is to incorporate new features into the regression function of the Kriging surrogate. To this end, the whole set of candidate function c_i is used to fit the data in a linear model as:

$$y(x) = \sum_{i=1}^{p} a_i b_i(\mathbf{x}) + \sum_{i=1}^{t} \beta_i c_i(\mathbf{x}) \tag{12.11}$$

where t is the number of candidate functions. The first part of Equation (12.11) is the regression function considered in the kriging surrogate and therefore, the unknown coefficients a can be determined independent of the β. In blind Kriging, the estimates of β can be considered as the weights associated with the candidate functions. One alternative for determining β is the conventional least-square solution. However, this may yield erroneous result specifically in cases where number of candidate feature is more than the number of samples available. As an alternative, a Gaussian prior distribution is introduced for β,

$$\beta \sim \mathcal{N}\left(0, \sigma_b^2\right) \tag{12.12}$$

Moreover, the choice of the correlation function is restricted to the product correlation form:

$$\psi(x, x') = \prod_{i=1}^{d} \psi_j\left(\left|x_j - x_j'\right|\right) \tag{12.13}$$

The variance-covariance matrix R can be constructed as:

$$R_j = U_j^{-1} \psi_j \left(U_j^{-1}\right)^T \tag{12.14}$$

From Equation (12.13) and Equation (12.14), it is evident that the number of considered features can be chosen per dimension and afterwards the full matrix R is obtained as:

$$\mathbf{R} = \bigotimes_{j=1}^{d} R_j \tag{12.15}$$

Once the correlation matrix R is constructed, the posterior of β is estimated as:

$$\hat{\beta} = \frac{\sigma_b^2}{\sigma^2} \mathbf{R} \mathbf{M}_c^T \Psi^{-1} (\mathbf{y} - \mathbf{M} a) \tag{12.16}$$

where $\mathbf{M}_c$ is the model matrix of all the candidate variable and $\mathbf{M}$ is the model matrix of all currently chosen variables and Ψ is the correlation matrix of the samples. The coefficient $\hat{\beta}$ obtained using the proposed approach quantifies the importance of the associated candidate feature. The advantage of the proposed approach, over other available alternatives, is mainly threefold. Firstly, the Bayesian approach proposed utilizes the already available data, making the procedure computationally efficient. Secondly, the variable selection method satisfies the hierarchy criteria. As per this effect, the lower order effects should be chosen before higher order effect. Last, but not the least, the heredity criteria is also satisfied. As per this criterion, an effect cannot be important unless its parent effect is also important.

12.2.4 Co-Kriging

The fourth variant of Kriging, proposed by Kennedy et al. (Kennedy and O'Hagan 2000), is known as Co-Kriging. Co-Kriging utilizes the correlation between fine and course model data to enhance the prediction accuracy. Unlike other variants of Kriging, Co-Kriging can be utilized for multi-fidelity analysis. Moreover, computational cost (in terms of CPU time) is significantly less for Co-Kriging, as compared to the other variants.

Suppose $X_c = \{x_c^1, \ldots, x_c^{nc}\}$ and $X_e = \{x_e^1, \ldots, x_e^{ne}\}$ be the low-fidelity and high-fidelity sets of sample points. The associate functions are denoted by $y_c = \{y_c^1, \ldots, y_c^{nc}\}$ and $y_e = \{y_e^1, \ldots, y_e^{ne}\}$. Creating a Co-Kriging model can be interpreted as constructing two Kriging models in sequence. In the first step, a Kriging model $Y_C(x)$ of the course data (X_c, y_c) is constructed. Subsequently, the second Kriging model $Y_D(x)$ is constructed on the residuals of the fine and course data (X_e, y_d), where

$$y_d = y_e - \rho\mu_c(X_e) \tag{12.17}$$

where, $\mu(X_e)$ is obtained using Equation (12.5). The parameter ρ is estimated as part of the MLE of the second Kriging model. It is to be noted that the choice of the correlation function and regression function of both the Kriging model can be adjusted separately. The resulting Co-Kriging model is represented using Equation (12.5), where $r(x)$ and Ψ are written as functions of two separate Kriging models:

$$r(x) = \left(\rho\sigma_c^2 r_c(x) \quad \rho\sigma_c^2 r_c(x, X_f) + \sigma_d^2 r_d(x)\right) \tag{12.18}$$

and

$$\Psi = \begin{pmatrix} \sigma_c^2\Psi_c & \rho\sigma_c^2\Psi_c(X_c, X_f) \\ 0 & \rho\sigma_c^2\Psi_c(X_f, X_f) + \sigma_d^2\Psi_d \end{pmatrix} \tag{12.19}$$

where $(F_c, \sigma_c, \Psi_c, M_c)$ and $(F_d, \sigma_d, \Psi_d, M_d)$ are obtained from $Y_C(x)$ and $Y_D(x)$. Similarly σ_c and σ_d are the process variance and $\Psi_c(\bullet, \bullet)$ and $\Psi_d(\bullet, \bullet)$ are correlation matrices of $Y_C(x)$ and $Y_D(x)$ respectively.

12.2.5 Stochastic Kriging

Although the interpolation property of Kriging is advantageous while dealing with deterministic simulation problems, it often yields erroneous results for problems involving uncertainties/noise. In order to address this issue, stochastic Kriging has been developed. In stochastic Kriging, the noise is modelled as a separate Gaussian process $\xi(x)$ with zero mean and covariance Σ. The stochastic Kriging predictor thus becomes:

$$\hat{y}(x) = M\alpha + r(x)\left(\Psi + \frac{1}{\sigma^2}\Sigma\right)^{-1}(\bar{y} - F\alpha) \tag{12.20}$$

where $\frac{1}{\sigma^2}\Sigma$ is a matrix resembling noise-to-signal ratios and $\bar{y}$ is a vector containing the average function values of the repeated simulations for each sample. Note that if the entries of Σ are zero, Equation (12.20) converges to Equation (12.5).

The covariance matrix Σ for noise can have various forms. In stochastic simulation, Σ is created based on repeated simulation. In the simplest form, Σ takes the following form:

$$\Sigma = \begin{pmatrix} \mathrm{var}(y^i) & \cdots & 0 \\ \vdots & \ddots & \vdots \\ 0 & \cdots & \mathrm{var}(y^i) \end{pmatrix} \tag{12.21}$$

where $\mathrm{var}(y^i)$ is the variance between the repeated simulation of data point i. On contrary, if the problem under consideration is deterministic (but noisy), Σ consists of scalar values 10^λ on its diagonal. The variable λ is estimated as part of the likelihood optimization of the Kriging.

12.2.6 Correlation function

All the variants of Kriging are dependent on the choice of correlation function to create an accurate surrogate. Various type of correlation function is available in literature (Rivest and Marcotte 2012, Biscay et al. 2013, Putter and Young 2001). In this work, we restrict ourselves to the stationary correlation function defined as:

$$\psi(x, x') = \prod_j \psi_j(\theta, x_i - x_i') \tag{12.22}$$

The correlation function defined in Equation (12.22) has two desirable properties. Firstly, the correlation function for multivariate functions can be represented as product of one dimensional correlation. Secondly, the correlation is stationary and depends only in the distance between two points. Various correlation functions that satisfies Equation (12.22) are available in literature. In this study, seven correlation functions, namely (a) exponential correlation function, (b) generalised exponential correlation function (c) Gaussian correlation function (d) linear correlation function (e) spherical correlation function (f) cubic correlation function and (g) spline correlation function have been investigated. The mathematical forms of all the correlation functions are provided below:

i) Exponential correlation function:

$$\psi_j(\theta; d_j) = \exp\left(-\theta_j |d_j|\right) \tag{12.23}$$

ii) Generalised exponential correlation function:

$$\psi_j\left(\theta;d_j\right) = \exp\left(-\theta_j\left|d_j\right|^{\theta_{n+1}}\right), \quad 0 < \theta_{n+1} \le 2 \tag{12.24}$$

iii) Gaussian correlation function:

$$\psi_j\left(\theta;d_j\right) = \exp\left(-\theta_j d_j^{\,2}\right) \tag{12.25}$$

iv) Linear correlation function:

$$\psi_j\left(\theta;d_j\right) = \max\left\{0, 1-\theta_j\left|d_j\right|\right\} \tag{12.26}$$

v) Spherical correlation function:

$$\psi_j\left(\theta;d_j\right) = 1 - 1.5\xi_j + 0.5\xi_j^2, \quad \xi_j = \min\left\{1, \theta_j\left|d_j\right|\right\} \tag{12.27}$$

vi) Cubic correlation function:

$$\psi_j\left(\theta;d_j\right) = 1 - 3\xi_j^2 + 2\xi_j^3, \quad \xi_j = \min\left\{1, \theta_j\left|d_j\right|\right\} \tag{12.28}$$

vi) Spline correlation function:

$$\psi_j\left(\theta;d_j\right) = \begin{cases} 1 - 5\xi_j^2 + 30\xi_j^3, & 0 \le \xi_j \le 0.2 \\ 1.25\left(1-\xi_j^3\right), & 0.2 \le \xi_j \le 1 \\ 0, & \xi_j > 1 \end{cases} \tag{12.29}$$

where $\xi_j = \theta_j\left|d_j\right|$

For all the correlation functions described above, $d_j = x_i - x_i'$.

12.2.7 Maximum likelihood estimation (MLE)

Another important aspect of all Kriging surrogate is determination of the hyperparameters governing the correlation functions. The conventional way of determining the hyperparameters is by employing the maximum likelihood estimate (MLE). There are several variants of likelihood that one can utilize. In this study, we limit our discussion to marginal likelihood and Pseudo likelihood only.

12.2.7.1 Marginal likelihood

The natural log of the marginal likelihood is given as:

$$\log\left(\mathcal{L}_{\text{marginal}}\right) = \frac{n}{2}\log\left(2\pi\right) + \frac{n}{2}\log\left(\sigma^2\right) + \frac{1}{2}\log\left(\left|\Psi\right|\right) + \frac{1}{2\sigma^2}\left(y - F\alpha\right)^T \Psi^{-1}\left(y - F\alpha\right) \tag{12.30}$$

Equation (12.30) is simplified by taking derivatives with respect to α and σ^2 and equating to zero. The quantity $\frac{n}{2}\log\left(\sigma^2\right)$ in Equation (12.30) denotes the quality of fit. Similarly, $\frac{1}{2}\log\left(\left|\Psi\right|\right)$ denotes the complexity penalty. Therefore, the marginal likelihood automatically balances flexibility and accuracy. It is to be noted that marginal likelihood depends on correct specification of Kriging model for the data and hence, may not be robust enough when the Kriging model is misspecified.

12.2.7.2 Pseudo likelihood

In order to address the robustness issue of marginal likelihood, the pseudo likelihood has emerged as an attractive alternative. The pseudo likelihood estimate is given as:

$$\log\left(\mathcal{L}_{\mathrm{PL}}\right) = \sum_{i=1}^{n} -\frac{1}{2}\log\left(\sigma_i^2\right) - \frac{\left(y_i - F\alpha - \mu^i\right)^2}{2\sigma_i^2} - \frac{1}{2}\log\left(2\pi\right) \qquad (12.31)$$

where

$$\mu^i = y^i - F\alpha - \frac{\Psi^{-1}y}{\Psi_{ii}^{-1}} \qquad (12.32)$$

and

$$\sigma_i^2 = \frac{1}{\Psi_{ii}^{-1}} \qquad (12.33)$$

Since pseudo likelihood is independent of the model selection and hence, yields a more robust solution.

12.3 Stochastic approach using Kriging model

The stochasticity in material properties of laminated composite shallow doubly curved shells, such as longitudinal elastic modulus, transverse elastic modulus, longitudinal shear modulus, transverse shear modulus, Poisson's ratio, mass density and geometric properties such as ply-orientation angle as input parameters are considered for the uncertain natural frequency analysis. In the present study, the frequency domain feature (first three natural frequencies) is considered as output. It is assumed that the distribution of randomness for input parameters exists within a certain band of tolerance with respect to their deterministic mean values following a uniform random distribution. In the present investigation, $\pm 10°$ for ply orientation angle with subsequent $\pm 10\%$ tolerance for material properties from deterministic mean value are considered for numerical illustration. For the purpose of comparative assessment of different Kriging model variants, both low and high dimensional input parameter space is considered to address the issue of dimensionality in surrogate modelling. For low dimensional input parameter space, stochastic variation of layer-wise ply orientation angles $\{\theta(\bar{\omega})\}$ are considered, while combined variation of all aforementioned layer-wise stochastic input parameters $\{g(\bar{\omega})\}$ are considered to explore the relatively higher dimensional input parameter space as follows:

***Case*-1** Variation of ply-orientation angle only: $\theta(\bar{\omega}) = \{\theta_1 \ \theta_2 \ \theta_3........\theta_i......\theta_l\}$

Case-2 Combined variation of ply orientation angle, elastic modulus (longitudinal and transverse), shear modulus (longitudinal and transverse), Poisson's ratio and mass density:

$$g\{\theta(\bar{\omega}), \rho(\bar{\omega}), G_{12}(\bar{\omega}), G_{23}(\bar{\omega}), E_1(\bar{\omega})\} = \{ \Phi_1(\theta_1...\theta_l), \Phi_2(E_{1(1)}...E_{1(l)}), \Phi_3(E_{2(1)}...E_{2(l)}),$$
$$\Phi_4(G_{12(1)}..G_{12(l)}), \Phi_5(G_{23(1)}..G_{23(l)}), \Phi_6(\mu_1...\mu_l), \Phi_7(\rho_1..\rho_l)\}$$

where θ_i, $E_{1(i)}$, $E_{2(i)}$, $G_{12(i)}$, $G_{23(i)}$, μ_i and ρ_i are the ply orientation angle, elastic modulus along longitudinal and transverse direction, shear modulus along longitudinal direction, shear modulus along transverse direction, Poisson's ratio and mass density, respectively and '*l*' denotes the number of layer in the laminate. $\bar{\omega}$ represents the stochastic character of the input parameters. In the present investigation a 4 layered composite laminate is considered having total 4 and 28 random input parameters for individual and combined variations respectively. Figure 12.1 shows a schematic representation of the stochastic system where *x* and *y(x)* are the collective input and output parameters with stochastic character respectively. Latin hypercube sampling (Mckay et al. 2000) is employed in this study for generating sample points to ensure the representation of all portions of the vector space. In Latin hypercube sampling, the interval of each dimension is divided into *m* non-overlapping intervals having equal probability considering a uniform distribution, so the intervals should have equal size. Moreover, the sample is chosen randomly from a uniform distribution of a point in each interval in each dimension and the random pair is selected considering equal likely combinations for the point from each dimension. Figure 12.2 represents the Kriging based uncertainty quantification algorithm wherein the actual finite element model of composite shell is effectively replaced by the computationally efficient Kriging models and subsequently comparative performance of different Kriging variants are judged on the basis of accuracy and computational efficiency.

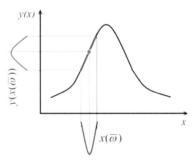

Fig. 12.1 Stochastic simulation model.

12.4 Results and discussion

In this section, the performance of the Kriging variants for stochastic free vibration analysis of laminated composite shells has been investigated. Two different

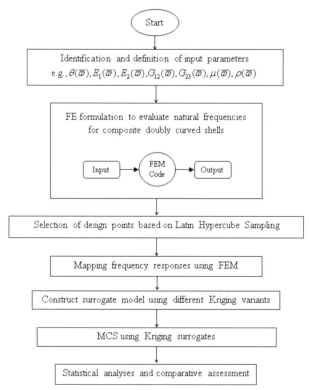

Fig. 12.2 Flowchart of stochastic natural frequency analysis using Kriging model variants.

cases of stochastic variations have been considered for individual and combined stochasticity having 4 and 28 random input parameters respectively, as discussed in the previous section. Apart from the performance of the Kriging variants, the effect of covariance functions has been illustrated. The two alternatives for computing the hyperparameters invovled in Kriging based surrogate, namely the maximum likelihood estimate and pseudo likelihood estimate, have also been illustrated. Additionally, stochastic Kriging has been utilized to investigate the effect of variance noise levels present in the system. For all the cases, results obtained are bechmarked against crude Monte Carlo simulation (MCS) results considering 10,000 realizations in each case. For full scale MCS, number of original FE analysis is the same as the sampling size.

12.4.1 Validation

A four layered graphite–epoxy symmetric angle-ply ($45°/–45°/–45°/45°$) laminated composite cantilever shallow doubly curved hyperbolic paraboloid ($R_x/R_y = –1$) shell has been considered for the analysis. The length, width and thickness of the composite laminate considered in the present analysis are 1 m, 1 mm and 5 mm, respectively. Material properties of graphite–epoxy composite considered with deterministic mean value as $E_1 = 138.0$ GPa, $E_2 = 8.96$ GPa, $v_{12} = 0.3$, $G_{12} = 7.1$ GPa, $G_{13} = 7.1$ GPa, $G_{23} = 2.84$ GPa, $\rho = 3202$ kg/m³. A typical discretization of (6×6) mesh on plan area with 36 elements and 133 nodes with natural coordinates of an isoparametric quadratic plate bending element has been considered for the present FEM approach. The finite element code is validated with the results available in the open literature as shown in Table 12.1 Qatu and Leissa (1991a). Convergence studies are performed using mesh division of (4 x 4), (6 x 6), (8 x 8), (10 x 10) and (12 x 12) wherein (6 x 6) mesh is found to provide best results with the least difference compared to benchmarking results. The marginal differences between the results by Qatu and Leissa (1991) and the present finite element approach can be attributed to the consideration of transverse shear deformation and rotary inertia and also to the fact that the Ritz method overestimates the structural stiffness of the composite plates.

Table 12.1 Non-dimensional fundamental natural frequencies $[\omega = \omega_n L^2\sqrt{(\rho/E_1 h^2)}]$ of three layered $[\theta/-\theta/\theta]$ graphite–epoxy twisted plates, L/b = 1, b/h = 20, ψ = 30°.

Ply-orientation Angle, θ	Present FEM (with mesh size)					Qatu and Leissa (1991a)
	4 x 4	6 x 6	8 x 8	10 x 10	12 x 12	
15°	0.8588	0.8618	0.8591	0.8543	0.8540	0.8759
30°	0.6753	0.6970	0.6752	0.6722	0.6717	0.6923
45°	0.4691	0.4732	0.4698	0.4578	0.4575	0.4831
60°	0.3189	0.3234	0.3194	0.3114	0.3111	0.3283

12.4.2 Comparative assessment of Kriging variants

In this section the performance of the kriging variants in stochastic free vibration analysis of FRP composite shells has been presented. For the ease of understanding and refering, following notations have been used hereafter:

a) KV1: Ordinary Kriging
b) KV2: Universal Kriging with pseudo likelihood estimate
c) KV3: Blind Kriging
d) KV4: Co-Kriging
e) KV5: Stochastic kriging with zero noise (Universal Kriging based on marginal likelihood estimator).

In addition to the above, individual and combined variation of input parameters as described in Section 12.3, are implied as first and second case throughout this chapter. For the first case (i.e., uncertainties in ply orientation angles only), the sample size is varied from 20 to 60 at an interval of 10. Figure 12.3 and Fig. 12.4 show the mean and standard deviation of error, corresponding to the first three natural frequencies. For all the three frequencies, KV5 yields the best results.

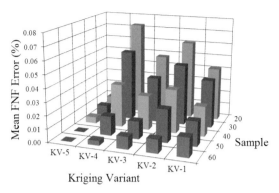

(a) Mean first natural frequency (FNF) error

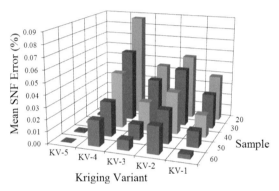

(b) Mean second natural frequency (SNF) error

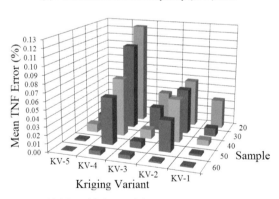

(c) Mean third natural frequency (TNF) error

Fig. 12.3 Error of the Kriging variants in predicting the mean error (%) of first three natural frequencies for the first case (individual variation).

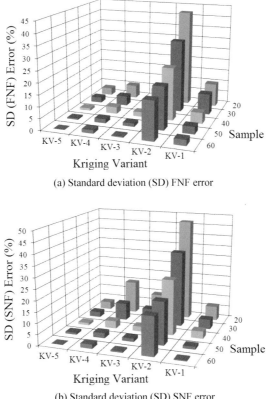

(a) Standard deviation (SD) FNF error

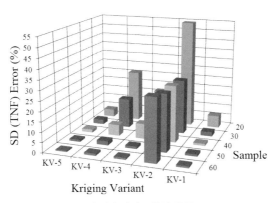

(b) Standard deviation (SD) SNF error

(c) Standard deviation (SD) TNF error

Fig. 12.4 Error of the Kriging variants in predicting the standard deviation error (%) of first three natural frequencies for the first case (individual variation).

Performance of the other Kriging variants varies from case to case. For instance, results obtained using KV4 outperforms all but KV5 in predicting the mean of the first natural frequency. However, for the mean of second natural frequencies, KV4 yields the worst results. Similarly the arguments hold for KV1 and KV3 as well. KV2, in most of the cases (specifically for standard deviation of response), yields erroneous results. This is probably due to the inability of the pseudo likelihood function to accurately predict the hyperparameters associated with the covariance function. Probability density function (PDF) and representative scatter plot of the first three natural frequencies, obtained using the kriging variants and crude MCS, are shown in Fig. 12.5 and Fig. 12.6, respectively. KV5 yields the best result followed by KV4 and KV3. Results obtained using KV2 is found to be eroneous. As already stated, this is due to erroneous hyperparameters obtained using the pseudo likelihood estimate.

For the second case (combined variation of all input parameters), the sample size is varied from 250 to 600 at an interval of 50. Figure 12.7 and 12.8 show the error in mean and standard deviation of the first three natural frequencies obtained using the five variants of Kriging. For mean natural frequencies, all but KV2 yields excellent results with extremely low error. However, for standard deviation of the first two natural frequencies ordinary Kriging (KV1) yields erroneous results. Similar to the previous case, KV3, KV4 and KV5 are found to yield excellent results with extremely low prediction error. Probability density function plot and representative scatter plot of the first three natural frequency, obtained using the kriging variants and crude MCS, are shown in Fig. 12.9 and Fig. 12.10 respectively. Apart from KV2, all the variants are found to yield excellent results.

It is worth mentioning here that the whole point of using a Kriging based uncertainty quantification approach is to achieve computational efficiency in terms of finite element simulation. For example, the probabilistic descriptions and statistical results of the natural frequencies presented in this study are based on 10,000 simulations. In case of crude MCS, same number of actual finite element simulations are needed to be carried out. However, in the present approach of Kriging based uncertainty quantification, the number of actual finite element simulations required is same as sample size to construct the Kriging models. Thus, if 50 and 550 samples are required to form the Kriging models for the individual and combined cases respectively, the corresponding levels of computational efficiency are about 200 times and 18 times with respect to crude MCS. Except the computational time associated with finite element simulation as discussed above, another form of significant computational time involved in the process of uncertainty quantification can be the building and prediction for the Kriging models. In general, the computational expenses are found to increase for higher dimension of the input parameter space. The second form of computational time can be different for different kriging variants. A comparative investigation on the computational times for different kriging model variants are provided next.

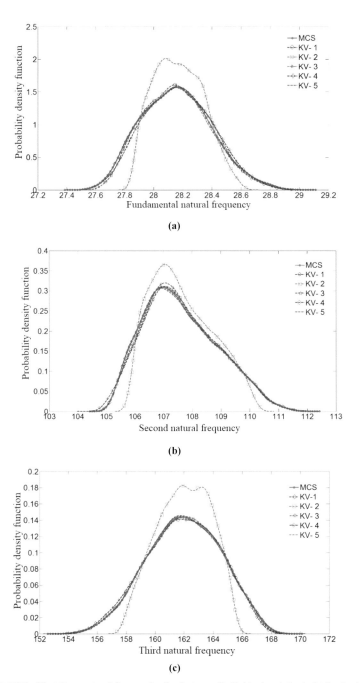

Fig. 12.5 PDF of first three natural frequencies for first case (individual variation) obtained using the Kriging variants (50 samples) and MCS.

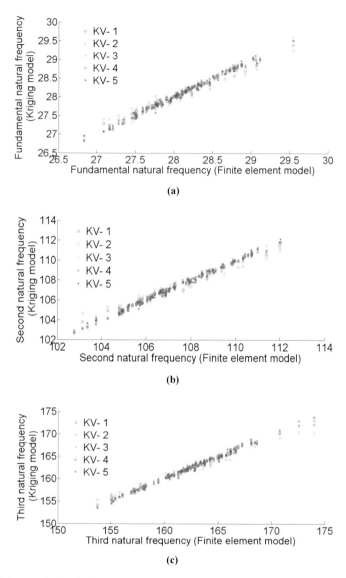

Fig. 12.6 Scatter plot for the first three natural frequencies for the first case (individual variation) obtained using 50 samples.

Table 12.2 reports the computational time required for Kriging model prediction by the five Kriging variants for both cases. Please note here that the time reported in this table is for building and predictions corresponding to different Kriging models and this time has no relation with the time required for finite element simulation of laminated composite shell. It is observed that for individual variation of input

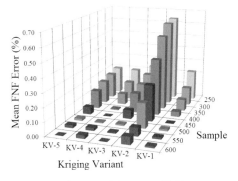

(a) Mean first natural frequency (FNF) error

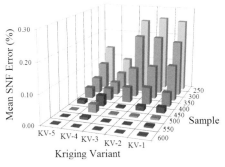

(b) Mean second natural frequency (SNF) error

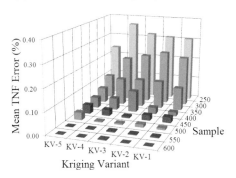

(c) Mean third natural frequency (TNF) error

Fig. 12.7 Error of the Kriging variants in predicting the mean of first three natural frequencies for the second case (combined variation).

parameters (4 input parameters), KV1 is the fastest followed by KV5 and KV4. However, for the combined variation case (28 input parameters), KV4 is much faster compared to the other variants. This is because unlike the other variants KV4 first generates a low fidelity solution from comparatively less number of

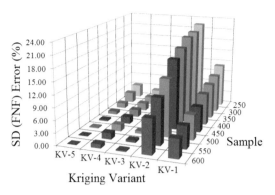

(a) Standard deviation (SD) FNF error

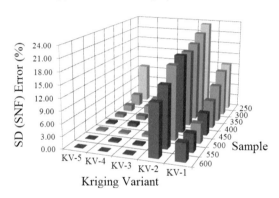

(b) Standard deviation (SD) SNF error

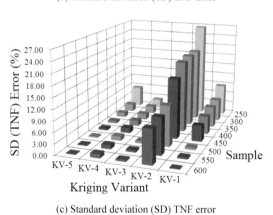

(c) Standard deviation (SD) TNF error

Fig. 12.8 Error of the Kriging variants in predicting the standard deviation of first three natural frequencies for the second case (combined variation).

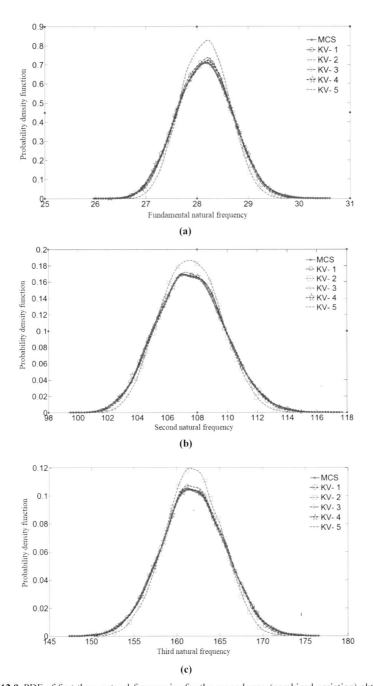

Fig. 12.9 PDF of first three natural frequencies for the second case (combined variation) obtained using 550 samples.

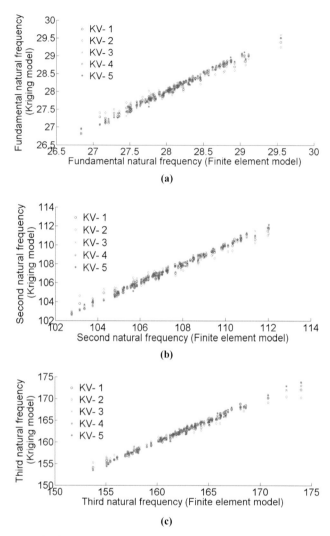

Fig. 12.10 Scatter plot for the first three natural frequencies for the second case (combined variation) obtained using 550 samples.

Table 12.2 CPU time of the Kriging variants.

Number of input parameters	Kriging variants				
	KV1	**KV2**	**KV3**	**KV4**	**KV5**
First case (Individual)	1.0511s	3.1391s	1.6714s	1.62s	1.2258s
Second case (Combined)	335.4105s	5205.1s	7330.1s	110.9570s	1031.3s

sample points and updates it by adding additional sample points. As a consequence, the matrix inversion involved in KV4 becomes less time consuming, making the overall procedure computationally efficient. However, this advantage of KV4 is only visible for large systems, involving large number of input variables (combined variation case). For smaller systems, inverting two matrices, instead of one, may not be advantageous as observed in the individual variation case.

12.4.3 Comparative assessment of various covariance functions

This section investigates the performance of the various covariance functions used in Kriging. To be specific, performance of the following seven covariance functions has been investigated:

a) COV1: Cubic covariance function
b) COV2: Exponential covariance function
c) COV3: Gaussian covariance function
d) COV4: Linear covariance function
e) COV5: Spherical covariance function
f) COV6: Spline covariance function
g) COV7: Generalised exponential covariance function.

The above mentioned covariance functions have been utilized in conjunction with ordinary Kriging (KV1); the reason being that ordinary Kriging does not have a regression part and hence the effect of the covariance function will be more prominent in such case. As per the study reported in previous subsection, the Kriging based surrogate is formulated with 50 sample points in the first case (individual variation). Figure 12.11 shows the PDF of response obtained using various covariance functions for the first case. It is observed that for all the three natural frequencies, Gaussian covariance (COV3) yields the best result followed by exponential (COV2) and generalised exponential (COV7) covariance function. The error (%) in mean and standard deviation of natural frequencies are shown in Fig. 12.12 and Fig. 12.13. For all the cases, Gaussian, expnential and generalised exponential covariance functions yields the best results. Interestingly, results obtained using cubic (COV1), linear (COV4), spherical (COV5) and spline (COV6) covariance functions are almost identical. Similar results have been observed for the second case (combined variation). However, due to paucity of space, the results for the second case have not been reported in this chapter.

12.4.4 Comparative assessment of various noise levels

In this section, the effect of noise, present within a system, has been simulated using the stochastic Kriging (KV5). The effect of noise in a system can be regarded as considering other sources of uncertainty besides conventional material and geometric uncertainties, *viz.,* error in modelling, human error and various other

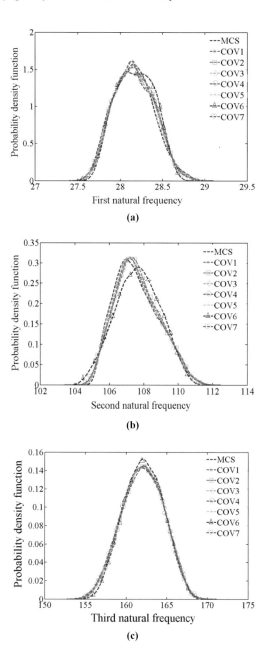

Fig. 12.11 PDF of the first three natual frequencies for the *first case* obtained using various covariance functions. For all the cases, Kriging based surrogate is formulated using 50 sample points. Ordinary Kriging (KV1) is used in conjunction with the covariance functions.

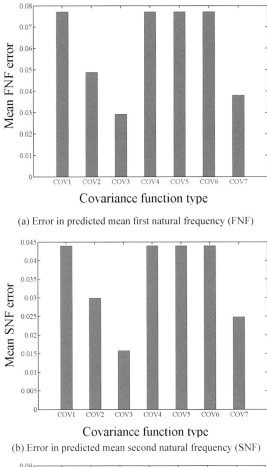

(a) Error in predicted mean first natural frequency (FNF)

(b) Error in predicted mean second natural frequency (SNF)

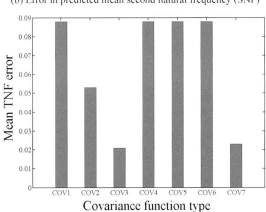

(c) Error in predicted mean third natural frequency (TNF)

Fig. 12.12 Error in predicted mean natural frequencies obtained using various covariance functions.

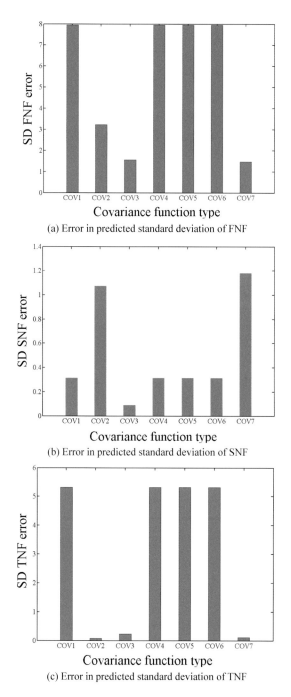

(a) Error in predicted standard deviation of FNF

(b) Error in predicted standard deviation of SNF

(c) Error in predicted standard deviation of TNF

Fig. 12.13 Error in predicted standard deviation of natural frequencies obtained using various covariance functions.

epistemic uncertainties involved in the system (Dey et al. 2016c). In the present study, gaussian white noise with specific level (p) has been introduced into the output responses as:

$$f_{ijN} = f_{ij} + p\xi_{ij} \tag{12.34}$$

where f_{ij} denotes the ith frequency in the jth sample in the design point set. The subscript N denotes the frequency in the presence of noise. ξ in Equation (12.34) denotes normal distributed random number with zero mean and unit variance. Typical results for the effect of noise have been reported for combined variation of input parameters. Figure 12.14 and Fig. 12.15 show the representative scatter plots and PDF of the first three natural frequencies corresponding to various noise levels. The distance of the points from a diagonal line increases with the increase of noise level in the scatter plots indicating lesser prediction accuracy of the Kriging

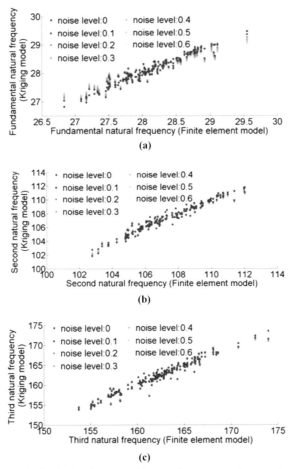

Fig. 12.14 Scatter plot for the first three natural frequencies corresponding to various noise levels.

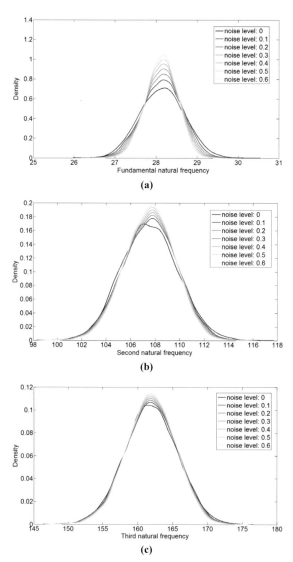

Fig. 12.15 Probability density function plot for the first three natural frequencies corresponding to various noise levels.

model. Significant change in the standard deviation of the frequencies is observed with changes in the noise level. However, the mean frequencies are found to be mostly insensitive to the noise.

12.5 Summary

This chapter presents a critical comparative assessment of five Kriging model variants in an exhaustive and comprehensive manner for quantifying uncertainty in the natural frequencies of composite doubly curved shells. Five kriging variants considered in this study are: Ordinary Kriging, Universal Kriging based on pseudo-likelihood estimator, Blind Kriging, Co-Kriging and Universal Kriging based on marginal likelihood estimator (Stochastic Kriging with zero noise). The comparative assessment has been carried out from the view point of accuracy and computational efficiency. Both low and high dimensional input parameter spaces have been considered in this investigation to explore the effect of dimensionality on different Kriging variants. Results have been presented for different sample sizes to construct the Kriging model variants. A comparative study on the performance of different covariant functions has also been carried out in conjunction to the present problem. Further, the effect of noise has been investigated considering Stochastic Kriging.

It is observed that Universal Kriging coupled with marginal likelihood estimate yields the best results for both low-dimensional and high-dimensional problems. Co-Kriging is found to be the fastest in terms of model bulding and prediction time for high dimensional input parameter space. In case of low dimensional input parameter space, ordinary kriging is relatively the fastest followed by co-kriging, blind kriging and universal kriging with pseudo likelihood estimate, respectively. However, the effect of relative differences in computational time becomes more crucial for large number of input parameters as considerable amount of time is required. Among various covariance functions investigated, it is observed that the Gaussian covariance function produces best accuracy. Stochastic kriging is an efficient tool for simulating the effect of noise present in a system. As evident from the results presented, stochastic Kriging yields highly accurate result for a system involving certain degree of inherent noise. Although this chapter focuses on stochastic natural frequency analysis of composite shells, the outcomes regarding comparative performance of the Kriging model variants may serve as a valuable reference for different computationally intensive problems in the broad field of science and engineering.

CHAPTER 13

A Comparative Study on Metamodel Based Stochastic Analysis of Composite Structures

13.1 Introduction

Various metamodel based uncertainty quantification algorithms for composite structures have been presented in the previous chapters of this book. Performance of the five different Kriging model variants was assessed in the last chapter. This chapter presents an exhaustive comparative investigation on different metamodels for critical comparative assessment of uncertainty in natural frequencies of composite plates on the basis of computational efficiency and accuracy. Both individual and combined variations of input parameters have been considered to account for the effect of low and high dimensional input parameter spaces in the surrogate based uncertainty quantification algorithms including the rate of convergence. Probabilistic characterization of the first three stochastic natural frequencies is carried out by using a finite element model that includes the effects of transverse shear deformation based on Mindlin's theory in conjunction with a layer-wise random variable approach. The results obtained by different metamodels have been compared with the results of direct Monte Carlo simulation (MCS) method for high fidelity uncertainty quantification. The crucial issue regarding influence of sampling techniques on the performance of metamodel based uncertainty quantification has been addressed as an integral part of this chapter.

The exhaustive utilization of computational power has favoured the development of very high-fidelity finite element models to deal with industrial problems. In spite of advances in capacity and speed of computer, the enormous computational cost of running complex, intricate scientific and engineering simulations makes it impractical to rely exclusively on simulation codes for the purpose of uncertainty quantification. Hence these high-fidelity models come with the drawback that they can be very-time consuming so that only a few runs of the model can be affordable. Thus these models are practically unusable in computationally intensive methods like traditional Monte Carlo simulation (MCS) based stochastic analysis that requires thousands of realizations to be carried out. In general, such complicated models can be considered as a system (often referred to I/O system), for which the output quantity of interest (O) is evaluated corresponding to a particular set of values for the input parameters (I). In case of analyses that require large number of model evaluation, it is a common practise to employ a computationally efficient surrogate or metamodel based approach, in which outputs are only evaluated for a limited set of algorithmically chosen input points and then an equivalent mathematical model is constructed to emulate the underlying mapping of the I/O system. The need of integrating the surrogate models and probabilistic approaches has significant demand for assessing the response characteristics of composite structures by accounting the uncertainties in the models as well as the random input parameters (e.g., geometrical parameters, fibre parameters and material properties) Baran et al. (2016). As discussed in the previous chapters, due to the dependency on a large number of parameters in complex production and fabrication processes of laminated composites, the system properties can be random in nature resulting inevitable uncertainty in the responses of such structures.

While adopting a surrogate based approach for uncertainty quantification, an obvious question that a designer may have is which technique is superior to the other and on what basis should the various surrogate modelling techniques be selected. Some studies demonstrate the application of one metamodeling technique or the other typically for a specific problem; however, the present chapter shows a comprehensive comparative study of the various metamodeling techniques in conjunction to composites. Multiple factors contribute to the success of a given metamodeling technique, ranging from the stochasticity and dimensionality of the problem to the associated data sampling technique and the internal parameter settings of the various modelling techniques. Overall, the knowledge of the performance of different metamodeling techniques with respect to different modelling criteria is of utmost importance to designers while choosing an appropriate technique for a particular application. A concise literature review on application of different surrogate modelling techniques is presented in the following paragraphs.

A preferable strategy for the analyses requiring repetitive model evaluation is to utilize approximation models which are often referred to as metamodels—"model of the model" (Kleijnen 1987) that effectively replace the expensive simulation model (Arregui–Mena et al. 2016) in a computationally efficient manner. Metamodeling techniques have been widely used for design evaluation and optimization in many engineering applications; a comprehensive review of metamodeling applications in mechanical and aerospace systems can be found in the paper by Simpson et al. (1997) and will therefore not be repeated here. For the interested readers, a review of metamodeling applications in optimization can be found in the articles by Barthelemy and Haftka (1993) and Sobieszczanski-Sobieski and Haftka (1997). A variety of surrogate modelling techniques exist wherein response surface methodology (Dey et al. 2015d, Myers and Montgomery 2002) and artificial neural network (ANN) methods (Smith 1993, Dey et al. 2016e) are found as the two well-known approaches for constructing simple and fast approximations of complex computer codes. An interpolation method known as Kriging is widely utilised for the design and analysis of computer experiments (Sacks et al. 1989b, Booker et al. 1999, Mukhopadhyay et al. 2017c).

The other promising statistical techniques, such as multivariate adaptive regression splines (MARS) (Friedman 1991, Mukhopadhyay 2018a) and radial basis function (RBF) approximations (Naskar et al. 2017, Dyn et al. 1986), moving least square (MLS) (Youn and Choi 2004, Soo et al. 2010), support vector regression (SVR) (Trevor et al. 2004, Dai et al. 2015) and polynomial neural network (PNN) (Dey et al. 2016c, Mellit et al. 2010) have also drawn significant attention of many researchers. Previously, Simpson et al. (1998) compared Kriging methods against polynomial regression models for the multidisciplinary design optimization of an aerospike nozzle involving three design variables while Giunta et al. (1998) compared Kriging models and polynomial regression models for a test problem. In contrast, Varadarajan et al. (2000) compared the ANN method with the polynomial regression model for an engine design problem involving nonlinear thermodynamic behaviour. Yang et al. (2000) compared four approximation methods such as, enhanced multivariate adaptive regression splines (MARS), stepwise regression, ANN, and the moving least square method for the construction of safety related functions in automotive crash analysis for a relatively small sample size.

In the literature, there are several successful applications of surrogate modelling techniques in the optimization of traditional composite laminates with straight fibers. Radial Basis Functions (Irisarri et al. 2011), second order polynomials (Rikards et al. 2006) and Neural Networks (Bisagni and Lanzi 2002) are found to be effective in reducing the time to find the maximum buckling load of a composite stiffened panel. Liu et al. (2000) used a cubic response surface combined with a two-level optimization technique to maximize the buckling load of a composite wing. Lee and Lin (2003) used trigonometric functions as the base functions to build a metamodel for the stacking sequence optimization of a composite propeller. Kalnins et al. (2010) compared the performance of Radial Basis Functions, multivariate adaptive regression splines and polynomials for optimization of the post-buckling characteristics of damaged composite stiffened structure. In another attempt, Lanzi and Giavotto (2006) compared the performance of Radial Basis Functions, Neural Networks, and Kriging metamodels in a multi-objective optimization problem for maximum post-buckling load and minimum weight of a composite stiffened panel. These methods are found to yield similar results and none of them is identified as being significantly superior. While there is a considerable amount of existing research on the use of metamodels for the constant stiffness composite design, only a few attempts look at their application in variable stiffness design. Among those worthy of mention are the following, the optimization of a variable stiffness laminate in vibration (Vandervelde and Milani 2009), the buckling load of a variable stiffness composite cylinder (Blom et al. 2010), and the simultaneous optimization of the buckling load and in-plane stiffness of a variable stiffness laminate ignoring the presence of defects, i.e., gaps and overlaps (Arian et al. 2012). Of late, Arian Nik et al. (2014, 2015) used the defect layer method and a Kriging metamodel to simultaneously maximize the buckling load and in-plane stiffness of a variable stiffness laminate with embedded defects. Dey et al. (2015a, 2017) have presented surrogate based stochastic investigation of composite structures including a comparative study on their performances.

The above-mentioned works are demonstrated as the potential method indicating that the surrogate model can be utilised as a beneficial tool for reduction of computational burden in optimization process. Based on literature review, it is found that in the following areas metamodeling can play a significant role: (a) *Model approximation.* Approximation of computation-intensive processes across the entire design space, or global approximation, is used to reduce computation costs, (b) *Design space exploration.* The design space is explored to enhance the engineers' understanding of the design problem by working on a cheap-to-run metamodel, (c) *Problem formulation.* Based on an enhanced understanding of a design optimization problem, the number and search range of design variables may be reduced; certain ineffective constraints may be removed; a single objective optimization problem may be changed to a multi-objective optimization problem or vice versa. Metamodel can assist the formulation of an optimization problem that is easier to solve or more accurate than otherwise, (d) *Optimization support.* Industry has various optimization needs, e.g., global optimization, multi-objective optimization, multidisciplinary design optimization, probabilistic optimization, and so on. Each type of optimization has its own challenges. Metamodeling can be applied and integrated to solve various types of optimization problems that involve computation-intensive functions. The literature review presented above reveals that there is no recommendation found regarding selection of surrogate model for analyses of composites and other applications. Furthermore, the performance of a surrogate model is described as problem dependent and the best surrogate model is unknown at the outset.

For surrogate model formation, few algorithmically chosen design points are evaluated using the expensive model/experiments. Finally on the basis of the information gathered through these design points over the design space, a fully functional metamodel is constructed. The "*Classic*" experimental designs are originated from the theory of Design of Experiments when physical experiments are conducted. These methods focus on planning experiments so that the random error in physical experiments has minimum influence on the approval or disapproval of a hypothesis. Widely used "classic" experimental designs include factorial or fractional factorial design (Myers and Montgomery 2002, Vianna et al. 2010, Chen 1995), central composite design (CCD) (Myers and Montgomery 2002),

Box-Behnken (Myers and Montgomery 1995), optimal design (Mukhopadhyay et al. 2015b) and Plackett-Burman designs (Myers and Montgomery 2002). Mukhopadhyay et al. (2015b) presented a comparative assessment of different designs of experiment methods in conjunction to a system identification problem using multi-objective optimization and suggested that the D-optimal design and CCD perform better compared to other considered designs of experiment methods. These classic methods tend to spread the sample points around boundaries of the design space and leave a few at the centre of the design space. As computer experiments involve mostly systematic error rather than random error as in physical experiments, Sacks et al. (1989a) stated that in the presence of systematic rather than random error, a good experimental design tends to fill the design space rather than to concentrate on the boundary. They also stated that "classic" designs, e.g., CCD and D-optimal designs can be inefficient or even inappropriate for deterministic computer codes. Jin et al. (2001) confirmed that a consensus among researchers was that experimental designs for deterministic computer analyses should be space filling. Koehler and Owen (1996) described several Bayesian and Frequentist "Space Filling" designs, including maximum entropy design by Currin et al. (1991), mean squared-error designs, minimax and maximin designs by Johnson et al. (1990), Latin Hypercube designs, orthogonal arrays, and scrambled nets. Four types of space filling sampling methods are relatively more often used in the literature. These are orthogonal arrays (Taguchi et al. 1993, Owen 1992, Hedayat et al. 1999), Latin Hypercube designs (Mckay et al. 1979, Iman and Conover 1980, Tang 1993, Park 1994, Ye et al. 2000), Hammersley sequences (Kalagnanam and Diweker 1997, Meckesheimer et al. 2002) and uniform designs (Fang et al. 2000). Hammersley sequences and uniform designs belong to a more general group called low discrepancy sequences (Chen et al. 2006) wherein Hammersley sampling is found to provide better uniformity than Latin Hypercube designs. A comparison of these sampling methods was carried out by Mukhopadhyay (2018a). If any knowledge of the space is available, these methods may be tailored to achieve higher efficiency. They may also play a more active role for iterative sampling-metamodeling processes. Mainly due to the difficulty of knowing the "appropriate" sampling size a priori, sequential and adaptive sampling has gained popularity in recent years. Lin (2004) proposed a sequential exploratory experiment design (SEED) method to sequentially generate new sample points. Jin et al. (2005) applied simulated annealing to quickly generate optimal sampling points. Sasena et al. (2002) used the Bayesian method to adaptively identify sample points that gave more information. Wang (2003) proposed an inheritable Latin Hypercube design for adaptive metamodeling. Samples were repetitively generated for fitting a Kriging model in a reduced space by Wang et al. (2004). Jin et al. (2002) compared a few different sequential sampling schemes and found that sequential sampling allows engineers to control the sampling process and it is generally more efficient than one-stage sampling. One can custom design the flexible sequential sampling schemes for specific design problems.

Metamodeling evolves from the classical Design of Experiments (DOE) theory, in which polynomial functions are used as response surfaces, or metamodels. Response surfaces are typically second-order polynomial models and therefore, they have limited capability to accurately model nonlinear functions of arbitrary shape. Obviously, higher-order response surfaces can be used to model a nonlinear design space. However, instabilities may arise, or it may be difficult to take enough sample points in order to estimate all of the coefficients in the polynomial equation particularly in high dimensions. Hence, many researchers advocate the use of a sequential response surface modelling approach using move limits or a trust region approach. Besides the commonly used polynomial functions, Sacks et al. (1989) and Pronzotto and Muller (2012) proposed the use of a stochastic model, called Kriging (Cressie 1988), to treat the deterministic computer response as a realization of a random function with respect to the actual system response. Neural networks have also been applied in generating the response surfaces for system approximation (Papadrakakis et al. 1998). Other types of models include radial basis functions (RBF) (Dyn et al. 1986, Fang and Horstemeyer 2006) multivariate adaptive regression splines (MARS) (Friedman 1991), least interpolating polynomials (De Boor and Ron 1990) and inductive learning (Langley and Simon 1995). A combination of polynomial functions and artificial neural networks has also been archived (Varadarajan et al. 2000). There is no conclusion about which model is definitely superior to the others. However, insights have been gained through a number of studies (Giunta and Watson 1998, Simpson et al. 2001). In recent years, Kriging models and related Guassian processes are intensively studied (Dey et al. 2015b, Qian et al. 2004, Martin and Simpson 2005, Li and Sudjianto 2012, Kleijnen and Van 2003, Daberkow and Mavris 2002). In general the Kriging models are more accurate for nonlinear problems but difficult to obtain and use because a global optimization process is applied to identify the maximum likelihood estimators. Kriging is also flexible in either interpolating the sample points or filtering noisy data. On the contrary, a polynomial model is easy to construct, clear on parameter sensitivity, and cheap to work with but is less accurate than the Kriging model (Jin et al. 2001). However, polynomial functions do not interpolate the sample points and are limited by the chosen function type. The RBF model, especially the multi-quadric RBF, can interpolate sample points and at the same time is easy to construct. It thus seems to reach a trade-off between Kriging and polynomials. Recently, a new model called Support Vector Regression (SVR) was used and tested (Clarke et al. 2005). SVR achieved high accuracy over all other metamodeling techniques including Kriging, polynomial, MARS, and RBF over a large number of test problems.

The Least Interpolating Polynomials use polynomial basis functions and also interpolate responses. They choose a polynomial basis function of "minimal degree" as described by Fang and Horstemeyer (2006) and hence are called "least interpolating polynomials". This type of metamodel deserves more study. In addition, Pérez et al. (2002) transformed the matrix of second-order terms of a quadratic polynomial model into the canonical form to reduce the number of terms. Messac and his team developed an extended RBF model (Mullur and Messac 2005) by adding extra terms to a regular RBF model to increase its flexibility, based on which an optimal model could be searched for. Turner and Crawford proposed a NURBS-based metamodel, which was applied only to low dimensional problems (Turner and Crawford 2005). If gradient information can be reliably and inexpensively obtained, it can be utilized in metamodeling (Morris et al. 1993). High dimensional model representation is found to be successfully applied in the problems related to optimization and system identification (Mukhopadhyay et al. 2015c, 2016d). A multipoint approximation (MPA) strategy has also received some attention (Wang et al. 1996, Rasmussen 1998, Shin and Grandhi 2001). MPA uses blending functions to combine multiple local approximations, and usually gradient information is used in metamodeling. Metamodels can also be constructed when design variables are modeled as fuzzy numbers (Madu 1995, Kleijnen 2004). Each metamodel type has its associated fitting method. For example, polynomial functions are usually fitted with the (weighted) least square method; the kriging method is fitted with the search for the Best Linear Unbiased Predictor (BLUP). Simpson et al. (1997) illustrated a detailed review on the equations and fitting methods for common metamodel types. In general computer experiments have very small random error which might be caused by the pseudorandom number generation or rounding (Kleijnen 2004). Giunta et al. (1994) found that numerical noises in computing the aerodynamic drag of High Speed Civil Transport (HSCT) caused many spurious local minima of the objective function. The problem was due to the discontinuous variations in calculating the drag by using the panel flow solver method. Madsen et al. (2000) stated that noises could come from the complex numerical modelling techniques. In case of physical or noisy computer experiments, it is found that Kriging and RBF are more sensitive to numerical noise than polynomial models (Jin et al. 2003). However, Kriging, RBF, and ANN could be modified to handle noises, assuming the signal to noise ratio is acceptable (Van Beers and Kleijnen 2004).

The different modelling methods and sampling techniques are summarized in Table 13.1. All of these techniques can be used to create approximations of existing computer analyses, and produce fast analysis modules for more efficient computation. These metamodeling techniques also yield insight into the functional relationship between input and output parameters. A designer's goal is usually to arrive at improved or robust solutions which are the values of design variables that best meet the design objectives. A search for these solutions usually relies on an optimization technique which generates and evaluates many potential solutions in the path toward design improvement; thus, fast analysis modules are an imperative.

In the later stages of design when detailed information about specific solutions is available, highly accurate analysis is essential. In the early stages of design, however, the focus is on generating, evaluating, and comparing potential conceptual configurations. The early stages of design are characterised by a large amount of information, often uncertain, which must be managed. To ensure the identification of a 'good' system configuration, a comprehensive search is necessary. In this case,

Table 13.1 Sampling Techniques and metamodeling methods.

1. Sampling techniques	a) Classic methods
	➢ Factorial design
	➢ Central composite
	➢ Box-behnken
	➢ Optimal designs
	➢ Plackett-burman
	b) Space-filling methods
	➢ Simple grids
	➢ Latin hypercube
	➢ Sobol sequence
	➢ Orthogonal arrays (Taguchi)
	➢ Hammersley sequence
	➢ Uniform designs
	➢ Minimax and maximin
	c) Hybrid methods
	d) Random or human selection
	e) Importance sampling
	f) Directional simulation
	g) Discriminative sampling
	h) Sequential or adaptive methods
2. Modelling methods	a) Polynomial regression (linear, quadratic, or higher)
	b) High dimensional model representation (HDMR) [cut-HDMR, RS-HDMR, GHDMR]
	c) Polynomial Chaos Expansion (PCE)
	d) Splines [linear, cubic, Non-Uniform Rational B-splines (NURBS)]
	e) Multivariate Adaptive Regression Splines (MARS)
	f) Gaussian process
	g) Kriging
	h) Radial Basis Functions (RBF)
	i) Least interpolating polynomials (Moving least square)
	j) Artificial Neural Network (ANN)
	k) Group Method of Data Handling - Polynomial Neural Network (GMDH - PNN)
	l) Knowledge base or decision tree
	m) Support vector machine (SVM)
	n) Weighted least squares regression
	o) Best linear unbiased predictor (BLUP)
	p) Multipoint approximation (MPA)
	q) Sequential or adaptive metamodeling
	r) Hybrid models

the trade-off between accuracy and efficiency may be appropriate. The creation of metamodels allows fast analysis, facilitating both comprehensive and efficient design space search at the expense of marginal loss of accuracy. Over the last few decades uncertainty quantification in complex structural systems including laminated composites has gained huge attention from the scientific community to realistically analyse and design the performance of the system (Dey et al. 2015a, 2016f). A careful review on the literature concerning uncertainty quantification of laminated composites reveals that there are distinctively three different approaches in probabilistic modelling of such structures: random variable approach (structural and material attributes are same throughout the composite including each layers for a particular sample of Monte Carlo simulation), layer-wise random variable approach (structural and material attributes are varied layer-wise for a particular sample of Monte Carlo simulation) and random field approach (structural and material attributes are varied spatially in all the dimensions for a particular sample of Monte Carlo simulation). Recently a non-probabilistic approach of fuzzy uncertainty propagation model has been proposed for composites that is applicable to the situation where explicit probability distribution of the material properties are not available (Dey et al. 2016d). A typical metamodel based algorithm for uncertainty quantification of a system is shown in Fig. 13.1. Performance assessment of different metamodels in uncertainty quantification of composite structures is particularly critical because of the fact that composite structures normally have a high dimensional input parameter space with complex and interdependent variables.

The present chapter focuses on stochastic structural dynamics of laminated composite plates by exhaustive comparative investigation of surrogate modelling

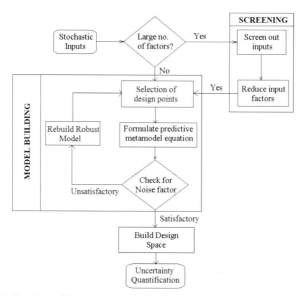

Fig. 13.1 Flowchart of Uncertainty Quantification (UQ) using metamodeling approach.

techniques for uncertainty quantification. To access the accuracy and computational efficiency obtained by different metamodels, a finite element model including the effects of transverse shear deformation based on Mindlin's theory in conjunction with a layer-wise random variable approach is employed to study the stochastic free vibration characteristics of graphite–epoxy composite cantilever plates. An eight noded isoparametric quadratic plate bending element with five degrees of freedom at each node is considered in the finite element formulation. Detail description of the finite element model for composite plates can be found in Chapter 3. Both individual and combined variation of stochastic input parameters are considered to account for the effect of dimensionality by employing the most prominent metamodeling techniques such as polynomial regression (PR), kriging, high dimensional model representation (HDMR), polynomial chaos expansion (PCE), artificial neural network (ANN), moving Least Square (MLS), support Vector Regression (SVR), multivariate adaptive regression splines (MARS), radial basis function (RBF) and polynomial neural network (PNN). For each of the surrogate modelling techniques, the rate of convergence with respect to traditional Monte Carlo simulation has been studied considering both low and high dimensional input parameter space. Different sampling techniques are used (namely 2^k factorial designs, central composite design, A-Optimal design, I-Optimal design, D-Optimal design, Taguchi's orthogonal array design, Box-Behnken design, Latin hypercube sampling and Sobol sequence) to construct the surrogate models. The sampling technique for a particular surrogate modelling method is chosen on the basis of available literature (as furnished in

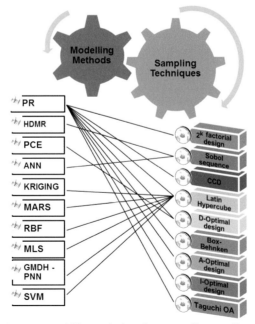

Fig. 13.2 Surrogate modelling methods and corresponding sampling techniques.

Fig. 13.2) to ensure best possible performance of each surrogate. As an integral part of this study, a comparative assessment of different design of experiment algorithms (2^k factorial designs, central composite design, A-Optimal design, I-Optimal design, D-Optimal design, Taguchi's orthogonal array design, Box-Behnken design) is presented considering the polynomial regression method. This chapter is organized hereafter as, Section 13.2: general overview and mathematical concepts of different metamodels considered in this study, Section 13.3: metamodel based stochastic free vibration analysis algorithm for laminated composite plates (detail theoretical formulation for the natural frequency analysis of composite plates can be found in Chapter 3), Section 13.4: results on comparative performance and discussion, Section 13.5: summary.

13.2 Mathematical formulation of metamodels

In general, the metamodels can be used as surrogates of the actual computationally expensive simulation or experimental model when a large number of evaluations are needed. The metamodels thus represent the results of the structural analysis (actual model evaluation) encompassing every possible combination of all input variables. From this, thousands of combinations of all design variables can be created and perform a pseudo analysis for each variable set, by simply adopting the corresponding predictive values. The formation of a metamodel is typically a three-step process. First step is selection of representative sample points (which are capable of acquiring information of the entire design space in an optimal manner), based on which the metamodel is constructed. In the second step, outputs or responses are evaluated corresponding to each sample point obtained. After obtaining the set of design points and corresponding responses, the last step is constructing the mathematical or statistical model to map input-output relationship. There exists both several sampling techniques (Myers and Montgomery 2002, Giunta et al. 2003, Santner et al. 2003, Koehler and Owen 1996) as well as metamodel formation methods (Slawomir and Yang 2011) as discussed in Section 13.1. One of the main concerns is selection of appropriate DOE method and metamodeling technique for a particular problem. All the sampling methods and metamodeling techniques have their unique properties and there exists no universal model that can be regarded as the best choice for all types of problems. Sampling method and metamodeling technique for a particular problem should be chosen depending on the complexity of the model, presence of noise in sampling data, nature and dimension (number) of input parameters, desired level of accuracy and computational efficiency. Before using a particular metamodeling technique it is essential to check it rigorously for its quality of fitting and prediction capability (Jin et al. 2001, Kim et al. 2009, Li et al. 2010). Description of the following metamodeling techniques can be found in the previous chapters of this book: polynomial regression (Chapter 4), kriging (Chapter 5), high dimensional model representation (Chapter 3), polynomial chaos expansion (Chapter 11), moving Least Square (Chapter 9), support Vector Regression (Chapter 8), multivariate adaptive regression splines (Chapter 10). Brief mathematical background of the remaining metamodeling techniques considered in this study (radial basis function, polynomial neural network and artificial neural network) is presented next.

13.2.1 Radial Basis Function (RBF)

Quadratic surrogates have the benefit of being easy to implement while still being able to model curvature of the underlying function. Another way to model curvature is to consider interpolating surrogates, which are linear combinations of nonlinear basis functions and satisfy the interpolation points. RBF is often used to perform the interpolation of scattered multivariate data (Buhmann 2000, Hardy 1971, Krishnamurthy 2003). The metamodel that appears in a linear combination of Euclidean distances can be expressed as

$$\hat{Y}(x) = \sum_{k=1}^{n} w_k \phi_k (X, x_k) \tag{13.1}$$

where, n is the number of sampling points, w_k is the weight determined by the least-squares method and $\phi_k(X, x_k)$ is the k-th basis function determined at the sampling point x_k. Various symmetric radial functions are used as the basis function. The radial function for RBF model can be expressed as (Maharshi et al. 2018),

$$R_f(x) = \exp\left(-\frac{(x-c)^T (x-c)}{r^2}\right) \qquad \text{(For Gaussian)} \tag{13.2}$$

$$R_f(x) = \sqrt{1 + \frac{(x-c)^T (x-c)}{r^2}} \qquad \text{(For multi-quadratic)} \tag{13.3}$$

$$R_f(x) = \frac{1}{\sqrt{1 + \frac{(x-c)^T (x-c)}{r^2}}} \qquad \text{(For inverse multi-quadratic)} \tag{13.4}$$

$$R_f(x) = \frac{1}{1 + \frac{(x-c)^T (x-c)}{r^2}} \qquad \text{(For Cauchy)} \tag{13.5}$$

The RBF method can be treated to be as an interpolator like Kriging. However, such an interpolation method has shortcomings in that the appearance of a metamodel varies significantly with the type of basis function and its internal parameters. In the present study, a Gaussian basis function is employed with the fixed parameter $r^2 = 1$. It should be noted that an RBF passes through all the sampling points exactly. This means that function values from the approximate function are equal to the true function values at the sampling points. This can be seen from the way that the coefficients are found. Therefore, it would not be possible to check RBF model fitness with ANOVA, which is a drawback of RBF.

13.2.2 Group Method of Data Handling—Polynomial Neural Network (GMDH-PNN)

In general, the Polynomial Neural Network (PNN) algorithm (Dey et al. 2016c, Mellit et al. 2010) is the advanced succession of the Group Method of Data Handling (GMDH) method wherein different linear, modified quadratic, cubic polynomials are used. By choosing the most significant input variables and polynomial order among various types of forms available, the best partial description (*PD*) can be obtained based on selection of nodes of each layer and generation of additional layers until the best performance is reached. Such methodology leads to an optimal PNN structure wherein the input-output data set can be expressed as

$$(X_i, Y_i) = (x_{1i}, x_{2i}, x_{3i}, \ldots\ldots x_{ni}, y_i) \text{ where } i = 1,2,3\ldots\ldots n \quad (13.6)$$

By computing the polynomial regression equations for each pair of input variable x_i and x_j and output Y of the object system which desires to modelling

$$Y = A + Bx_i + Cx_j + Dx_i^2 + Ex_j^2 + Fx_ix_j \text{ where } i, j = 1,2,3\ldots\ldots n \quad (13.7)$$

where A, B, C, D, E, F are the coefficients of the polynomial equation. This provides $n(n-1)/2$ high-order variables for predicting the output Y in place of the original n variables $(x_1, x_2, \ldots\ldots, x_n)$. After finding these regression equations from a set of input-output observations, we then find out which ones to save. This gives the best predicted collection of quadratic regression models. We now use each of the quadratic equations that we have just computed and generate new independent observations that will replace the original observations of the variables $(x_1, x_2, \ldots\ldots, x_n)$. From these new independent variables we will combine them exactly as we did before. That is, we compute all of the quadratic regression equations of Y versus these new variables. This will provide a new collection of $n(n-1)/2$ regression equation for predicting Y from the new variables, which in turn are estimates of Y from above equations. Now the best of the new estimates is selected to generate new independent variables from the selected equations to replace the old, and combine all pair of these new variables. This process is continued until the regression equations begin to have a poorer predictability power than did the previous ones. In other words, it is the time when the model starts to become overfitted. The estimated output $\hat{Y}$, Y can be further expressed as

$$\hat{Y} = \hat{f}(x_1, x_2, x_3, \ldots\ldots x_n)$$

$$= A_0 + \sum_{i=1}^{n} B_i x_i + \sum_{i=1}^{n}\sum_{j=1}^{n} C_{ij} x_i x_j + \sum_{i=1}^{n}\sum_{j=1}^{n}\sum_{k=1}^{n} D_{ijk} x_i x_j x_k + \ldots\ldots \quad (13.8)$$

$$\text{where } i, j, k = 1,2,3\ldots\ldots n$$

where $X(x_1, x_2, \ldots\ldots, x_n)$ is the input variables vector and $P(A_0, B_i, C_{ij}, D_{ijk}, \ldots\ldots)$ is vector of coefficients or weight of the Ivakhnenko polynomials. Components of the input vector X can be independent variables, functional forms or finite difference terms. This algorithm allows to find simultaneously the structure of model and model system output on the values of most significant inputs of the system. The following steps are to be performed for the framework of the design procedure of PNN (Oh et al. 2003):

Step 1: *Determination of input variables:* Define the input variables as $x_i = 1,2,3,......n$ related to output variable Y. If required, the normalization of input data is also completed.

Step 2: *Create training and testing data*: Create the input-output data set (n) and divide into two parts, namely, training data (n_{train}) and testing data (n_{test}) where $n = n_{train} + n_{test}$. The training data set is employed to construct the PNN model including an estimation of the coefficients of the partial description of nodes situated in each layer of the PNN. Next, the testing data set is used to evaluate the estimated PNN model.

Step 3: *Selection of structure*: The structure of PNN is selected based on the number of input variables and the order of *PD* in each layer. Two kinds of PNN structures, namely a basic PNN and a modified PNN structure are distinguished. The basic taxonomy for the architectures of PNN structure is furnished in Fig. 13.3.

Step 4: *Determination of number of input variables and order of the polynomial*: Determine the regression polynomial structure of a *PD* related to PNN structure. The input variables of a node from n input variables $x_1, x_2, x_3,......x_n$ are selected. The total number of *PD*s located at the current layer differs according to the number of the selected input variables from the nodes of the preceding layer. This results in $k = n!/(n! - r!)r!$ nodes, where r is the number of the chosen input variables. The choice of the input variables and the order of a *PD* itself help to select the best model with respect to the characteristics of the data, model design strategy, nonlinearity and predictive capability.

Step 5: *Estimation of coefficients of PD*: The vector of coefficients A_i is derived by minimizing the mean squared error between Y_i and $\hat{Y}_i$.

$$PI = \frac{1}{n_{train}} \sum_{i=1}^{n_{train}} (Y_i - \hat{Y}_i)^2 \tag{13.9}$$

where *PI* represents a criterion which uses the mean squared differences between the output data of original system and the output data of the model. Using the training data subset, this gives rise to the set of linear equations

$$Y = \sum_{i=1}^{n} X_i \, A_i \tag{13.10}$$

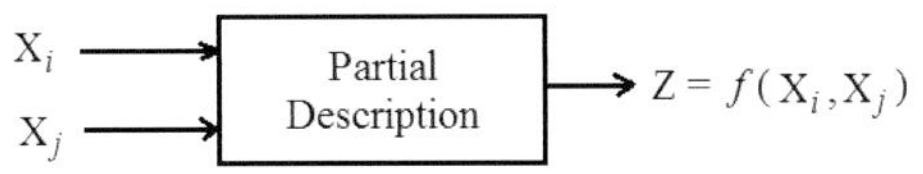

Fig. 13.3 Taxonomy for architectures of PNN.

The coefficients of the *PD* of the processing nodes in each layer are derived in the form

$$A_i = [X_i^T \, X_i]^{-1} \, X_i^T \, Y \tag{13.11}$$

where, $Y = [y_1, y_1, y_1, y_1, \ldots\ldots, y_{n_{train}}]^T$

$X_i = [x_{1i}, x_{2i}, x_{3i}\ldots\ldots X_{ki}\ldots\ldots X_{train\ i}]^T$

$X_{ki}^T = [x_{ki1} \; x_{ki2}\ldots\ldots x_{kin}\ldots x_{ki1}^m \; x_{ki2}^m \ldots\ldots\ldots x_{kin}^m]^T$

$A_i = [A_{0i} \; A_{1i} \; A_{2i}\ldots\ldots A_{n'i}]^T$

with the following notations i as the node number, k as the data number, n_{train} as the number of the training data subset, n as the number of the selected input variables, m as the maximum order, and n' as the number of estimated coefficients. This procedure is implemented repeatedly for all nodes of the layer and also for all layers of PNN starting from the input layer and moving to the output layer.

Step 6: *Selection of PDs with the best predictive capability*: Each *PD* is estimated and evaluated using both the training and testing data sets. Then we compare these values and choose several PDs, which give the best predictive performance for the output variable. Usually a predetermined number *W* of PDs is utilized.

Step 7: *Check the stopping criterion*: The stopping condition indicates that a sufficiently good PNN model is accomplished at the previous layer, and the modelling can be terminated. This condition reads as $PI_j > PI^*$ where PI_j is a minimal identification error of the current layer whereas PI^* denotes a minimal identification error that occurred at the previous layer.

Step 8: *Determination of new input variables for the next layer*: If PI_j (the minimum value in the current layer) has not been satisfied (so the stopping criterion is not satisfied), the model has to be expanded. The outputs of the preserved *PD*s serve as new inputs to the next layer.

13.2.3 Artificial Neural Network (ANN)

The fundamental processing element of ANN is an artificial neuron (or simply a neuron). A biological neuron receives inputs from other sources, combines them, generally performs a non-linear operation on the result, and then outputs the final result (Manohar and Divakar 2005). In the present study, the stochastic natural frequencies can be determined due to variability of input parameters. The ability of the ANNs, to recognize and reproduce the cause-effect relationships through training for the multiple input-output systems makes them efficient to represent even the most complex systems (Prareek et al. 2002). The main advantages of ANN as compared to response surface method (RSM) include:

a) ANN does not require any prior specification of suitable fitting function, and
b) It also has a universal approximation capability to approximate almost all kinds of non-linear functions including quadratic functions, whereas RSM is generally useful for quadratic approximations (Desai et al. 2008).

A multi-layer perceptron (MLP) based feed-forward ANN, which makes use of the back propagation learning algorithm, was applied for computational modelling. The network consists of an input layer, one hidden layer and an output layer. Each neuron acts firstly as a adding junction, summing together all incoming values. After that, it is filtered through an activation transfer function, the output of which is forwarded to the next layer of neurons in the network. The hyperbolic tangent was used as the transfer function for the input and hidden layer nodes. The reason behind employing the transfer function as logistic function or hyperbolic tangent (*tanh*) can be described as the logistic function generates the values nearer to zero if the argument of the function is substantially negative. Hence, the output of the hidden neuron can be made close to zero, and thus lowering the learning rate for all subsequent weights. Thus, it will almost stop learning. The *tanh* function, in the similar fashion, can generate a value close to −1.0, and thus will maintain learning. The algorithm used to train ANN in this study is quick propagation (QP). This algorithm is belonging to the gradient descent back-propagation. It has been reported in the literature that quick propagation learning algorithm can be adopted for the training of all the ANN models. The performance of the ANNs are statistically measured by the root mean squared error (*RMSE*), the coefficient of determination (R^2) and the absolute average deviation (AAD) obtained as follows:

$$RMSE = \sqrt{\frac{1}{n}\sum_{i=1}^{n}(Y_i - Y_{id})^2} \tag{13.12}$$

$$R^2 = 1 - \frac{\sum_{i=1}^{n}(Y_i - Y_{id})^2}{\sum_{i=1}^{n}(Y_{id} - Y_m)^2} \tag{13.13}$$

$$AAD = \left[\frac{1}{n}\sum_{i=1}^{n}\left|\frac{(Y_i - Y_{id})}{Y_{id}}\right|\right] \times 100 \tag{13.14}$$

where n is the number of points, Y_i is the predicted value, Y_{id} is the actual value, and Y_m is the average of the actual values.

13.3 Metamodel based stochastic natural frequency analysis

The stochasticity in layer-wise material properties of laminated composite plates, such as longitudinal elastic modulus, transverse elastic modulus, longitudinal shear modulus, transverse shear modulus, Poisson's ratio, mass density and geometric properties such as ply-orientation angle are considered as input parameters. In the present study, frequency domain feature (first three natural frequencies) is considered as output. It is assumed that the distribution of randomness of input parameters exists within a certain band of tolerance with their central deterministic mean values following a uniform random distribution. Both individual (ply-orientation angle) and combined layer-wise variation of input parameters are considered to account for the effect of low and high dimensional input parameter space in the surrogate based uncertainty quantification algorithms as follows

a) Variation of ply-orientation angle only: $\theta(\bar{\omega}) = \{\theta_1 \ \theta_2 \ \theta_3 ....\theta_i......\theta_l\}$

b) Combined variation of ply orientation angle, elastic modulus (longitudinal and transverse), shear modulus (longitudinal and transverse), Poisson's ratio and mass density:

$$g\{\theta(\bar{\omega}), \rho(\bar{\omega}), G_{12}(\bar{\omega}), G_{23}(\bar{\omega}), E_1(\bar{\omega})\} = \{ \ \Phi_1(\theta_1...\theta_l), \Phi_2(E_{1(1)}...E_{1(l)}), \Phi_3(E_{2(1)}...E_{2(l)}),$$
$$\Phi_4(G_{12(1)}..G_{12(l)}), \ \Phi_5(G_{23(1)}..G_{23(l)}), \ \Phi_6(\mu_1...\mu_l), \Phi_7(\rho_1..\rho_l)\}$$

where θ_i, $E_{1(i)}$, $E_{2(i)}$, $G_{12(i)}$, $G_{23(i)}$, μ_i and ρ_i are the ply orientation angle, elastic modulus along longitudinal and transverse direction, shear modulus along longitudinal direction, shear modulus along transverse direction, Poisson's ratio and mass density, respectively and '*l*' denotes the number of layer in the laminate. In the present study, $\pm 5°$ for ply orientation angle with subsequent $\pm 10\%$ tolerance for material properties from deterministic mean value are considered following standard industry practise for presenting results. The sampling technique for a particular surrogate modelling method is chosen on the basis of available literature to ensure best possible performance of each surrogate as furnished in Fig. 13.2. Figure 13.4 presents the working flowchart of stochastic natural frequency analysis using surrogate models.

13.4 Results and discussion

The previous investigations in the field of laminated composites have focused on the deterministic aspect of different static and dynamic responses over the last few decades (Qatu and Leissa 1991b, Splichal et al. 2015, Daniel 2016, Ochoa and Reddy 1992, Toranabene et al. 2016, Fantuzzi et al. 2015, Pandya and Kant 1988, Kant and Swaminathan 2000, Tornabene et al. 2011, 2013, Viola et al. 2013, Tornabene et al. 2014a, 2014b, 2015a, 2015b). A relatively new area of research is the quatification of uncertainty in laminated composite structures (Srinamula and Chryssanthopoulos 2009). The amount of research carried out in the field of uncertainty quantification of composite structures is insufficient owing to the computational intensiveness of such analyses. However, the stage of research on application of metamodels to achive computational efficeincy in the uncertainty quantification of composites is still in its infancy. As discussed in the introduction section, all the investigations on metamodel based uncertainty quantification of laminated composites are

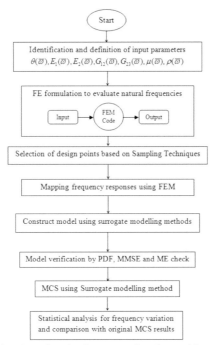

Fig. 13.4 Flowchart of stochastic surrogate based natural frequency analysis.

performed using a single metamodel. Thus there exists a strong rationale among the scientific community to investigate the relative performance of different metamodels, which is the focus of the present study. In the present chapter, a three layered graphite-epoxy symmetric angle-ply (45°/–45°/45°) laminated composite cantilever plate is considered to investigate the comparative performance of different metamodels on the basis of accuracy and computational efficiency. The length, width and thickness of the composite laminate considered in the present analysis are 1 m, 1 m and 5 mm, respectively. Material properties of graphite–epoxy composite is considered with deterministic value as $E_1 = 138.0$ GPa, $E_2 = 8.96$ GPa, $G_{12} = 7.1$ GPa, $G_{13} = 7.1$ GPa, $G_{23} = 2.84$ GPa, $\mu = 0.3$, $\rho = 3202$ kg/m³. An eight noded isoparametric quadratic plate bending element is considered for the present FEM approach. For full scale MCS, the number of original finite element analysis is same as the sampling size. In general, the performance function for complex composite structures is not available as an explicit function of the random design variables. The considered metamodels are employed to find the predictive and representative surrogates relating the first three natural frequencies to a number of input variables on a comparative basis. Thus the metamodels are used to determine the first three natural frequencies corresponding to given values of stochastic input variables, instead of time-consuming and computationally intensive finite element analysis. Table 13.2 presents the finite element mesh convergence study for non-dimensional fundamental natural frequencies of three layered graphite–epoxy

Table 13.2 Convergence study for non-dimensional fundamental natural frequencies $[\omega = \omega_n\, L^2 \sqrt{(\rho/E_1 t^2)}]$ of three layered ($\theta°/-\theta°/\theta°$) graphite-epoxy untwisted composite plates, $a/b = 1$, $b/t = 100$, considering $E_1 = 138$ GPa, $E_2 = 8.96$ GPa, $G_{12} = 7.1$ GPa, $v_{12} = 0.3$.

Ply angle, θ	Methods	Mesh sizes				Qatu and Leissa (1991a)
		4×4	6×6	8×8	10×10	
0°	Present FEM	1.0112	1.0133	1.0107	1.0040	1.0175
	ANSYS	1.0111	1.0130	1.0101	1.0035	
45°	Present FEM	0.4591	0.4603	0.4603	0.4604	0.4613
	ANSYS	0.4588	0.4600	0.4598	0.4696	
90°	Present FEM	0.2553	0.2567	0.2547	0.2542	0.2590
	ANSYS	0.2550	0.2565	0.2545	0.2541	

untwisted composite plates validated with the results obtained from ANSYS as well as Quatu and Leissa (1991a). Validation of the developed deterministic finite element code with the results of commercial packages like ANSYS caters to more confidence in the present analysis. Other than validation of the deterministic finite element formulation by computer code, ANSYS is also employed to validate the stochastic model of three layered angle-ply (45°/–45°/45°) composite cantilever plates corresponding to individual and combined variation of input parameters following a similar algorithm as presented Chapter 3. A stochastic convergence study is carried out with respect to mesh sizes (4×4), (6×6), (8×8), (10×10) and (12×12) as furnished in Fig. 13.5. To enumerate best predictive mesh convergence, a (6×6) mesh size is considered in the present comparative study corresponding to individual and combined variation of input parameters.

A comparative assessment of different design of experiment methods has been carried out in conjunction with the polynomial regression method. Figure 13.6 presents the error in percentage of mean and standard deviation of the first three natural frequencies for polynomial regression based stochastic analysis using different design of experiment algorithms with respect to MCS results for individual variation of ply orientation angle $\{\theta(\bar{\omega})\}$ and combined variation $\{\theta(\bar{\omega}),\ E_1(\bar{\omega}),\ E_2(\bar{\omega}),\ G_{12}(\bar{\omega}),\ G_{23}(\bar{\omega}),\ \mu(\bar{\omega}),\ \rho(\bar{\omega})\}$. D-optimal design method is observed to be the most computationally efficient and accurate compared to other design of experiment algorithms. The scatter plot and probability density function plot for first three natural frequencies corresponding to the combined variation of input parameters are furnished in Fig. 13.7 considering polynomial regression using the D-optimal design method along with traditional Monte Carlo simulation results. The figures corroborate excellent capability of the D-optimal design based polynomial regression method in prediction as well as characterizing the probabilistic features for the first three natural frequencies.

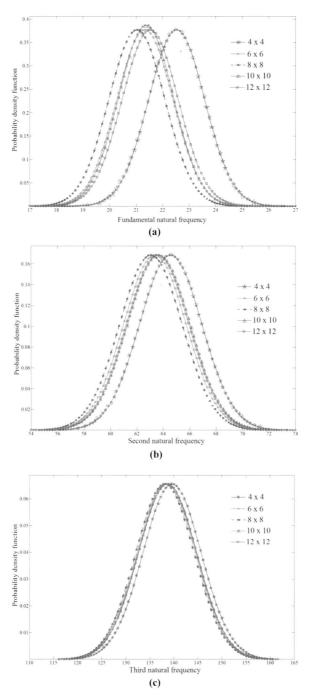

Fig. 13.5 Finite element mesh convergence study using ANSYS for combined stochasticity.

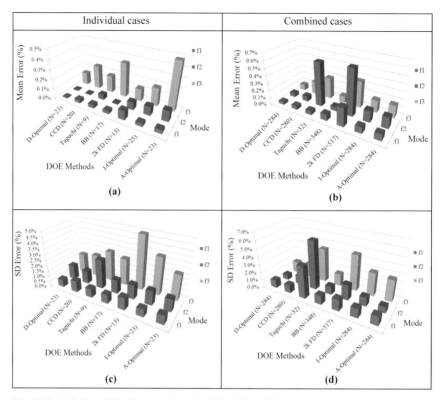

Fig. 13.6 (a–d) Error (%) of mean and standard deviation of first three natural frequencies between polynomial regression method with different design of experiment methods and MCS results for individual variation of ply orientation angle and combined variation (f_1 f_2 and f_3 denote first three modes of vibration).

Figure 13.8 presents the percentage error of mean and standard deviation of the first three natural frequencies between the considered surrogate modelling methods and MCS results with respect to different sample sizes for individual variation of ply orientation angle [$\theta(\bar{\omega})$], while Fig. 13.9 indicates the error in percentage for mean and standard deviation of the first three natural frequencies between surrogate modelling methods and MCS results with respect to different sample sizes for combined variation of all stochastic input parameters {$\theta(\bar{\omega})$, $E_1(\bar{\omega})$, $E_2(\bar{\omega})$, $G_{12}(\bar{\omega})$, $G_{23}(\bar{\omega})$, $\mu(\bar{\omega})$, $\rho(\bar{\omega})$}. In general, for all cases, the sparsity of the first three natural frequencies for combined variation of input parameters are found to be higher than that of individual variation of input parameter, as expected. As the sample size increases, the percentage of error of mean and standard deviation of the first three natural frequencies between surrogate modelling methods and MCS results are found to reduce irrespective of modelling methods. An exhaustive study is carried out to enumerate the best minimum sample size required to construct the metamodel for all tested modelling methods corresponding to suitable sampling

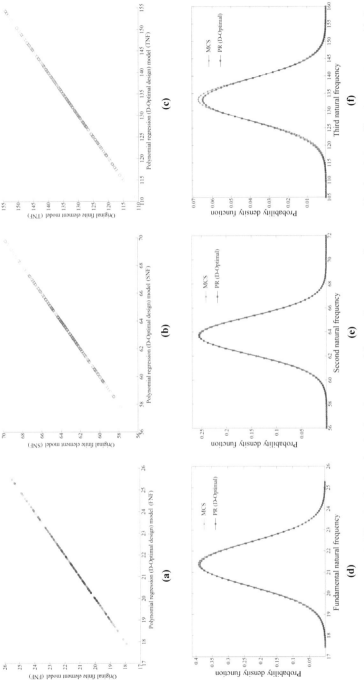

Fig. 13.7 Scatter diagram and probability density function for first three natural frequencies corresponding to combined variation of input parameters considering polynomial regression using D-optimal design method.

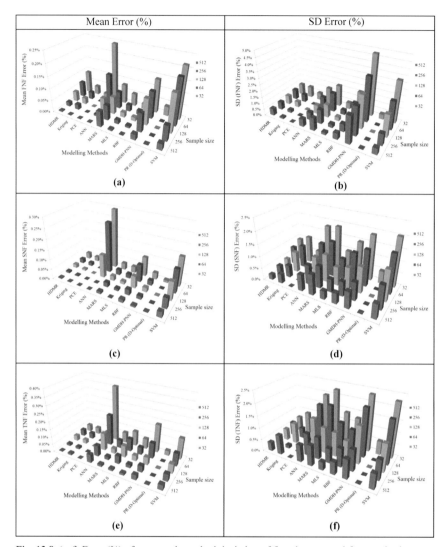

Fig. 13.8 (a–f) Error (%) of mean and standard deviation of first three natural frequencies between surrogate modelling methods and MCS results with respect to different sample sizes for individual variation of ply orientation angle [$\theta(\bar{\omega})$] for angle-ply (45°/–45°/45°) composite plates.

techniques. Polynomial regression with D-optimal design method is found to require least number of samples for suitable fitment of surrogates corresponding to individual as well as combined variation cases. In contrast, Group method of data handling—Polynomial neural network (GMDH-PNN) method and Support Vector Regression (SVR) are observed to require the maximum number of sample for individual variation while Artificial neural network (ANN) method is found to

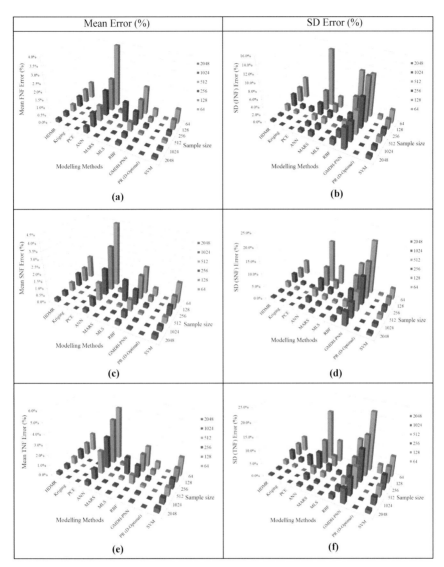

Fig. 13.9 (a–f) Error (%) of mean and standard deviation of first three natural frequencies between surrogate modelling methods and MCS results with respect to different sample sizes for combined variation $\{\theta(\bar{\omega}),\ E_1(\bar{\omega}),\ E_2(\bar{\omega}),\ G_{12}(\bar{\omega}),\ G_{23}(\bar{\omega}),\ \mu(\bar{\omega}),\ \rho(\bar{\omega})\}$ for angle-ply (45°/–45°/45°) composite plates.

need the maximum number of samples for combined variation compared to other tested modelling methods. Table 13.3 presents the minimum number of samples required for different tested metamodeling methods to obtain reasonable accuracy in terms of mean and standard deviation for the layer-wise stochastic analysis of

Table 13.3 Minimum sample size required for different metamodeling methods.

Sl. No.	Metamodeling methods	Minimum number of samples required for model formation	
		Individual variation	Combined variation
1.	High dimensional model representation (HDMR)	256	512
2.	Kriging method	128	256
3.	Polynomial chaos expansion (PCE)	64	128
4.	Artificial neural network (ANN)	256	2048
5.	Multivariate adaptive regression splines (MARS)	64	128
6.	Moving Least Square (MLS)	128	512
7.	Radial basis function (RBF)	64	1024
8.	Group method of data handling - Polynomial neural network (GMDH-PNN)	512	1024
9.	Polynomial regression (by D-optimal)	32	64
10.	Support Vector Regression (SVR)	512	1024

composite plate for both individual and combined variation. A clear idea about the performance of different metamodeling techniques from the viewpoint of computational efficiency can be perceived for both low and relatively higher dimensional input parameter space. The probability distributions obtained by using ten different metamodeling methods along with traditional MCS for first three natural frequencies corresponding to individual (only ply angle) and combined variation of input parameters are shown in Fig. 13.10 and Fig. 13.11, respectively. From the viewpoint of accuracy in probabilistic characterization with respect to traditional MCS, performances are comparatively worse for ANN and SVM in case of individual stochasticity and SVM and PCE in case of combined stochasticity respectively. Other metamodels are found to obtain satisfactory results, polynomial regression based on D-optimal design being the best. ANN performs better for the higher dimensional input parameter space (combined case), even though it requires more samples compared to most of the other methods. However, it can be noted that the results presented in Fig. 13.10 and Fig. 13.11 are obtained using the corresponding sample size provided in Table 13.3, which is finalized on the basis of error analysis for mean and standard deviation. The trade-off between desired level of accuracy and computational efficiency should be judged based on specific requirements for a particular problem. The results presented in this chapter along with the in-depth previous comparative investigations on response surface method (Mukhopadhyay 2015b) and kriging model variants (as presented in Chapter 12) can provide a reasonably composed guideline for choosing sampling method and surrogate modelling technique for future applications. However, it should always

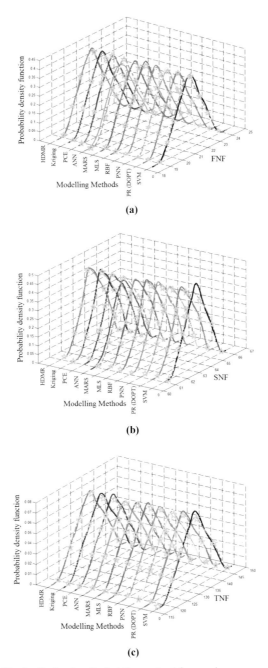

(a)

(b)

(c)

Fig. 13.10 Probability density function for first three natural frequencies corresponding to individual variation of input parameters (colour code: ▬▬▬ MCS and specification of other colours are indicated in the figures).

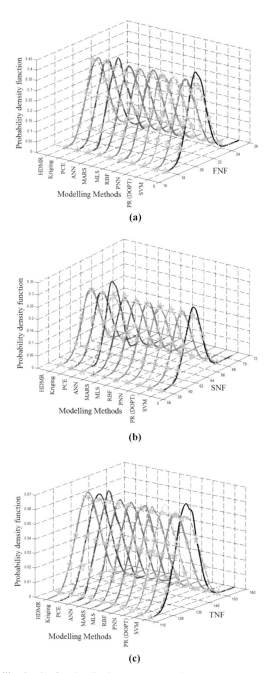

Fig. 13.11 Probability density function for first three natural frequencies corresponding to combined variation of input parameters (colour code: ——— MCS and specification of other colours are indicated in the figures).

be noted that surrogate modelling being a problem-specific technique, it is quite difficult to identify a single surrogate model that works best for all problems. Thus future researches are necessary to investigate the comparative performances of different surrogates for other types of problems in structural mechanics from different angles such as non-linearity, dimension of input parameter space and the effect of correlation among them, noise, etc. The present chapter will serve as an important reference for such future investigations.

13.5 Summary

This chapter presents a critical comparative assessment of different metamodels (such as polynomial regression, kriging, high dimensional model representation, polynomial chaos expansion, artificial neural network, moving least square, support vector regression, multivariate adaptive regression splines, radial basis function and polynomial neural network) for stochastic natural frequency analysis of composite laminates from the viewpoint of accuracy (with respect to traditional Monte Carlo simulation) and computational efficiency. First three stochastic natural frequencies of a laminated composite plate are considered for individual and combined variation of layer-wise random input parameters. A comparative investigation is presented on different design of experiment methods (such as 2^k factorial designs, central composite design, A-Optimal design, I-Optimal, D-Optimal, Taguchi's orthogonal array design, Box-Behnken design) in conjunction with the polynomial regression technique. D-optimal design is found to obtain the most satisfactory results compared to others. For each of the metamodeling techniques, the rate of convergence with respect to traditional Monte Carlo simulation has been studied considering both low and high dimensional input parameter space. Probabilistic descriptions of the natural frequencies obtained on the basis of different metamodeling techniques are presented along with direct Monte Carlo simulation results.

Polynomial regression with D-optimal design method is found to be most computationally cost effective for suitable fitment of surrogates corresponding to individual as well as combined variation of input parameters. Group method of data handling—polynomial neural network (GMDH-PNN) method and support vector regression (SVR) are observed to be least computationally efficient for individual variation while the artificial neural network (ANN) method is found to be the most computationally expensive for combined variation compared to other metamodels. From the viewpoint of accuracy in probabilistic characterization with respect to traditional MCS, performances are comparatively worse for ANN and SVM in case of individual stochasticity while SVM and PCE shows relatively less accuracy in case of combined stochasticity. On the basis of the stochastic results presented in this chapter, a clear idea about the performance of different metamodeling techniques from the viewpoint of accuracy and computational efficiency can be perceived for both low and relatively higher dimensional input parameter space.

Overall, this book deals with the aspect of different forms of uncertainty associated with composites as well as critical evaluation of the efficient metamodel-based uncertainty propagation approaches. For modelling uncertainty, a layer-wise random variable based approach has been adopted in this book. Further investigations are needed on development of metamodeling approaches for composites considering random field based modelling and multi-scale progressive damage analyses in a stochastic framework. This book can serve as a valuable reference in such future endeavours.

Bibliography

Adali, S. and I.U. Cagdas. 2011. Failure analysis of curved composite panels based on first-ply and buckling failures. *Procedia Engineering* 10: 1591–1596.

Adhikari, S. and K.H. Haddad. 2014. A spectral approach for fuzzy uncertainty propagation in finite element analysis. *Fuzzy Sets and Systems* 243: 1–24.

Afeefa, S, W.G. Abdelrahman, T. Mohammad and S. Edward. 2008. Stochastic finite element analysis of the free vibration of laminated composite plates. *Computational Mechanics* 41: 495–501.

Aguib, S., A. Nour, H. Zahloul, G. Bossis, Y. Chevalier and P. Lançon. 2014. Dynamic behavior analysis of a magnetorheological elastomer sandwich plate. *International Journal of Mechanical Sciences* 87: 118–136.

Ahmadi, M., F. Vahabzadeh, B. Bonakdarpour, E. Mofarrah and M. Mehranian. 2005. Application of the central composite design and response surface methodology to the advanced treatment of olive oil processing wastewater using Fenton's peroxidation. *Journal of Hazardous Materials*, 123(1-3): 187–195.

Al-Assaf, Y. and H. El Kadi. 2001. Fatigue life prediction of unidirectional glass fiber/epoxy composite laminae using neural networks. *Composite Structures* 53(1): 65–71.

Al-Assaf, Y. and H. El Kadi. 2007. Fatigue life prediction of composite materials using polynomial classifiers and recurrent neural networks. *Composite Structures* 77(4): 561–569.

Alibeigloo, A. and M. Alizadeh. 2015. Static and free vibration analyses of functionally graded sandwich plates using state space differential quadrature method. *European Journal of Mechanics - A/Solids* 54: 252–266.

Alkhateb, H., A. Al-Ostaz and K.I. Alzebdeh. 2009. Developing a stochastic model to predict the strength and crack path of random composites. *Composites Part B: Engineering* 40(1): 7–16.

Allahyari, H., I.M. Nikbin, S. Rahimi and A. Allahyari. 2018. Experimental measurement of dynamic properties of composite slabs from frequency response. *Measurement* 114: 150–161.

Allegri, G., S. Corradi and M. Marchetti. 2006. Stochastic analysis of the vibrations of an uncertain composite truss for space applications. *Composites Science and Technology* 66: 273–282.

Altenbach, H. 1998. Theories for laminated and sandwich plates. *Mech Compos Mater* 34(3): 243–52.

Altmann, F., J.U. Sickert, V. Mechtcherine and M. Kaliske. 2012. A fuzzy-probabilistic durability concept for strain-hardening cement-based composites (SHCCs) exposed to chlorides: Part 1: Concept development. *Cement and Concrete Composites* 34(6): 754–762.

An, X., B.C. Khoo and Y. Cui. 2017. Nonlinear aeroelastic analysis of curved laminated composite panels. *Composite Structures* 179: 377–414.

Andrzej, T., P. Ratko and K. Predrag. 2011. Influence of transverse shear on stochastic instability of symmetric cross-ply laminated plates. *Probabilistic Engineering Mechanics* 26: 454–460.

Angelikopoulos, P., C. Papadimitriou and P. Koumoutsakos. 2015. X-TMCMC: Adaptive kriging for Bayesian inverse modeling. *Computer Methods in Applied Mechanics and Engineering* 289: 409–428.

Ankenmann, B., B.L. Nelson and J. Staum. 2010. Stochastic kriging for simulation metamodeling. *Operations Research* 58(2): 371–382.

Anlas, G. and G. Göker. 2001. Vibration analysis of skew fibre-reinforced composite laminated plates. *Journal of Sound and Vibration* 242(2): 265–276.

ANSYS, Release 14.5, ANSYS Inc. 2012.

António, C.C. and L.N. Hoffbauer. 2007. Uncertainty analysis based on sensitivity applied to angle-ply composite structures. *Reliability Engineering & System Safety* 92(10): 1353–1362.

António, C.C. and L.N. Hoffbauer. 2013. Uncertainty assessment approach for composite structures based on global sensitivity indices. *Composite Structures* 99: 202–212.

Anuja, G. and R. Katukam. 2015. Parametric studies on the cutouts in heavily loaded aircraft beams. *Materials Today: Proceedings* 2(4-5): 1568–1576.

Arian Nik, M., K. Fayazbakhsh, D. Pasini and L. Lessard. 2012. Surrogate-based multiobjective optimization of a composite laminate with curvilinear fibers. *Compos. Struct.* 94: 2306–23.

Arian Nik, M., K. Fayazbakhsh, D. Pasini and L. Lessard. 2014. Optimization of variable stiffness composites with embedded defects induced by Automated Fiber Placement. *Composite Structures* 107: 160–6.

Arjangpay, A., A. Darvizeh, M. Yarmohammad Tooski and R. Ansari. 2018. An experimental and numerical investigation on low velocity impact response of a composite structure inspired by dragonfly wing configuration. *Composite Structures* 184: 327–336.

Arregui-Mena, J.D., L. Margetts and P.M. Mummery. 2016. Practical application of the stochastic finite element method. *Archives of Computational Methods in Engineering* 23(1): 171–190.

Artero-Guerrero, J.A., J. Pernas-Sánchez, J. Martín-Montal, D. Varas and J. López-Puente. 2017. The influence of laminate stacking sequence on ballistic limit using a combined Experimental/FEM/Artificial Neural Networks (ANN) methodology. *Composite Structures*.

Arunraj, N.S., S. Mandal and J. Maiti. 2013. Modeling uncertainty in risk assessment: an integrated approach with fuzzy set theory and Monte Carlo simulation. *Accident Analysis & Prevention* 55: 242–255.

Aslan, N. 2008. Application of response surface methodology and central composite rotatable design for modeling and optimization of a multi-gravity separator for chromite concentration. *Powder Technology* 185(1): 80–86.

Au, S.K. and J.L. Beck. 1999. A new adaptive importance sampling scheme for reliability calculations. *Structural Safety* 21: 135–138.

Aydogdu, M. and M.C. Ece. 2006. Buckling and vibration of non-ideal simply supported rectangular isotropic plates. *Mech. Res. Commun* 33: 532–40.

Aydogdu, M. 2009. A new shear deformation theory for laminated composite plates. *Composite Structures* 89(1): 94–101.

Babuška, I., M. Motamed and R. Tempone. 2014. A stochastic multiscale method for the elastodynamic wave equation arising from fiber composites. *Computer Methods in Applied Mechanics and Engineering* 276: 190–211.

Babuška, I. and S. Silva Renato. 2014. Dealing with uncertainties in engineering problems using only available data. *Computer Methods in Applied Mechanics and Engineering* 270: 57–75.

Bahadori, R., H. Gutierrez, S. Manikonda and R. Meinke. 2018. A mesh-free Monte-Carlo method for simulation of three-dimensional transient heat conduction in a composite layered material with temperature dependent thermal properties. *International Journal of Heat and Mass Transfer* 119: 533–541.

Baharlou, B. and A.W. Leissa. 1987. Vibration and buckling of generally laminated composite plates with arbitary edge conditions. *Int. J. Mechanical Sciences* 29(8): 545–555.

Bakshi, K. and D. Chakravorty. 2014. First ply failure study of thin composite conoidal shells subjected to uniformly distributed load. *Thin-Walled Structures* 76: 1–7.

Balokas, G., S. Czichon and R. Rolfes. 2017. Neural network assisted multiscale analysis for the elastic properties prediction of 3D braided composites under uncertainty. *Composite Structures*.

Banat, D. and R.J. Mania. 2018. Progressive failure analysis of thin-walled fibre metal laminate columns subjected to axial compression. *Thin-Walled Structures* 122: 52–63.

Banerjee, J.R. 2001a. Explicit analytical expressions for frequency equation and mode shapes of composite beams. *International Journal of Solids and Structures* 38(14): 2415–2426.

Banerjee, J.R. 2001b. Frequency equation and mode shape formulae for composite Timoshenko beams. *Composite Structures* 51(4): 381–388.

Baran, I., K. Cinar, N. Ersoy, R. Akkerman and J.H. Hattel. 2016. A review on the mechanical modeling of composite manufacturing processes. *Archives of Computational Methods in Engineering*, DOI 10.1007/s11831–016–9167–2.

Barthelemy, J.-F. M. and R.T. Haftka. 1993. Approximation concepts for optimum structural design—a review. *Structural Optimization* 5(3): 129–144.

Bathe, K.J. 1990. Finite Element Procedures in Engineering Analysis, *PHI*, New Delhi.

Batou, A. and C. Soize. 2013. Stochastic modeling and identification of an uncertain computational dynamical model with random fields properties and model uncertainties. *Archive of Applied Mechanics* 83(6): 831–848.

Bayer, V. and C. Bucher. 1999. Importance sampling for first passage problems of nonlinear structures. *Probab. Eng. Mech.* 14: 27–32.

Bayraktar, H. and S.F. Turalioglu. 2005. A Kriging-based approach for locating a sampling site—in the assessment of air quality. *Stochastic Environmental Research and Risk Assessment* 19(4): 301–305.

Bert, C.W. and V. Birman. 1988. Parametric instability of thick orthotropic circular cylindrical shells. *Acta Mechanica* 71: 61–76.

Bert, C.W. 1991. Literature review: research on dynamic behavior of composite and sandwich plates – V: Part II. *Shock Vib Digest* 23: 9–21.

Beslin, O. and J.L. Guyader. 1996. The use of an "ectoplasm" to predict free vibrations of plates with cut-outs. *Journal of Sound and Vibrations*, 191(5): 935–954.

Bezerra, E.M., A.C. Ancelotti, L.C. Pardini, J.A.F.F. Rocco, K. Iha and C.H.C. Ribeiro. 2007. Artificial neural networks applied to epoxy composites reinforced with carbon and E-glass fibers: Analysis of the shear mechanical properties. *Materials Science and Engineering: A* 464(1): 177–185.

Bhat, M.R. 2015. Probabilistic stress variation studies on composite single lap joint using Monte Carlo simulation. *Composite Structures* 121: 351–361.

Bisagni, C. 1999. Experimental buckling of thin composite cylinders in compression. *AIAA Journal* 37(2): 276–278.

Bisagni, C. and L. Lanzi. 2002. Post-buckling optimisation of composite stiffened panels using neural networks, *Compos Struct*. 58: 237–47.

Biscay, L.R, D.G. Camejo, J.–M. Loubes and A.L. Muñiz. 2013. Estimation of covariance functions by a fully data-driven model selection procedure and its application to Kriging spatial interpolation of real rainfall data. *Stat. Methods Appt.* 23: 149–174.

Blachut, J. 2004. Buckling and first ply failure of composite toroidal pressure hull. *Computers & Structures* 82(23): 1981–1992.

Blom, A.W., P.B. Stickler and Z. Gürdal. 2010. Optimization of a composite cylinder under bending by tailoring stiffness properties in circumferential direction. *Compos. Part B Eng.* 41: 157–65.

Bohlooli, H., A. Nazari, G. Khalaj, M.M. Kaykha and S. Riahi. 2012. Experimental investigations and fuzzy logic modeling of compressive strength of geopolymers with seeded fly ash and rice husk bark ash. *Composites Part B: Engineering* 43(3): 1293–1301.

Bolotin, V.V. 1964. The dynamic stability of elastic system, San Francisco: Holden-Day.

Booker, A.J., J.E. Jr. Dennis, P.D. Frank, D.B. Serafini, V. Torczon and M.W. Trosset. 1999. A rigorous framework for optimization of expensive functions by surrogates. *Struct. Optim.* 17: 1–13.

Borkowski, J.J. 1995. Spherical prediction-variance properties of central composite and Box-Behnken designs. *Technometrics* 37: 399–410.

Bowles, D.E. and S.S. Tompkins. 1989. Prediction of coefficients of thermal expansion for unidirectional composite. *J. Compos. Mater.* 23: 370–381.

Brampton, C.J., D.N. Betts, C.R. Bowen and H.A. Kim. 2013. Sensitivity of bistable laminates to uncertainties in material properties, geometry and environmental conditions. *Composite Structures* 102: 276–286.

Breitkpf, P., H. Naceur, A. Rassineux and P. Villon. 2005. Moving least squares response surface approximation: formulation and metal forming applications. *Computers and Structures* 83: 1411–1428.

Bruno, D., S. Lato and R. Zinno. 1993. Nonlinear analysis of doubly curved composite shells of bimodular material. *Composites Engineering* 3(5): 419–435.

Buhmann, M.D. 2000. Radial basis functions. *Acta Numerica* 1–38.

Bui, T.Q., M.N. Nguyen and C. Zhang. 2011. An efficient mesh-free method for vibration analysis of laminated composite plates. *Comput. Mech.* 48: 175–93.

Carpenter, W.C. 1993. Effect of design selection on response surface performance. NASA Contractor Report 4520.

Carr, J.R. 1990. UVKRIG: A FORTRAN-77 program for Universal Kriging. *Computers & Geosciences* 16(2): 211–236.

Carrera, E. 2003. Theories and finite elements for multilayered plates and shells: a unified compact formulation with numerical assessment and benchmarking. *Archives of Computational Methods in Engineering* 10(3): 215–296

Carrera, E. and S. Brischetto. 2009. A survey with numerical assessment of classical and refined theories for the analysis of sandwich plates. *Appl. Mech. Rev.* 62: 1–17.

Carrera, E., F.A. Fazzolari and L. Demasi. 2011. Vibration analysis of anisotropic simply supported plates by using variable kinematic and Rayleigh–Ritz method. *J. Vib. Acoust.* 133: 1–16.

Carrere, N., Y. Rollet, F.H. Leroy and J.F. Maire. 2009. Efficient structural computations with parameters uncertainty for composite applications. *Composites Science and Technology* 69: 1328–1333.

Chakrabarti, A. and A. Sheikh. 2004. Vibration of laminate-faced sandwich plate by a new refined element, *J. Aerosp. Eng.* 17(3): 123–134.

Chakrabarti, A., A.H. Sheikh, M. Griffith and D.J. Oehlers. 2013. Dynamic response of composite beams with partial shear interactions using a higher order beam theory. *Journal of Structural Engineering* 139(1): 47–56.

Chakraborty, D. 2005. Artificial neural network based delamination prediction in laminated composites. *Materials & Design* 26(1): 1–7.

Chakravorty, D., J.N. Bandyopadhyay and P.K. Sinha. 1995. Free vibration analysis of point supported laminated composite doubly curved shells—A finite element approach. *Computers & Structures* 54(2): 191–198.

Chalak, H.D., A. Chakrabarti, A.H. Sheikh and M.A. Iqbalb. 2015. Stability analysis of laminated soft core sandwich plates using higher order zig-zag plate theory. *Mechanics of Advanced Materials and Structures* 22: 897–907.

Chandrashekhar, M. and R. Ganguli. 2010. Nonlinear vibration analysis of composite laminated and sandwich plates with random material properties. *International Journal of Mechanical Sciences* 52(7): 874–891.

Chandrashekhara, K. 1989. Free vibrations of anisotropic laminated doubly curved shells. *Computers & Structures* 33: 435–440.

Chang, Y.-T., S-J. Ho and B.-S. Chen. 2014. Robust stabilization design of nonlinear stochastic partial differential systems: Fuzzy approach. *Fuzzy Sets and Systems* 248: 61–85.

Chang, R.R. 2000. Experimental and theoretical analyses of first-ply failure of laminated composite pressure vessels. *Composite Structures* 49(2): 237–243.

Charmpis, D.C., G.I. Schueller and M.F. Pellissetti. 2007. The need for linking micromechanics of materials with stochastic finite elements: a challenge for materials science. *Comput. Mater. Sci.* 41(1): 27–37.

Chaudhuri, R.A. and K.R. Abuarja. 1988. Extract solution of shear flexible doubly curved antisymmetric angle ply shells. *Int. J. Engineering Sciences* 26(6): 587–604.

Chen, J.-F., E.V. Morozov and K. Shankar. 2014. Simulating progressive failure of composite laminates including in-ply and delamination damage effects. *Composites Part A: Applied Science and Manufacturing* 61: 185–200.

Chen, L.-W and L.-Y. Chen. 1989. Thermal buckling behaviour of laminated composite plates with temperature-dependent properties. *Compos Struct.* 13(4): 275–87.

Chen, L.W. and J.Y. Yang. 1990. Dynamic stability of laminated composite plates by the finite element method. *Computers & Structures* 36(5): 845–851.

Chen, N.-Z. and S.C. Guedes. 2008. Spectral stochastic finite element analysis for laminated composite plates. *Computer Methods in Applied Mechanics and Engineering* 197(51): 4830–4839.

Chen, V.C.P., K.L. Tsui, R.R. Barton and M. Meckesheimer. 2006. A review on design, modeling and applications of computer experiments. *IIE Transactions* 38: 273–291.

Chen, W. 1995. A Robust Concept Exploration Method for Configuring Complex System, Ph.D. Dissertation Thesis, Mechanical Engineering, Georgia Institute of Technology, Atlanta, GA.

Chen, X., B.L. Nelson and K.-K. Kim. 2012. Stochastic kriging for conditional value-at-risk and its sensitivities. pp 1–12. *In: Proc. Title Proc. 2012 Winter Simul. Conf. IEEE.*

Chen, X., B.E. Ankenman and B.L. Nelson. 2013. Enhancing stochastic kriging metamodels with gradient estimators. *Oper. Res.* 61: 512–528.

Chen, X. and K.-K. Kim. 2014. Stochastic kriging with biased sample estimates. *ACM Trans Model Comput. Simul.* 24: 1–23.

Chen, X. and Z. Qiu. 2018. A novel uncertainty analysis method for composite structures with mixed uncertainties including random and interval variables. *Composite Structures* 184: 400–410.

Chen, X.L., G.R. Liu and S.P. Lim. 2003. An element free Galerkin method for the free vibration analysis of composite laminates of complicated shape. *Composite Structures* 59(2): 279–289.

Chen, Y., S. Hou, K. Fu, X. Han and L. Ye. 2017. Low-velocity impact response of composite sandwich structures: Modelling and experiment. *Composite Structures* 168: 322–334.

Cheng, J. and X. Ru-cheng. 2007. Probabilistic free vibration analysis of beams subjected to axial loads. *Advances in Engineering Software* 38(1): 31–38.

Cheng, X., J. Zhang, J. Bao, B. Zeng, Y. Cheng and R. Hu. 2018. Low-velocity impact performance and effect factor analysis of scarf-repaired composite laminates. *International Journal of Impact Engineering* 111: 85–93.

Cherkassky, V. and Y. Ma. 2004. Practical selection of SVM parameters and noise estimation for SVM regression. *Neural Network* 17: 113–26.

Chiachio, M., J. Chiachio and G. Rus. 2012. Reliability in composites—a selective review and survey of current development. *Composites Part B: Engineering* 43(3): 902–913.

Chiang, M.Y.M., X. Wang, C.R. Schultheisz and J. He. 2005. Prediction and three-dimensional Monte-Carlo simulation for tensile properties of unidirectional hybrid composites. *Composites Science and Technology* 65(11): 1719–1727.

Cho, M. and J.S. Kim. 2000. A post-process method for laminated shells with a doubly curved nine-noded finite element. *Composites Part B: Engineering* 31(1): 65–74.

Choi, H. and M. Kang. 2014. Optimal sampling frequency for high frequency data using a finite mixture model. *Journal of Korean Statistical Society* 43(2): 251–262.

Choi, I.H. 2017. Low-velocity impact response analysis of composite pressure vessel considering stiffness change due to cylinder stress. *Composite Structures* 160: 491–502.

Choi, I.H. 2018. Finite element analysis of low-velocity impact response of convex and concave composite laminated shells. *Composite Structures* 186: 210–220.

Choi, S. and R.V. Grandhi and R.A. Canfield. 2004. Structural reliability under non-gaussian stochastic behavior. *Computers & Structures* 82: 1113–1121.

Chow, S.T., K.M. Liew and K.Y. Lam. 1992. Transverse vibration of symmetrically laminated rectangular composite plates. *Composite Structures* 20(4): 213–226.

Chowdhury, R. and B.N. Rao. 2009. Assessment of high dimensional model representation techniques for reliability analysis. *Probabilistic Engineering Mechanics* 24: 100–115.

Chowdhury, R. and S. Adhikari. 2012. Fuzzy parametric uncertainty analysis of linear dynamical systems: A surrogate modeling approach. *Mechanical Systems and Signal Processing* 32: 5–17.

Chowdhury, R., B.N. Rao and A.M. Prasad. 2009. High dimensional model representation for structural reliability analysis. *Communications in Numerical Methods in Engineering* 25(4): 301–337.

Chowdhury, N.T., J. Wang, W.K. Chiu and W. Yan. 2016. Matrix failure in composite laminates under tensile loading. *Composite Structures* 135: 61–73.

Clarke, S.M., J.H. Griebsch and T.W. Simpson. 2005. Analysis of support vector regression for approximation of complex engineering analyses, transactions of ASME. *Journal of Mechanical Design* 127(6): 1077–1087.

Clarson, B.L. and R.D. Ford. 1962. The response of a typical aircraft structure to jet noise. *Journal of the Royal Aeronautical Society* 66: 31–40.

Clemens, M. and J. Seifert. 2015. Dimension reduction for the design optimization of large scale high voltage devices using co-kriging surrogate modeling. *IEEE Trans Magn.* 51: 1–4.

Cook, R.D., D.S. Malkus and M.E. Plesha. 1989. *Concepts and Applications of Finite Element Analysis.* New York, John Wiley and Sons.

Corr, R.B. and A. Jennings. 1976. A simultaneous iteration algorithm for symmetric eigenvalue problems. *Int. J. for Numerical Methods in Engineering* 10(3): 647–663.

Couckuyt, I., A. Forrester, D. Gorissen, F. De Turck and T. Dhaene. 2012. Blind kriging: implementation and performance analysis. *Adv. Eng. Softw.* 49: 1–13.

Craig, J.A. 1978. *D-optimal Design Method: Final Report and User's Manual.* USAF Contract F33615–78–C–3011, FZM–6777, General Dynamics, Forth Worth Div.

Craven, P. and G. Wahba. 1979. Smoothing noisy data with spline functions, *Numer. Math.* 31: 377–403.

Cresssie, N. 1988. Spatial prediction and ordinary Kriging. *Mathematical Geology* 20(4): 405–421.

Cressie, N.A.C. 1990. The Origins of Kriging. *Mathematical Geology* 22: 239–252.

Cressie, N.A.C. 1993. *Statistics for Spatial Data: Revised Edition*, Wiley, New York.

Crino, S. and D.E. Brown. 2007. Global optimization with multivariate adaptive regression splines, IEEE transactions on systems. *Man and Cybernetics Part B: Cybernetics* 37(2).

Currin, C., T.J. Mitchell, M.D. Morris and D. Ylvisaker. 1991. Bayesian prediction of deterministic functions, with applications to the design and analysis of computer experiments. *Journal of American Statistical Association* 86(416): 953–963.

Daberkow, D.D. and D.N. Mavris. 2002. An investigation of metamodeling techniques for complex systems design. pp. 2002–5457. *In: 9th AIAA/ISSMO Symposium on Multidisciplinary Analysis and Optimization*, Atlanta, Georgia, USA, September 4–6, AIAA.

Dai, H., B. Zhang and W. Wei. 2015. A multi wavelet support vector regression method for efficient reliability assessment. *Reliability Engineering and System Safety* 136: 132–139.

Dai, K.Y., G.R. Liu, K.M. Lim and X.L. Chen. 2004. A mesh-free method for static and free vibration analysis of shear deformable laminated composite plates. *Journal of Sound and Vibration* 269: 633–652.

Daniel, I.M. 2016. Yield and failure criteria for composite materials under static and dynamic loading. *Progress in Aerospace Sciences* 81: 18–25.

De Boor, C. and A. Ron. 1990. On multivariate polynomial interpolation. *Constructive Approximation* 6: 287–302.

De Lima, A.M.G., A.W. Faria and D.A. Rade. 2010. Sensitivity analysis of frequency response functions of composite sandwich plates containing viscoelastic layers. *Composite Structures* 92(2): 364–376.

Debski, H., P. Rozylo and A. Gliszczynski. 2018. Effect of low-velocity impact damage location on the stability and post-critical state of composite columns under compression. *Composite Structures* 184: 883–893.

Degrauwe, D., G. Lombaert and G.D. Roeck G.D.2010. Improving interval analysis in finite element calculations by means of affine arithmetic. *Comput. Struct.* 88(3-4): 247–254.

Dehkordi, M.B., S.M.R. Khalili and E. Carrera. 2016. Non-linear transient dynamic analysis of sandwich plate with composite face-sheets embedded with shape memory alloy wires and flexible core- based on the mixed LW (layer-wise)/ESL (equivalent single layer) models. *Composites Part B: Engineering* 87: 59–74.

DeMunck, M., D. Moens, W. Desmet and D. Vandepitte. 2008. A response surface based optimization algorithm for the calculation of fuzzy envelope FRFs of models with uncertain properties. *Comput. Struct.* 86(10): 1080–1092.

Denga, Y., Y. Chen, Y. Zhang and S. Mahadevan. 2012. Fuzzy dijkstra algorithm for shortest path problem under uncertain environment. *Applied Soft Computing* 12(3): 1231–1237.

Desai, K.M., S.A. Survase, P.S. Saudagar, S.S. Lele and R.S. Singhal. 2008. Comparison of artificial neural network (ANN) and response surface methodology (RSM) in fermentation media optimization: case study of fermentative production of scleroglucan. *Biochemical Engineering Journal* 41(3): 266–273.

Deutsch, C.V. 1996. Correcting for negative weights in Ordinary Kriging. *Computers & Geosciences* 22(7): 765–773.

Deutsch, F. 2000. *Best Approximation in Inner Product Space.* Springer, New York.

Dey, P. and M.K. Singha. 2006. Dynamic stability analysis of composite skew plates subjected to periodic in-plane load. *Thin-Walled Structures* 44: 937–942.

Dey, S. and A. Karmakar. 2013. A comparative study on free vibration analysis of delaminated torsion stiff and bending stiff composite shells. *Journal of Mechanical Science and Technology* 27(4): 963–972.

Dey, S. and A. Karmakar. 2012a Free vibration analyses of multiple delaminated angle-ply composite conical shells—A finite element approach. *Composite Structures* 94(7): 2188–2196.

Dey, S. and A. Karmakar. 2012b. Finite element analyses of bending stiff composite conical shells with multiple delamination. *Journal of Mechanics of Materials and Structures* 7(2): 213–224.

Dey, S. and A. Karmakar. 2012c. Natural frequencies of delaminated composite rotating conical shells - A finite element approach. *Finite Elements in Analysis and Design* 56: 41–51.

Dey, S., T. Mukhopadhyay, S.K. Sahu and S. Adhikari. 2018a. Stochastic dynamic stability analysis of composite curved panels subjected to non-uniform partial edge loading. *European Journal of Mechanics/A Solids* 67: 108–122.

Dey, S., T. Mukhopadhyay, S. Naskar, T.K. Dey, H.D. Chalak and S. Adhikari. 2018b. Probabilistic characterization for dynamics and stability of laminated soft core sandwich plates. *Journal of Sandwich Structures & Materials*, DOI: 10.1177/1099636217694229.

Dey, S., T. Mukhopadhyay and S. Adhikari. 2017. Metamodel based high-fidelity stochastic analysis of composite laminates: A concise review with critical comparative assessment. *Composite Structures* 171: 227–250.

Dey, S., T. Mukhopadhyay, H.H. Khodaparast and S. Adhikari. 2016a. A response surface modelling approach for resonance driven reliability based optimization of composite shells. *Periodica Polytechnica - Civil Engineering* 60(1): 103–111.

Dey, S., T. Mukhopadhyay, A. Spickenheuer, S. Adhikari and G. Heinrich. 2016b. Bottom up surrogate based approach for stochastic frequency response analysis of laminated composite plates. *Composite Structures* 140: 712–727.

Dey, S., S. Naskar, T. Mukhopadhyay, U. Gohs, S. Sriramula, S. Adhikari and G. Heinrich. 2016c. Uncertain natural frequency analysis of composite plates including effect of noise—a polynomial neural network approach. *Composite Structures* 143: 130–142

Dey, S., T. Mukhopadhyay, H.H. Khodaparast and S. Adhikari. 2016d. Fuzzy uncertainty propagation in composites using Gram-Schmidt polynomial chaos expansion. *Applied Mathematical Modelling* 40(7-8): 4412–4428.

Dey, S., T. Mukhopadhyay, A. Spickenheuer, U. Gohs and S. Adhikari. 2016e. Uncertainty quantification in natural frequency of composite plates—an artificial neural network based approach. *Advanced Composites Letters* 25(2): 43–48.

Dey, S., T. Mukhopadhyay, S.K. Sahu and S. Adhikari. 2016f. Effect of cutout on stochastic natural frequency of composite curved panels. *Composites Part B: Engineering* 105: 188–202.

Dey, S., T. Mukhopadhyay and S. Adhikari. 2015a. Stochastic free vibration analysis of angle-ply composite plates—a RS-HDMR approach. *Composite Structures* 122: 526–536.

Dey, S., T. Mukhopadhyay and S. Adhikari. 2015b. Stochastic free vibration analyses of composite doubly curved shells—a kriging model approach. *Composites Part B: Engineering* 70: 99–112.

Dey, S., T. Mukhopadhyay, H.H. Khodaparast and S. Adhikari. 2015c. Stochastic natural frequency of composite conical shells. *Acta Mechanica* 226(8): 2537–2553.

Dey, T.K., T. Mukhopadhyay, A. Chakrabarti and U.K. Sharma. 2015d. Efficient lightweight design of FRP bridge deck. *Proceedings of the Institution of Civil Engineers—Structures and Buildings* 168(10): 697–707.

Dey, S., T. Mukhopadhyay, S.K. Sahu, G. Li, H. Rabitz and S. Adhikari. 2015e. Thermal uncertainty quantification in frequency responses of laminated composite plates. *Composites Part B: Engineering* 80: 186–197.

Dey, S., T. Mukhopadhyay, H.H. Khodaparast, P. Kerfriden and S. Adhikari. 2015f. Rotational and ply-level uncertainty in response of composite shallow conical shells. *Composite Structures* 131 594–605.

Diamond, P. 1989. Fuzzy Kriging. *Fuzzy Sets and Systems* 33(3): 315–332.

Díaz-Madroñero, M., D. Peidro and J. Mula. 2014. A fuzzy optimization approach for procurement transport operational planning in an automobile supply chain. *Applied Mathematical Modelling* 38(23): 5705–5725.

Dimitrov, N., P. Friis-Hansen and C. Berggreen. 2013. Reliability analysis of a composite wind turbine blade section using the model correction factor method: numerical study and validation. *Applied Composite Materials* 20: 17–39.

Dimopoulos, C.A. and C.J. Gantes. 2015. Numerical methods for the design of cylindrical steel shells with unreinforced or reinforced cutouts. *Thin-Walled Structures* 96: 11–28.

Dixit, V., N. Seshadrinath and M.K. Tiwari. 2016. Performance measures based optimization of supply chain network resilience: A NSGA-II + Co-Kriging approach. *Computers & Industrial Engineering* 93: 205–214.

Dong, H. and J. Wang. 2015. A criterion for failure mode prediction of angle-ply composite laminates under in-plane tension. *Composite Structures* 128: 234–240.

Douville, M.A. and P. Le Grognec. 2013. Exact analytical solutions for the local and global buckling of sandwich beam-columns under various loadings. *International Journal of Solids and Structures* 50(16-17): 2597–2609.

Duc N.D. and P.H. Cong. 2015. Nonlinear thermal stability of eccentrically stiffened functionally graded truncated conical shells surrounded on elastic foundations. *European Journal of Mechanics—A/Solids* 50: 120–131.

Dvorak, G. J. and Norman Laws. 1986. Analysis of first ply failure in composite laminates. *Engineering Fracture Mechanics* 25(5-6): 763–770.

Dyn, N., D. Levin and S. Rippa. 1986. Numerical procedures for surface fitting of scattered data by radial basis functions. *SIAM J. Scientific and Statistical Computing* 7: 639–659.

Eftekhari, M., M. Mahzoon and S. Ziaie Rad. 2011. An evolutionary search technique to determine natural frequencies and mode shapes of composite Timoshenko beams. *Mechanics Research Communications* 38(3): 220–225.

Eiblmeier, J. and J. Loughlan. 1997. The influence of reinforcement ring width on the buckling response of carbon fibre composite panels with circular cut-outs. *Composite Structures* 38(1-4): 609–622.

El Kadi and Y. Al-Assaf. 2002. Prediction of the fatigue life of unidirectional glass fiber/epoxy composite laminae using different neural network paradigms. *Composite Structures* 55(2): 239–246.

Elmalich, D. and O. Rabinovitch. 2012. A high-order finite element for dynamic analysis of soft-core sandwich plates. *J. Sandw. Struct. Mater.* 14(5): 525–55.

Elsayed, K. 2015. Optimization of the cyclone separator geometry for minimum pressure drop using Co-Kriging. *Powder. Technol.* 269: 409–424.

Emery, X. 2005. Simple and ordinary multigaussian kriging for estimating recoverable reserves. *Mathematical Geology* 37(3): 295–319.

Evan-Iwanowski, R.M. 1965. On the parametric response of structures. *Applied Mechanics Review* 18: 699–702.

Fang, C. and G.S. Springer. 1993. Design of composite laminates by a Monte Carlo method. *Composite Materials* 27(7): 721–753.

Fang, H. and M.F. Horstemeyer. 2006. Global response approximation with radial basis functions. *Journal of Engineering Optimization* 38(4): 407–424.

Fang, K.T., D.K.J. Lin P. Winker and Y. Zhang. 2000. Uniform design: theory and application. *Technometrics* 39(3): 237–248.

Fang, S.E. and R. Perera. 2009. A response surface methodology based damage identification technique. *Smart Mater. Struct.* 18, doi:10.1088/0964-1726/18/6/065009.

Fantuzzi, N., M. Bacciocchi, F. Tornabene, E. Viola and A.J.M. Ferreira. 2015. Radial basis functions based on differential quadrature method for the free vibration analysis of laminated composite arbitrarily shaped plates. *Composites Part B: Engineering* 78: 65–78.

Fantuzzi, N. and F. Tornabene. 2016. Strong formulation isogeometric analysis (SFIGA) for laminated composite arbitrarily shaped plates. *Composites Part B: Engineering* 96: 173–203.

Farooq, U. and M. Peter. 2014. Ply level failure prediction of carbon fibre reinforced laminated composite panels subjected to low velocity drop-weight impact using adaptive meshing techniques. *Acta Astronautica* 102: 169–177.

Fayazbakhsh, K., M. Arian Nik, D. Pasini and L. Lessard. 2013. Defect layer method to capture effect of gaps and overlaps in variable stiffness laminates made by automated fiber placement. *Compos. Struct*. 97: 245–51.

Fazzolari, F.A. 2014. A refined dynamic stiffness element for free vibration analysis of cross-ply laminated composite cylindrical and spherical shallow shells. *Composites Part B: Engineering* 62: 143–158.

Fedorov, V.V. 1989. Kriging and other estimators of spatial field characteristics (with special reference to environmental studies). *Atm. Environment* (1967), 23(1): 175–184.

Ferreira, A.J.M, C.M.C Roque, E. Carrera, M. Cinefra and O. Polit. 2011. Two higher order Zig-Zag theories for the accurate analysis of bending, vibration and buckling response of laminated plates by radial basis functions collocation and a unified formulation. *J. Compos. Mater.* 45: 2523–36.

Ferreira, A.J.M., C.M.C. Roque, R.M.N. Jorge and E.J. Kansa. 2005. Static deformations and vibration analysis of composite and sandwich plates using a layerwise theory and multiquadrics discretizations. *Eng Anal Bound Elem* 29: 1104–14.

Ferreira, A.J.M. and G.E. Fasshauer. 2007. Analysis of natural frequencies of composite plates by an RBF-pseudo spectral method. *Composite Structures* 79: 202–210.

Ferreira, A.J.M., G.E. Fasshauer, R.C. Batra and J.D. Rodrigues. 2008. Static deformations and vibration analysis of composite and sandwich plates using a layerwise theory and RBF-PS discretizations with optimal shape parameter. *Composite Structures* 86: 328–43.

Fletcher, R. 1989. *Practical Methods of Optimization*. John Wiley & Sons, New York.

Forrester, A. and A. Keane. 2008. *Engineering Design via Surrogate Modelling: A Practical Guide*. John Wiley & Sons.

Friedman, J.H. 1991. Multivariate adaptive regression splines. *The Annals of Statistics* 19(1): 1–67.

Friswell, M.I., O. Bilgen, S.F. Ali, G. Litak and S. Adhikari. 2015. The effect of noise on the response of a vertical cantilever beam energy harvester. *ZAMM - Journal of Applied Mathematics and Mechanics* 95(5): 433–443.

Gadade, A.M., A. Lal and B.N. Singh. 2016a. Accurate stochastic initial and final failure of laminated plates subjected to hygro-thermo-mechanical loadings using Puck's failure criteria. *International Journal of Mechanical Sciences* 114: 177–206.

Gadade, A.M., A. Lal and B.N. Singh. 2016b. Finite element implementation of Puck's failure criterion for failure analysis of laminated plate subjected to biaxial loadings. *Aerospace Science and Technology* 55: 227–241.

Ganapathi, M., P. Boisse and D. Solaut. 1999. Non-linear dynamic stability analysis of composite laminates under periodic in-plane loads. *International Journal for Numerical Methods in Engineering* 46: 943–956.

Ganapathi, M. and D.P. Makhecha. 2001. Free vibration analysis of multi-layered composite laminates based on an accurate higher-order theory. *Composites Part B: Engineering* 32(6): 535–543.

Ganesan, R. and V.K. Kowda. 2005. Free vibration of composite beam-columns with stochastic material and geometric properties subjected to random axial loads. *Journal of Reinforced Plastics and Composites* 24(1): 69–91.

Gao, Y. and S. Tong. 2016. Composite adaptive fuzzy output feedback dynamic surface control design for stochastic large-scale nonlinear systems with unknown dead zone. *Neurocomputing* 175: Part A, 55–64.

Gao, W.N. Zhang, and J. Ji. 2009. A new method for random vibration analysis of stochastic truss structures. *Finite Elements in Analysis and Design* 45(3): 190–199.

Gao, Y.H. and Y.F. Xing. 2017. The multiscale asymptotic expansion method for three-dimensional static analyses of periodical composite structures. *Composite Structures* 177: 187–195.

Garg, A.K., R.K. Khare and T. Kant. 2006. Free vibration of skew fiber-reinforced composite and sandwich laminates using a shear deformable finite element model. *J. Sandwich Structures and Materials*, 8: 33–53, http://dx.doi.org/10.1177/1099636206056457.

Gaspar, B., A.P. Teixeira and S.C. Guedes. 2014. Assessment of the efficiency of Kriging surrogate models for structural reliability analysis. *Probabilistic Engineering Mechanics* 37: 24–34.

Ghafari, E. and J. Rezaeepazhand. 2016. Vibration analysis of rotating composite beams using polynomial based dimensional reduction method. *International Journal of Mechanical Sciences* 115: 93–104.

Ghaffari, A., H. Abdollahi, M.R. Khoshay and B.I. Soltani. 2008. Performance comparison of neural networks. *Environmental Science & Technology* 42(21): 7970–7975.

Ghanem, R.G. and P.D. Spanos. 2002. *Stochastic Finite Elements—A Spectral Approach*. Revised, Dover Publications Inc., NY.

Ghavanloo, E. and S.A. Fazelzadeh. 2013. Free vibration analysis of orthotropic doubly curved shallow shells based on the gradient elasticity. *Composites Part B: Engineering* 45(1): 1448–1457.

Ghiasi, Y. and V. Nafisi. 2015. The improvement of strain estimation using universal kriging. *Acta Geod Geophys* 50: 479–490.

Giunta, G., F. Biscani, S. Belouettar and E. Carrera. 2011. Hierarchical modelling of doubly curved laminated composite shells under distributed and localised loadings, *Composites Part B: Engineering* 42(4): 682–691.

Giunta, A.A., J.M. Dudley, R. Narducci, B. Grossman, R.T. Haftka, W.H. Mason and L.T. Watson. 1994. Noisy aerodynamic response and smooth approximations in HSCT design. *In*: 5th AIAA/ USAF/NASA/ISSMO Symposium on Multidisciplinary Analysis and Optimization, Vol. 2, AIAA, Panama City, FL.

Giunta, A.A, V. Balabanov, D. Haim, B. Grossman, W.H. Mason and L.T. Watson. 1996. Wing design for high-speed civil transport using DOE methodology, USAF/NASA/ISSMO Symposium. *AIAA* Paper 96–4001.

Giunta, A. A. and L.T. Watson. 1998. A comparison of approximation modeling techniques: polynomial versus interpolating models. *In*: *Proceedings of the 7th AIAA/USAF/NASA/ISSMO Symposium on Multidisciplinary Analysis & Optimization*. Vol. 1, American Institute of Aeronautics and Astronautics, Inc., St. Louis, MO, September 2–4, AIAA–98–4758.

Giunta, A.A., S.F. Wojtkiewicz and M.S. Eldred. 2003. Overview of modern design of experiments methods for computational simulations. *American Institute of Aeronautics and Astronautics*. Paper AIAA 2003–0649.

Giunta, A., L.T. Watson and J. Koehler. 1988. *A Comparison of Approximation Modeling Techniques: Polynomial Versus Interpolating Models*. Proc. 7th AIAA/USAF/NASA/ISSMO.

Goyal, V.K. and R.K. Kapania. 2008. Dynamic stability of uncertain laminated beams subjected to subtangential loads. *Int. J. Solids Struct.* 45(10): 2799–817.

Grover, N., D.K. Maiti and B.N. Singh. 2013. A new inverse hyperbolic shear deformation theory for static and buckling analysis of laminated composite and sandwich plates. *Composite Structures* 95: 667–675.

Guan, G.-F., H.-T. Wang and X. Wei. 2014. Multi-input multi-output random vibration control of a multi-axis electro-hydraulic shaking table. *Journal of Vibration and Control*. DOI: 10.1177/1077546314521444.

Guilleminot, J., C. Soize, D. Kondo and C. Binetruy. 2008. Theoretical framework and experimental procedure for modelling mesoscopic volume fraction stochastic fluctuations in fiber reinforced composites. *Int. J. of Solids and Structures* 45(21): 5567–5583.

Gulshan, T.M.N.A., A. Chakrabartiand and M. Talha. 2014. Bending analysis of functionally graded skew sandwich plates with through-the thickness displacement variations. *Journal of Sandwich Structures and Materials* 16(2): 210–248.

Günay, M.G. and T. Timarci. 2017. Static analysis of thin-walled laminated composite closed-section beams with variable stiffness. *Composite Structures* 182: 67–78.

Gunn, S.R. 1997. *Support Vector Machines for Classification and Regression, Technical Report, Image Speech and Intelligent Systems Research Group*. University of Southampton, UK.

Gupta, A. and M. Talha. 2016. An assessment of a non-polynomial based higher order shear and normal deformation theory for vibration response of gradient plates with initial geometric imperfections. *Composites Part B: Engineering* 107: 141–161.

Habibi, M., L. Laperrière and H.M. Hassanabadi. 2018. Influence of low-velocity impact on residual tensile properties of nonwoven flax/epoxy composite. *Composite Structures* 186: 175–182.

Haddad, K.H., Y. Govers, S. Adhikari, M. Link, M.I. Friswell, J.E. Mottershead and J. Sienz. 2014. Fuzzy Model Updating and its Application to the DLR AIRMOD Test Structure. USD 2014, Leuven, Belgium, 15–17 September 2014, 4631–4644.

Haldar, A. and S. Mahadevan. 1993. *Probabilistic Structural Mechanics Handbook*. Chapman & Hall.

Hanss, M. and K. Willner. 2000. A fuzzy arithmetical approach to the solution of finite element problems with uncertain parameters. *Mech. Research Comm.* 27(3): 257–272.

Hanss, M. 2002. The transformation method for the simulation and analysis of systems with uncertain parameters. *Fuzzy Sets Syst.* 130(3): 277–289.

Hanss, M. 2005. *Applied Fuzzy Arithmetic—An Introduction with Engineering Applications*, Springer Publication, ISBN 3–540–24201–5, New York.

Hardy, R.L. 1971. Multiquadratic equations of topography and other irregular surfaces. *J. Geophys.* 76: 1905–15.

Hasim, K.A. 2017. Isogeometric static analysis of laminated composite plane beams by using refined zigzag theory. *Composite Structures.*

Hastie, T., S. Rosset, R. Tibshirani and J. Zhu. 2004. The entire regularization path for the support vector machine. *Journal of Machine Learning Research* 5: 1391–1415.

Hazimeh, R., G. Challita, K. Khalil and R. Othman. 2015. Finite element analysis of adhesively bonded composite joints subjected to impact loadings. *International Journal of Adhesion and Adhesives* 56: 24–31.

He, Q., J. Wang, Y. Liu, D. Dai and F. Kong. 2012. Multiscale noise tuning of stochastic resonance for enhanced fault diagnosis in rotating machines. *Mechanical Systems and Signal Processing* 28: 443–457.

He, W., J. Liu, S. Wang and D. Xie. 2018. Low-velocity impact response and post-impact flexural behaviour of composite sandwich structures with corrugated cores. *Composite Structures.*

Hedayat, A.S., N.J.A. Sloane and J. Stufken. 1999. Orthogonal Arrays: Theory and Applications, Springer, New York.

Hengl, T, G.B.M. Heuvelink and D.G. Rossiter. 2007. About regression-kriging: from equations to case studies. *Comput. Geosci.* 33: 1301–1315.

Hertog, D.D., J.P.C. Kleijnen and A.Y.D. Siem. 2006. The correct kriging variance estimated by bootstrapping. *Journal of the Operational Research Society* 57(4): 400–409.

Honda, S. and Y. Narita. 2012. Natural frequencies and vibration modes of laminated composite plates reinforced with arbitrary curvilinear fiber shape paths. *Journal of Sound and Vibration* 331(1): 180–191.

Hota, S.S. and P. Padhi. 2007. Vibration of plates with arbitrary shapes of cutouts. *Journal of Sound and Vibration* 302(4-5): 1030–1036.

Hou, H. and G. He. 2018. Static and dynamic analysis of two-layer Timoshenko composite beams by weak-form quadrature element method. *Applied Mathematical Modelling* 55: 466–483.

Hu, H.T. and H.W. Peng. 2013. Maximization of fundamental frequency of axially compressed laminated curved panels with cutouts. *Composites Part B: Engineering* 47: 8–25.

Hu, X.X., T. Sakiyama, H. Matsuda and C. Morita 2002. Vibration of twisted laminated composite conical shells. *International Journal of Mechanical Sciences* 44(8): 1521–1541.

Huang, C., H. Zhang and S.M. Robeson. 2016. Intrinsic random functions and universal kriging on the circle. *Statistics & Probability Letters* 108: 33–39.

Huang, M. and T. Sakiyama. 1999. Free vibration analysis of rectangular plates with variously shaped holes. *Journal of Sound and Vibration* 226(4): 769–786.

Huang, X., J. Chen and H. Zhu. 2016. Assessing small failure probabilities by AK–SS: an active learning method combining kriging and subset simulation. *Structural Safety* 59: 86–95.

Huber, K.P., M.R. Berthold and H. Szczerbicka. 1996. Analysis of simulation models with fuzzy graph based metamodeling. *Performance Evaluation* 27-28: 473–490.

Hung, Y. 2011. Penalized blind kriging in computer experiments. *Stat Sin* 21: 1171–1190.

Hwu, C., H.W. Hsu and Y.H. Lin 2017. Free vibration of composite sandwich plates and cylindrical shells. *Composite Structures* 171: 528–537.

Iman, R.L. and W.J. Conover. 1980. Small sensitivity analysis techniques for computer models with an application to risk assessment. *Communication Statistics—Theory and Methods* A9(17): 1749–1842.

Irisarri, F.X., F. Laurin, F.H. Leroy and J.F. Maire. 2011. Computational strategy for multiobjective optimization of composite stiffened panels. *Compos. Struct.* 93, 1158–67.

Ishikawa, T., K. Koyama and S. Kobayaski. 1978. Thermal expansion coefficients of unidirectional composites. *J. Compos. Mater.* 12: 153–168.

Iurlaro, L., M. Gherlone, M.D. Sciuva and A. Tessler. 2013. Assessment of the refined zigzag theory for bending, vibration, and buckling of sandwich plates: a comparative study of different theories. *Composite Structures* 106: 777–92.

Iwatsubo, T., M. Saigo and Y. Sugiyama. 1973. Parametric instability of clamped-clamped and clamped-simply supported columns under periodic axial load. *Journal of Sound and Vibration* 30: 65–77.

Jagtap, K.R., A. Lal and B.N. Singh. 2011. Stochastic nonlinear free vibration analysis of elastically supported functionally graded materials plate with system randomness in thermal environment. *Composite Structures* 93(12): 3185–3199.

Jagtap, K.R., S.Y. Ghorpade, A. Lal and B.N. Singh. 2017. Finite element simulation of low velocity impact damage in composite laminates. *Materials Today: Proceedings* 4(2): 2464–2469.

Jayatheertha, C., J.P.H. Webber and S.K. Morton. 1996. Application of artificial neural networks for the optimum design of a laminated plate. *Computers & Structures* 59(5): 831–845.

Jenq, S.T., G.C. Hwang and S.M. Yang. 1993. The effect of square cut-outs on the natural frequencies and mode shapes of GRP cross-ply laminates.*Composites Science and Technology* 47(1): 91–101.

Jeong, S., M. Mitsuhiro and Y. Kazuomi. 2005. Efficient optimization design method using Kriging model. *Journal of Aircraft* 42(2): 413–420.

Jiang, D., Y. Li, Q. Fei and S. Wu. 2015. Prediction of uncertain elastic parameters of a braided composite. *Composite Structures* 126: 123–131.

Jiang, L. and H. Hu. 2017. Low-velocity impact response of multilayer orthogonal structural composite with auxetic effect. *Composite Structures* 169: 62–68.

Jin, R., W. Chen and A. Sudjianto. 2005. An efficient algorithm for constructing optimal design of computer experiments. *Journal of Statistical Planning and Inferences* 134(1): 268–287.

Jin, R., X. Du and W. Chen. 2003. The use of metamodeling techniques for optimization under uncertainty. *Structural and Multidisciplinary Optimization* 25(2): 99–116.

Jin, R., W. Chen and A. Sudjianto. 2002. On sequential sampling for global metamodeling for in engineering design. *In*: *ASME 2002 Design Engineering Technical Conferences and Computer and Information in Engineering Conference*. Montreal, Canada, September 29-October 2, DETC2002/DAC–34092.

Jin, R., W. Chen and T.W. Simpson. 2001. Comparative studies of metamodeling techniques under multiple modeling criteria. *Structural and Multidisciplinary Optimization* 23(1): 1–13.

Johnson, M.E., L.M. Moore and D. Ylvisaker. 1990. Minimax and maximin distance designs. *Journal of Statistical Planning and Inferences* 26(2): 131–148.

Jones, R.M. 1975. *Mechanics of Composite Materials*. McGraw-Hill Book Co., NY.

Jones, R.M. 1975. *Mechanics of Composite Materials*. Washington, D.C.: McGraw-Hill, Scripta.

Joseph, V.R., Y. Hung and A. Sudjianto. 2008. Blind kriging: a new method for developing metamodels. *J. Mech. Des.* 130:031102.

Kalagnanam, J.R. and U.M. Diwekar. 1997. An efficient sampling technique for off-line quality control. *Technometrics* 39(3): 308–319.

Kalnins, K., R. Rikards, J. Auzins and C. Bisagni, H. Abramovich and R. Degenhardt. 2010. Metamodeling methodology for postbuckling simulation of damaged composite stiffened structures with physical validation. *Int. J. Struct. Stab. Dyn.* 10: 705–16.

Kam, T.Y. and E.S. Chang 1997. Reliability formulation for composite laminates subjected to first-ply failure. *Composite Structures* 38(1-4): 447–452.

Kam, T.Y. and T.B. Jan. 1995. First-ply failure analysis of laminated composite plates based on the layerwise linear displacement theory. *Composite Structures* 32(1-4): 583–591.

Kam, T.Y., H.F. Sher, T.N. Chao and R.R. Chang. 1996. Predictions of deflection and first-ply failure load of thin laminated composite plates via the finite element approach. *International Journal of Solids and Structures* 33(3): 375–398.

Kam, T.Y., Y.W. Liu and F.T. Lee. 1997. First-ply failure strength of laminated composite pressure vessels. *Composite Structures* 38(1-4): 65–70.

Kam, T.Y. and F.M. Lai. 1999. Experimental and theoretical predictions of first-ply failure strength of laminated composite plates. *International Journal of Solids and Structures* 36(16): 2379–2395.

Kamiński, B. 2015. A method for the updating of stochastic kriging metamodels. *Eur. J. Oper. Res.* 247: 859–866.

Kaminski, M. 2013. The Stochastic Perturbation Method for Computational Mechanics. John Wiley & Sons.

Kang, S.-C., H.-M. Koh and J.F. Choo. 2010. An efficient response surface method using moving least squares approximation for structural reliability analysis. *Probabilistic Engineering Mechanics* 25: 365–371.

Kant, T. 1993. A critical review and some results of recently developed refined theories of fiber-reinforced laminated composites and sandwiches. *Composite Structures* 23(4): 293–312.

Kant, T. and K. Swaminathan. 2000. Estimation of transverse/interlaminar stresses in laminated composites—a selective review and survey of current developments. *Composite Structures* 49(1): 65–75.

Karbhari, V.M. and S. Matthias. 2007. Fuzzy logic based approach to FRP retrofit of columns. *Composites Part B: Engineering* 38(5-6): 651–673.

Karmakar, A. and P.K. Sinha. 2001. Failure analysis of laminated composite pretwisted rotating plates. *J. Reinforced Plastics and Composites* 20: 1326–1357.

Karsh, P.K., T. Mukhopadhyay and S. Dey. 2018. Spatial vulnerability analysis for the first ply failure strength of composite laminates including effect of delamination. *Composite Structures* 184: 554–567.

Karsh, P.K., T. Mukhopadhyay and S. Dey. 2018. Stochastic dynamic analysis of twisted functionally graded plates. *Composites Part B: Engineering* 147: 259–278.

Kayikci, R. and F.O. Sonmez. 2012. Design of composite laminates for optimum frequency response. *Journal of Sound and Vibration* 331(8): 1759–1776.

Kennedy, M. and A. O'Hagan. 2000. Predicting the output from a complex computer code when fast approximations are available. *Biometrika* 87: 1–13.

Kepple, J., M. Herath, G. Pearce, G. Prusty, R. Thomson and R. Degenhardt. 2015. Improved stochastic methods for modelling imperfections for buckling analysis of composite cylindrical shells. *Engineering Structures* 100: 385–398.

Kersaudy, P., B. Sudret, N. Varsier, O. Picon and J. Wiart. 2015. A new surrogate modeling technique combining Kriging and polynomial chaos expansions—Application to uncertainty analysis in computational dosimetry. *Journal of Computational Physics* 286: 103–117.

Khandelwal, R.P., A. Chakrabarti and P. Bhargava. 2013. Vibration and buckling analysis of laminated sandwich plate having soft core. *Int. J. Struct. Stab. Dyn.* 13(8): 20–31.

Khashaba, U.A. and R. Othman. 2017. Low-velocity impact of woven CFRE composites under different temperature levels. *International Journal of Impact Engineering.*

Khdeir, A.A. and J.N. Reddy. 1999. Free vibrations of laminated composite plates using second-order shear deformation theory. *Computers & Structures* 71(6): 617–626.

Khodaparast, H.H., J.E. Mottershead and K.J. Badcock. 2011. Interval model updating with irreducible uncertainty using the Kriging predictor. *Mechanical Systems and Signal Processing* 25(4): 1204–1226.

Khuri, A.I. and S. Mukhopadhyay. 2010. Response surface methodology, John Wiley & Sons. *Inc. WIREs Comp. Stat.* 2: 128–149.

Kim, B.S., Y.B. Lee and D.H. Choi. 2009. Comparison study on the accuracy of metamodeling technique for non-convex functions. *Journal of Mechanical Science and Technology* 23(4): 1175–1181.

Kim, N., C.K. Jeon and J. Lee. 2013. Dynamic stability analysis of shear-flexible composite beams, *Archive of Applied Mechanics* 83(5): 685–707.

Kim, J.-K., C.-S. Kim and D.-Y. Song. 2003. Strength evaluation and failure analysis of unidirectional composites using Monte-Carlo simulation. *Materials Science and Engineering: A* 340(1): 33–40.

Kisa, M. 2004. Free vibration analysis of a cantilever composite beam with multiple cracks. *Composites Science and Technology* 64(9): 1391–1402.

Kishor D.K., R. Ganguli and S. Gopalakrishnan. 2011. Uncertainty analysis of vibrational frequencies of an incompressible liquid in a rectangular tank with and without a baffle using polynomial chaos expansion. *Acta Mechanica* 220(1-4): 257–273.

Kishore, M.D.V., Hari, B.N. Singh and M.K. Pandit. 2011. Nonlinear static analysis of smart laminated composite plate. *Aerospace Science and Technology* 15(3): 224–235.

Kleijnen, J.P.C. 1987. *Statistical Tools for Simulation Practitioners*. NY: Marcel Dekker.

Kleijnen, J.P.C. and W. Van Beers. 2003. Kriging for interpolation in random simulation. *Journal of the Operational Research Society* 54: 255–262.

Kleijnen, J.P.C. 2004. Design and Analysis of Monte Carlo Experiments, *In*: J.E. Gentle, W. Haerdle, and Y. Mori (eds.). *Handbook of Computational Statistics: Concepts and Fundamentals*. Springer-Verlag, Heidelberg, Germany.

Kochmann, D.M. and W.J. Drugan. 2009. Dynamic stability analysis of an elastic composite material having a negative-stiffness phase. *Journal of the Mechanics and Physics of Solids* 57(7): 1122–1138.

Koehler, J.R. and A. Owen. 1996. Computer experiments. pp. 261–308. *In*: S. Ghosh and C.R. Rao (eds.). *Handbook of Statistics*, Elsevier Science, New York.

Koji, S. 2013. Kawai Soshi, Alonso Juan J., Dynamic adaptive sampling based on Kriging surrogate models for efficient uncertainty quantification, Proceeding of 54th AIAA/ASME/ASCE/AHS/ASC Structures, Structural Dynamics, and Materials Conference, April 8–11, Boston, Massachusetts, USA.

Kollár, L.P. and G.S. Springer. 2009. *Mechanics of Composite Structures*. Cambridge University Press.

Koziel, S., A. Bekasiewicz, I. Couckuyt and T. Dhaene. 2014. Efficient Multi-Objective Simulation-Driven Antenna Design Using Co-Kriging. *IEEE Trans Antennas Propag* 62: 5900–5905.

Koziel, S. and X.-S.Yang (eds.). Computational Optimization, Methods and Algorithms, Springer, ISBN: 978-3-642-20858-4 (Print) 978-3-642-20859-1.

Krige, D.G. 1951. A Statistical Approach to Some Basic Mine Valuation Problems on the Witwatersrand. *J. Chem. Metall. Min. Soc. South Africa* 52: 119–139.

Krishnamurthy, T. 2003. Response surface approximation with augmented and compactly supported radial basis functions. The 44th AIAA/ASME/ASCE/AHS/ASC structures. *Structural Dynamics.* and materials conference. Norfolk, VA.

Kulkarni, S.D. and S. Kapuria .2008. Free vibration analysis of composite and sandwich plates using an improved discrete Kirchoff quadrilateral element based on third order zigzag theory. *Comput. Mech.* 42: 803–824.

Kumar, A., A. Chakrabarti, P. Bhargava and R. Chowdhury. 2015. Probabilistic failure analysis of laminated sandwich shells based on higher order zigzag theory. *Journal of Sandwich Structures and Materials* 17(5): 546–561.

Kumar, C.S., V. Arumugam, R. Sengottuvelusamy, S. Srinivasan and H.N. Dhakal. 2017b. Failure strength prediction of glass/epoxy composite laminates from acoustic emission parameters using artificial neural network. *Applied Acoustics* 115: 32–41.

Kumar, R.S. 2013. Analysis of coupled ply damage and delamination failure processes in ceramic matrix composites. *Acta Materialia* 61(10): 3535–3548.

Kumar, S.D., J. Magesh and V. Subramanian. 2017a. Tuning of bandwidth by superposition of bending and radial resonance modes in bilayer laminate composite. *Materials & Design* 122: 315–321.

Kumar, Y.V.S. and A. Srivastava. 2003. First ply failure analysis of laminated stiffened plates. *Composite Structures* 60(3): 307–315.

Kurşun, A., M. Şenel and H. M. Enginsoy. 2015. Experimental and numerical analysis of low velocity impact on a preloaded composite plate. *Advances in Engineering Software* 90: 41–52.

Kuttenkeuler, J. 1999. A finite element based modal method for determination of plate stiffnesses considering uncertainties. *Journal of Composite Materials* 33(8): 695–711.

Kwon, H. and S. Choi. 2015. A trended Kriging model with R^2 indicator and application to design optimization. *Aerospace Science and Technology* 43: 111–125.

Lal, A. and B.N. Singh. 2010. Stochastic free vibration of laminated composite plates in thermal environments. *Journal of Thermoplastic Composite Materials* 23(1): 57–77.

Lal, A., B.N. Singh and S. Kale. 2011. Stochastic post buckling analysis of laminated composite cylindrical shell panel subjected to hygrothermomechanical loading. *Composite Structures* 93: 1187–1200.

Lal, A. and B.N. Singh. 2011. Effect of random system properties on bending response of thermo-mechanically loaded laminated composite plates. *Applied Mathematical Modelling* 35(12): 5618–5635.

Lallemand, B., G. Plessis, T. Tison and P. Level. 1999. Neumann expansion for fuzzy finite element analysis. *Eng. Comput.* 16(5): 572–583.

Lan, X., Z. Feng and F. Lv. 2014. Stochastic principal parametric resonances of composite laminated beams,.*Shock and Vibration*. doi: 10.1155/2014/617828.

Lancaster, P. and K. Salkauskas. 1981. Surfaces generated by moving least squares methods. *Mathematics of Computation* 37(155): 141–158.

Langley, P. and H.A. Simon. 1995. Applications of machine learning and rule induction. *Communications of the ACM* 38(11): 55–64.

Lanhe, W., L. Hua and W. Daobin. 2005. Vibration analysis of generally laminated composite plates by the moving least squares differential quadrature method. *Composite Structures* 68: 319–330.

Lanzi, L. and V. Giavotto. 2006. Post-buckling optimization of composite stiffened panels: computations and experiments. *Compos. Struct.* 73: 208–20.

Lee, H.P., S.P. Lim and S.T. Chow. 1987. Free vibration of composite rectangular plates with rectangular cutouts. *Composite Structures* 8: 63–81.

Lee, H.P. and S.P. Lim. 1992. Free vibration of isotropic and orthotropic square plates with square cut outs subjected to in-plane forces. *Computers & Structures* 43(3): 431–437.

Lee, K.H. and D.H. Kang. 2006. A robust optimization using the statistics based on kriging metamodel. *Journal of Mechanical Science and Technology* 20(8): 1169–1182

Lee, J., Z. Urdal and O.H. Riffin 1995. Postbuckling of laminated composites with delaminations. *AIAA Journal* 33(10): 1963–1970.

Lee, S.Y. and D.S. Chung. 2010. Finite element delamination model for vibrating composite spherical shell panels with central cutouts. *Finite Elements in Analysis and Design* 46(3): 247–256.

Lee, Y.J and C.C. Lin. 2003. Regression of the response surface of laminated composite structures. *Compos Struct.* 62: 91–105.

Lee, Jaehong and Seung-Eock Kim. 2002. Free vibration of thin-walled composite beams with I-shaped cross-sections. *Composite Structures* 55(2): 205–215.

Lee, S.-P., J.-W. Jin and K.-W. Kang. 2014. Probabilistic analysis for mechanical properties of glass/epoxy composites using homogenization method and Monte Carlo simulation. *Renewable Energy* 65: 219–226.

Lee, S., T. Park and G.Z. Voyiadjis. 2002. Free vibration analysis of axially compressed laminated composite beam-columns with multiple delaminations. *Composites Part B: Engineering* 33(8): 605–617.

Lee, S., T. Park and G.Z. Voyiadjis. 2003. Vibration analysis of multi-delaminated beams. *Composites Part B: Engineering* 34(7): 647–659.

Leissa, A.W. and Y. Narita. 1989. Vibration studies for simply supported symmetrically laminated rectangular plates. *Composite Structures* 12: 113–132.

Leissa, A.W. and Y. Narita. 1984. Vibrations of corner point supported shallow shells of rectangular planform. *Earthquake Engng Struct. Dynam.* 12: 651–661.

Leissa, A.W. and Y. Narita. 1984. Vibrations of completely free shallow shells of rectangular planform. *Journal of Sound and Vibration* 96(2): 207–218.

Li, G., S.W. Wang, C. Rosenthal and H. Rabitz. 2001. High dimensional model representations generated from low dimensional data samples. I. mp-Cut-HDMR. *J. Math. Chem.* 30: 1–30.

Li, G., S.W. Wang, H. Rabitz, S. Wang and P.R. Jáffe. 2002. Global uncertainty assessments by high dimensional model representations (HDMR). *Chem. Eng. Sci.* 57: 4445–4460.

Li, G., M. Artamonov, H. Rabitz, S.W. Wang, P.G. Georgopoulos and M. Demiralp. 2003. High dimensional model representations generated from low order terms – Ip-RS-HDMR. *J. Comput. Chem.* 24: 647–656.

Li, G., J. Schoendorf and T.S. Ho. 2004. Multicut-HDMR with an application to an ionospheric model, *J. Comp. Chem.* 25: 1149–1156.

Li, G., J. Hu, S.W. Wang, P.G. Georgopoulos, J. Schoendorf and H. Rabitz . 2006. Random sampling-high dimensional model representation (RS-HDMR) and orthogonality of its different order component functions. *J. Phys. Chem. A* 110: 2474–2485.

Li, G. and H. Rabitz. 2012. General formulation of HDMR component functions with independent and correlated variables. *J. Math. Chem.* 50: 99–130.

Li, L., T. Romary and J. Caers. 2015. Universal kriging with training images. *Spat Stat* 14: 240–268.

Li, R., A. Sudjianto. 2012. Analysis of computer experiments using penalized likelihood in gaussian kriging models. *Technometrics* 47(2): 111–120.

Li, Y., S. Ng, M. Xie and T.Goh. 2010. A systematic comparison of metamodeling techniques for simulation optimization in decision support systems. *Applied Soft Computing* 10(4): 1257–1273.

Li, D.H., R.P. Wang, R.L. Qian, Y. Liu and G.H. Qing. 2016b. Static response and free vibration analysis of the composite sandwich structures with multi-layer cores. *International Journal of Mechanical Sciences* 111: 101–115.

Li, J., X. Tian, Z. Han and Y. Narita. 2016a. Stochastic thermal buckling analysis of laminated plates using perturbation technique. *Composite Structures* 139: 1–12.

Li, M. and H. Fan. 2018. Multi-failure analysis of composite Isogrid stiffened cylinders. *Composites Part A: Applied Science and Manufacturing.*

Li, X. and C. Guedes Soares. 2015. Spectral finite element analysis of in-plane free vibration of laminated composite shallow arches. *Composite Structures* 132: 484–494.

Liao, B.B. and P.F. Liu. 2017. Finite element analysis of dynamic progressive failure of plastic composite laminates under low velocity impact. *Composite Structures* 159: 567–578.

Lichtenstern, A. 2013. Kriging Methods in Spatial Statistics. Bachelor Thesis, Technische Universitat Munchen.

Liew, K.M., C.M. Lim and L.S. Ong. 1994. Vibration of pretwisted cantilever shallow conical shells. *I. J. Solids and Structures* 31: 2463–74.

Liew, K.M. and C.W. Lim. 1995. Vibratory characteristics of general laminates, I: Symmetric trapezoids. *Journal of Sound and Vibration* 183(4): 615–642.

Liew, K.M. 1996. Solving the vibration of thick symmetric laminates by Reissner/ Mindlin plate theory and the p-Ritz method. *Journal of Sound and Vibration* 198(3): 343–360.

Liew, K.M. and Y.Q. Huang. 2003. Bending and buckling of thick symmetrical rectangular laminates using the moving least squares differential quadrature method. *Int. J. Mech. Sci.* 45: 95–114.

Liew, K.M., J. Wang, T.Y. Ng and M.J. Tan. 2004. Free vibration and buckling analyses of shear-deformable plates based on FSDT mesh-free method. *Journal of Sound and Vibration* 276: 997–1017.

Lin, C.C. and Y.J. Lee. 2004. Stacking sequence optimization of laminated composite structures using genetic algorithm with local improvement. *Compos. Struct.* 63: 339–45.

Lin, Y. 2004. An efficient robust concept exploration method and sequential exploratory experimental design, Ph.D. Dissertation Thesis, Mechanical Engineering, Georgia Institute of Technology, Atlanta, 780.

Liu, B. and R.T. Haftka and M.A. Akgün. 2000. Two-level composite wing structural optimization using response surfaces. *Struct Multidiscipl Optim.* 20: 87–96.

Liu, J., Y.S. Cheng, R.F. Li and F.T.K. Au. 2010. A semi-analytical method for bending, buckling, and free vibration analyses of sandwich panels with square-honeycomb cores. *Int. J. Struct. Stab. Dyn.* 10(1): 127–51.

Liu, L., L.P. Chua and D.N. Ghista. 2007. Mesh-free radial basis function method for static, free vibration and buckling analysis of shear deformable composite laminates. *Composite Structures* 78: 58–69.

Liu, P.F. and J.Y. Zheng. 2006. A Monte Carlo finite element simulation of damage and failure in SiC/Ti–Al composites. *Materials Science and Engineering: A* 425(1): 260–267.

Liu, P.F., B.B. Liao, L.Y. Jia and X.Q. Peng. 2016. Finite element analysis of dynamic progressive failure of carbon fiber composite laminates under low velocity impact. *Composite Structures* 149: 408–422.

Liu, Q. 2015. Analytical sensitivity analysis of frequencies and modes for composite laminated structures. *International Journal of Mechanical Sciences* 90: 258–277.

Loja, M.A.R., J.I. Barbosa and C.M. Mota Soares. 2015. Dynamic behaviour of soft core sandwich beam structures using kriging-based layerwise models. *Composite Structures* 134: 883–894.

Longbiao, Li. 2017. Damage and failure of fiber-reinforced ceramic-matrix composites subjected to cyclic fatigue, dwell fatigue and thermomechanical fatigue. *Ceramics International* 43(16): 13978–13996.

Love, A. E.H. 1888. The small free vibrations and deformation of a thin elastic shell. *Philosophical Transactions of the Royal Society of London. A* 179: 491–546.

Luersen, M.A., C.A. Steeves and P.B. Nair. 2015. Curved fiber paths optimization of a composite cylindrical shell via Kriging-based approach. *Journal of Composite Materials* 49(29): 3583–3597.

Lugovy, M., N. Orlovskaya, V. Slyunyayev, E. Mitrentsis, M. Neumann, C.G. Aneziris, H. Jelitto, G.A. Schneider and J. Kuebler. 2017. Comparative study of static and cyclic fatigue of zrb 2-sic ceramic composites. *Journal of the European Ceramic Society*.

Luo, H., Y. Yan, T. Zhang, Z. He and S. Wang. 2017. Progressive failure numerical simulation and experimental verification of carbon-fiber composite corrugated beams under dynamic impact. *Polymer Testing* 63: 12–24.

Mace, B., K.Worden and G. Manson. 2005. Uncertainty in structural dynamics. *J. Sound Vib.* 288(3): 423–429.

Madsen, J.I., W. Shyy and R. Haftka. 2000. Response surface techniques for diffuser shape optimization. *AIAA Journal* 38(9): 1512–1518.

Madu, C.N. 1995. A fuzzy theoretic approach to simulation metamodeling. *Appl. Math. Lett.* 8(6): 35–41.

Mahadevan, S., X. Liu and Q. Xiao. 1997. A probabilistic progressive failure model for composite laminates. *Journal of Reinforced Plastics and Composites* 16(11): 1020–1038.

Mahata, A., T. Mukhopadhyay and S. Adhikari. 2016. A polynomial chaos expansion based molecular dynamics study for probabilistic strength analysis of nano-twinned copper. *Materials Research Express* 3: 036501.

Maharshi, K., T. Mukhopadhyay, B. Roy, L. Roy and S. Dey. 2018. Stochastic dynamic behaviour of hydrodynamic journal bearings including the effect of surface roughness. *International Journal of Mechanical Sciences* 142-143: 370–383.

Mahdi, A.N., K. Fayazbakhsh, D. Pasini and L. Lessard. 2014. A comparative study of metamodeling methods for the design optimization of variable stiffness composites. *Composite Structures* 107: 494–501.

Mahmoudkhani, S., H. Haddadpour and M. Navazi Hossein. 2013. Free and forced random vibration analysis of sandwich plates with thick viscoelastic cores. *Journal of Vibration and Control* 19(14): 2223–2240.

Maimı, P., P.P. Camanho, J.A. Mayugo and A. Turon. 2011. Matrix cracking and delamination in laminated composites. Part I: Ply constitutive law, first ply failure and onset of delamination. *Mechanics of Materials* 43(4): 169–185.

Makhecha, D.P., M. Ganapathi and B.P. Patel. 2002. Vibration and damping analysis of laminated/ sandwich composite plates using higher-order theory. *J. Reinf. Plast Comp.* 21(6): 559–75.

Malekzadeh, P., F. Bahranifard and S. Ziaee. 2013. Three-dimensional free vibration analysis of functionally graded cylindrical panels with cut-out using Chebyshev–Ritz method, *Composite Structures* 105: 1–13.

Malik, M.H. and A.F.M. Arif. 2013. ANN prediction model for composite plates against low velocity impact loads using finite element analysis. *Composite Structures* 101: 290–300.

Mallela, U.K. and A. Upadhyay. 2016. Buckling load prediction of laminated composite stiffened panels subjected to in-plane shear using artificial neural networks. *Thin-Walled Structures* 102: 158–164.

Mallick, P.K. 2007. Fiber - Reinforced Composites: Materials, Manufacturing, and Design, Third Edition , CRC Press.

Mallikarjuna and T. Kant. 1993. A critical review and some results of recently developed refined theories of fiber-reinforced laminated composites and sandwiches. *Composite Structures* 23: 293–312.

Manan, A. and J. Cooper. 2009. Design of composite wings including uncertainties: a probabilistic approach, *J. Aircr.* 46(2): 601–7.

Manan, A. and J.E. Cooper. 2010. Prediction of uncertain frequency response function bounds using polynomial chaos expansion. *Journal of Sound and Vibration* 329(16): 3348–3358.

Mandal, A., C. Ray and S. Haldar. 2017. Free vibration analysis of laminated composite skew plates with cut-out. *Archive of Applied Mechanics* 87(9): 1511–1523.

Manohar, B. and S. Divakar. 2005. An artificial neural network analysis of porcine pancreas lipase catalysed esterification of anthranilic acid with methanol. *Process Biochemistry* 40(10): 3372–3376.

Mantari, J.L., A.S. Oktem and C. Guedes Soares. 2012. Bending and free vibration analysis of isotropic and multilayered plates and shells by using a new accurate higher-order shear deformation theory. *Composites Part B: Engineering* 43(8): 3348–3360.

Mantari, J.L., E.M. Bonilla and C. Guedes Soares. 2014. A new tangential-exponential higher order shear deformation theory for advanced composite plates. *Composites Part B: Engineering* 60: 319–328.

Mantari, J.L. 2016. A simple polynomial quasi-3D HSDT with four unknowns to study FGPs. Reddy's HSDT assessment. *Composite Structures* 137: 114–120.

Mao, Z. and M. Todd. 2013. Statistical modeling of frequency response function estimation for uncertainty quantification. *Mechanical Systems and Signal Processing* 38(2): 333–345.

Marano, G.C. and R. Greco. 2011. Optimization criteria for tuned mass dampers for structural vibration control under stochastic excitation. *Journal of Vibration and Control* 17(5): 679–688.

Martin, J.D. and T.W. Simpson. 2004. On using Kriging models as probabilistic models in design. *SAE Transactions Journal of Materials & Manufacturing* 5: 129–139.

Martin, J.D. and T.W. Simpson. 2005. Use of Kriging models to approximate deterministic computer models. *AIAA Journal* 43(4): 853–863.

Martins, A.T., Z. Aboura, W. Harizi, A. Laksimi and K. Khellil. 2018. Analysis of the impact and compression after impact behavior of tufted laminated composites. *Composite Structures* 184: 352–361.

Mata-Díaz, A., J. López-Puente, D. Varas, J. Pernas-Sánchez and J.A. Artero-Guerrero. 2017. Experimental analysis of high velocity impacts of composite fragments. *International Journal of Impact Engineering* 103: 231–240.

Matheron, G. 1963. Principles of geostatistics. *Economic Geology* 58(8): 1246–1266.

Matheron, G.F.P.M. Traité de géostatistique appliquée, Editions Technip, France, 1962–63. (fundamental tools of linear geostatistics: variography, variances of estimation and dispersion, and kriging).

Matías, J.M. and W. González-Manteiga. 2005. Regularized kriging as a generalization of simple, universal, and bayesian kriging. *Stoch. Environ. Res. Risk Assess.* 20: 243–258.

Matlab. 2013. Version 8.2.0.701 (R2013b), MathWorks Inc.

McKay, M.D., R.J. Bechman and W.J. Conover. 1979. A comparison of three methods for selecting values of input variables in the analysis of output from a computer code. *Technometrics* 21(2): 239-245.

McKay, M.D., R.J. Beckman and W.J. Conover. 2000. A comparison of three methods for selecting values of input variables in the analysis of output from a computer code. *Technometrics* 42(1): 55–61.

Meckesheimer, M., A.J. Booker, R.R. Barton and T.W. Simpson. 2002. Computationally inexpensive metamodel assessment strategies. *AIAA Journal* 40(10): 2053–2060.

Mehar, K., S.K. Panda and T.R. Mahapatra. 2017. Theoretical and experimental investigation of vibration characteristic of carbon nanotube reinforced polymer composite structure. *International Journal of Mechanical Sciences* 133: 319–329.

Mehrez, L., A. Doostan, D. Moens and D. Vandepitte. 2012a. Stochastic identification of composite material properties from limited experimental databases, Part I: Experimental database construction. *Mechanical Systems and Signal Processing* 27: 471–483.

Mehrez, L., A. Doostan, D. Moens and D. Vandepitte. 2012b. Stochastic identification of composite material properties from limited experimental databases, Part II: Uncertainty modelling. *Mechanical Systems and Signal Processing* 27: 484–498.

Meirovitch, L. 1992. *Dynamics and Control of Structures*. J. Wiley & Sons, New York.

Mellit, A., M. Drif and A. Malek. 2010. EPNN-based prediction of meteorological data for renewable energy systems. *Revue des Energies Renouvelables* 13(1): 25–47.

Meng-Kao, Y. and T.K. Yao. 2004. Dynamic instability of composite beams under parametric excitation. *Composites Science and Technology* 64: 1885–1893.

Metya, S., T. Mukhopadhyay, S. Adhikari and G. Bhattacharya. 2017. System reliability analysis of soil slopes with general slip surfaces using multivariate adaptive regression splines. *Computers and Geotechnics* 87: 212–228

Michael, J.B. and R.D. Norman. 1974. On minimum-point second-order designs. *Technometrics* 16(4): 613–616.

Mindlin, R.D. 1951. Influence of rotatory inertia and shear on flexural motions of isotropic, elastic plates. *J. Appl. Mech.* 18: 31.

Mitchell, T.J. 1974. An algorithm for the construction of D-optimal experimental designs. *Technometrics* 16(2): 203–210.

Moens, D. and M. Hanss. 2011. Non-probabilistic finite element analysis for parametric uncertainty treatment in applied mechanics: Recent advances. *Finite Elements in Analysis and Design* 47(1): 4–16.

Möller, B. and M. Beer. 2004. *Fuzzy Randomness - Uncertainty in Civil Engineering and Computational Mechanics*. Springer, Berlin.

Mondal, S., A.K. Patra, S. Chakraborty and N. Mitra. 2015. Dynamic performance of sandwich composite plates with circular hole/cut-out: A mixed experimental–numerical study. *Composite Structures* 131: 479–489.

Montgomery, D.C. 1991. *Design and Analysis of Experiments*. J. Wiley and Sons, N.J.

Moore, R.E. 1966. *Interval Analysis*. Prentice-Hall, Englewood Cliffs, NJ, USA.

Moorthy, J. and J.N. Reddy. 1990. Parametric instability of laminated composite plates with transverse shear deformation. *International Journal of Solids and Structures* 26(7): 801–811.

Moreno-García, P., J.V. Araújo dos Santos and H. Lopes. 2014. A new technique to optimize the use of mode shape derivatives to localize damage in laminated composite plates. *Composite Structures* 108: 548–554.

Morris, M.D., T.J. Mitchell and D. Ylvisaker. 1993. Bayesian design and analysis of computer experiments: use of derivatives in surface prediction. *Technometrics* 35(3): 243–255.

Morse, L., Z.S. Khodaei and M.H. Aliabadi. 2018. Reliability based impact localization in composite panels using Bayesian updating and the Kalman filter. *Mechanical Systems and Signal Processing* 99: 107–128.

Muc, A. and P. Kędziora. 2001. A fuzzy set analysis for a fracture and fatigue damage response of composite materials. *Composite Structures* 54(2-3): 283–287.

Muc, A. and P. Romanowicz. 2017. Effect of notch on static and fatigue performance of multilayered composite structures under tensile loads. *Composite Structures* 178: 27–36.

Mukherjee, D., B.N. Rao and A.M. Prasad. 2011. Global sensitivity analysis of unreinforced masonry structure using high dimensional model representation. *Engineering Structures* 33: 1316–1325.

Mukhopadhyay, T. 2018a. A multivariate adaptive regression splines based damage identification methodology for web core composite bridges including the effect of noise. *Journal of Sandwich Structures & Materials*. DOI: 10.1177/1099636216682533.

Mukhopadhyay, T., S. Adhikari and A. Batou. 2018b. Frequency domain homogenization for the viscoelastic properties of spatially correlated quasi-periodic lattices. *International Journal of Mechanical Sciences*, DOI: 10.1016/j.ijmecsci.2017.09.004.

Mukhopadhyay, T., A. Mahata, S. Adhikari and M. Asle Zaeem. 2018. Probing the shear modulus of two-dimensional multiplanar nanostructures and heterostructures. *Nanoscale* 10: 5280–5294.

Mukhopadhyay, T. and S. Adhikari. 2017a. Stochastic mechanics of metamaterials. *Composite Structures* 162: 85–97.

Mukhopadhyay, T., A. Mahata, S. Adhikari and M. Asle Zaeem. 2017b. Effective elastic properties of two dimensional multiplanar hexagonal nano-structures, 2D Materials, 4025006.

Mukhopadhyay, T., S. Chakraborty, S. Dey, S. Adhikari and R. Chowdhury. 2017c. A critical assessment of Kriging model variants for high-fidelity uncertainty quantification in dynamics of composite shells. *Archives of Computational Methods in Engineering* 24(3): 495–518.

Mukhopadhyay, T. and S. Adhikari. 2017d. Effective in-plane elastic moduli of quasi-random spatially irregular hexagonal lattices. *International Journal of Engineering Science* 119: 142–179.

Mukhopadhyay, T., A. Mahata, S. Adhikari and M. Asle Zaeem. 2017e. Effective mechanical properties of multilayer nano-heterostructures. *Scientific Reports* 7: 15818.

Mukhopadhyay, T., S. Naskar, S. Dey and S. Adhikari. 2016a. On quantifying the effect of noise in surrogate based stochastic free vibration analysis of laminated composite shallow shells. *Composite Structures* 140: 798–805.

Mukhopadhyay, T. and S. Adhikari. 2016b. Effective in-plane elastic properties of auxetic honeycombs with spatial irregularity. *Mechanics of Materials* 95: 204–222.

Mukhopadhyay, T. and S. Adhikari. 2016c. Equivalent in-plane elastic properties of irregular honeycombs: An analytical approach. *International Journal of Solids and Structures* 91: 169–184.

Mukhopadhyay, T., R. Chowdhury and A. Chakrabarti. 2016d. Structural damage identification: A random sampling-high dimensional model representation approach. *Advances in Structural Engineering* 19(6): 908–927.

Mukhopadhyay, T. and S. Adhikari. 2016e. Free vibration analysis of sandwich panels with randomly irregular honeycomb core. *Journal of Engineering Mechanics* 142(11): 06016008.

Mukhopadhyay, T., A. Mahata, S. Dey and S. Adhikari. 2016f. Probabilistic analysis and design of HCP nanowires: an efficient surrogate based molecular dynamics simulation approach. *Journal of Materials Science & Technology* 32(12): 1345–1351.

Mukhopadhyay, T., T.K. Dey, S. Dey and A. Chakrabarti. 2015a. Optimization of fiber reinforced polymer web core bridge deck—a hybrid approach. *Structural Engineering International* 25(2): 173–183.

Mukhopadhyay, T., T.K. Dey, R. Chowdhury and A. Chakrabarti. 2015b. Structural damage identification using response surface based multi-objective optimization: a comparative study. *Arabian Journal for Science and Engineering* 40(4): 1027–1044.

Mukhopadhyay, T., T.K. Dey, R. Chowdhury, A. Chakrabarti and S. Adhikari. 2015c. Optimum design of FRP bridge deck: an efficient RS-HDMR based approach. *Structural and Multidisciplinary Optimization* 52(3): 459–477.

Mullur A.A. and A. Messac. 2005. Extended radial basis functions: more flexible and effective metamodeling. *AIAA Journal* 43(6): 1306–1315.

Myers, R.H. and D.C. Montgomery. 2002. *Response Surface Methodology: Process and Product Optimization Using Designed Experiments*, 2nd edn, Wiley, New York.

Myers, R.H., D.C. Montgomery and C.M. Anderson-Cook. 2016. Response Surface Methodology: Process and Product Optimization Using Designed Experiments, Wiley-Blackwell; 4th Revised ed. edition.

Nakagiri, S., H. Tatabatake and S. Tani. 1990. Uncertain eigen value analysis of composite laminated plates by SFEM. *Compos Struct.* 14: 9–12.

Nanda, N. and S. Kapuria. 2015. Spectral finite element for wave propagation analysis of laminated composite curved beams using classical and first order shear deformation theories. *Composite Structures* 132: 310–320.

Nanda, N., S. Kapuria and S. Gopalakrishnan. 2014. Spectral finite element based on an efficient layerwise theory for wave propagation analysis of composite and sandwich beams. *Journal of Sound and Vibration* 333(14): 3120–3137.

Narita, Y. and A.W. Leissa. 1992. Frequencies and mode shapes of cantilevered laminated composite plates. *Journal of Sound and Vibration* 154(1): 161–172.

Narita, Y. 2001. Closure to discussion of combinations for the free-vibration behavior of anisotropic rectangular plates under general edge conditions. *J. Appl. Mech.* 68(4): 685.

Naskar S., T. Mukhopadhyay, S. Sriramula and S. Adhikari. 2017. Stochastic natural frequency analysis of damaged thin-walled laminated composite beams with uncertainty in micromechanical properties. *Composite Structures* 160: 312–334.

Naskar, S., T. Mukhopadhyay and S. Sriramula. 2018. Probabilistic micromechanical spatial variability quantification in laminated composites. Composites Part B: Engineering, DOI: 10.1016/j. compositesb.2018.06.002.

Nayak, A.K., S.S.J. Moy and R.A. Shenoi. 2002. Free vibration analysis of composite sandwich plates based on Reddy's higher-order theory. *Composites Part B: Engineering* 33(7): 505–519.

Nechak, L., F. Gillot, S. Besset and J.J. Sinou. 2015. Sensitivity analysis and Kriging based models for robust stability analysis of brake systems. *Mechanics Research Communications* 69: 136–145.

Nejad, F.B., A. Rahai and A. Esfandiari. 2005. A structural damage detection method using static noisy data. *Engineering Structures* 27: 1784–1793

Neogi, S.D., A. Karmakar and D. Chakravorty. 2017. Finite element analysis of laminated composite skewed hypar shell roof under oblique impact with friction. *Procedia Engineering* 173: 314–322.

Nguyen Khuong, D. and H. Nguyen-Xuan. 2015. An isogeometric finite element approach for three-dimensional static and dynamic analysis of functionally graded material plate structures. *Composite Structures* 132: 423–439.

Noor, A.K., W.S. Burton and C.W. Bert. 1996. Computational models for sandwich panels and shells. *Appl. Mech. Rev.* 49(3): 155–99.

Oberkampf, W.L., S.M. DeL, B.M. Rutherford, K.V. Diegert and K.F. Alvin. 2000. Estimation of total uncertainty in computational simulation. *Sandia National Laboratories, SAND2000-0824, Albuquerque NM.*

Oberkampf, W.L., J.C. Helton and K. Sentz. 2001. Mathematical representation of uncertainty. *In: AIAA Non-Deterministic Approaches Forum* 1645: 16–19.

Ochoa, O.O. and J.N. Reddy. 1992. *Finite Element Analysis of Composite Laminates.* Springer Netherlands.

Oh, D.H. and L. Librescu. 1997. Free vibration and reliability of composite cantilevers featuring uncertain properties. *Reliability Engineering and System Safety* 56: 265–272.

Oh, S.K., W. Pedrycz and B.J. Park. 2003. Polynomial neural networks architecture, Analysis and design. *Comput. Electr. Eng.* 29(6): 703–725.

Oktem, A.S. and C. Guedes Soares. 2011. Boundary discontinuous Fourier solution for plates and doubly curved panels using a higher order theory. *Composites Part B: Engineering* 42(4): 842–850.

Olea, R.A. 2011. Optimal contour mapping using Kriging. *J. Geophys. Res.* 79: 695–702.

Omre, H. and K.B. Halvorsen. 1989. The bayesian bridge between simple and universal kriging. *Math Geol.* 21: 767–786.

Onkar, A.K. and D. Yadav. 2005. Forced nonlinear vibration of laminated composite plates with random material properties. *Composite Structures* 70: 334–342.

Ostachowicz, W.M. and S. Kaczmarczyk. 2001. Vibrations of composite plates with SMA fibres in a gas stream with defects of the type of delamination. *Composite Structures* 54(2): 305–311.

Owen, A. 1992. Orthogonal arrays for computer experiments, integration, and visualization. *Statistical Sinica* 2: 439–452.

Padmanabhan, S.K. and R. Pitchumani. 1999. Stochastic analysis of isothermal cure of resin systems. *Polymer Composites* 20(1): 72–85.

Panda, H.S., S.K. Sahu, P.K. Parhi and A.V. Asha. 2014. Vibration of woven fiber composite doubly curved panels with strip delamination in thermal field. *Journal of Vibration and Control.* DOI: 10.1177/1077546313520024.

Pandey, R., K.K. Shukla and A. Jain. 2008. Thermo-elastic stability analysis of laminated composite plates, an analytical approach. *Commun. Nonlinear. Sci. Numer. Simulat.* 14(4): 1679–99.

Pandit, M.K., B.N. Singh and A.H. Sheikh. 2008. Buckling of laminated sandwich plates with soft core based on an improved higher order zigzag theory. *Thin-Walled Structures* 46(11): 1183–1191.

Pandit, M.K., A.H. Sheikh and B.N. Singh. 2008. An improved higher order zigzag theory for the static analysis of laminated sandwich plate with soft core. *Finite Elements in Analysis and Design* 44(9-10): 602–610.

Pandya, B.N. and T. Kant. 1988. Finite element analysis of laminated composite plates using a higher-order displacement model. *Composites Science and Technology* 32(2): 137–155.

Papadrakakis, M., M. Lagaros and Y. Tsompanakis. 1998. Structural optimization using evolution strategies and neural networks. *Computer Methods in Applied Mechanics and Engineering* 156(1-4): 309–333.

Pareek, V.K., M.P. Brungs, A.A. Adesina and R. Sharma. 2002. Artificial neural network modeling of a multiphase photodegradation System. *Journal of Photochemistry and Photobiology A* 149: 139–146.

Park, I. and R.V. Grandhi. 2014. A Bayesian statistical method for quantifying model form uncertainty and two model averaging techniques. *Reliability Engineering & System Safety* 129: 46–56.

Park, J.S. 1994. Optimal latin-hypercube designs for computer experiments. *Journal of Statistical Planning Inference* 39: 95–111.

Park, J.S., C.G. Kim and C.S. Hong. 1995. Stochastic finite element method for laminated composite structures. *Journal of Reinforced Plastics and Composites* 14(7): 675–693.

Park, T., S.Y. Lee and G.Z. Voyiadjis. 2009. Finite element vibration analysis of composite skew laminates containing delaminations around quadrilateral cutouts. *Composites Part B: Engineering* 40(3): 225–236.

Park, H. 2017. Investigation on low velocity impact behavior between graphite/epoxy composite and steel plate. *Composite Structures* 171: 126–130.

Park, L. and U. Lee. 2015. Spectral element modeling and analysis of the transverse vibration of a laminated composite plate. *Composite Structures* 134: 905–917.

Park, L. and U. Lee. 2017. A generic type of frequency-domain spectral element model for the dynamics of a laminated composite plate. *Composite Structures* 172: 83–101.

Parthasarthy, G., N. Ganesan and C.V.R. Reddy. 1986. Study of unconstrained layer damping treatments applied to rectangular plates having central cutouts. *Computers & Structures* 23(3): 433–443.

Pascual, B. and S. Adhikari. 2012. Combined parametric-nonparametric uncertainty quantification using random matrix theory and polynomial chaos expansion. *Computers and Structures* 112-113(12): 364–379.

Patel, S. and C. Guedes Soares. 2017. System probability of failure and sensitivity analyses of composite plates under low velocity impact. *Composite Structures* 180: 1022–1031.

Patel, S.N., P.K. Datta and A.H. Sheikh. 2009. Parametric study on the dynamic instability behavior of laminated composite stiffened plate. *Journal of Engineering Mechanics* 135(11): 1331–1341.

Pawar, P.M., S. Nam Jung and B.P. Ronge. 2012. Fuzzy approach for uncertainty analysis of thin walled composite beams. *Aircraft Engineering and Aerospace Technology* 84(1): 13–22.

Pedronia, N., E. Zioa, E. Ferrariob, A. Pasanisic and M. Coupletc. 2013. Hierarchical propagation of probabilistic and non-probabilistic uncertainty in the parameters of a risk model. *Computers & Structures* 126: 199–213.

Perdikaris, P., D.Venturi, J.O. Royset and G.E. Karniadakis. 2015. Multi-fidelity modelling via recursive co-kriging and Gaussian-Markov random fields. *Proc. Math Phys. Eng. Sci.* 471: 20150018.

Pérez, V.M., J.E. Renaud and L.T. Watson. 2002. Adaptive experimental design for construction of response surface approximations. *AIAA Journal* 40(12): 2495–2503.

Peter, J. and M. Marcelet. 2008. Comparison of surrogate models for turbomachinery design. *WSEAS Transactions on Fluid Mechanics* 3(1): 10–17.

Pigoli, D., A. Menafoglio and P. Secchi. 2016. Kriging prediction for manifold-valued random fields. *Journal of Multivariate Analysis* 145: 117–131.

Piovan, M.T., J.M. Ramirez and R. Sampaio. 2013. Dynamics of thin-walled composite beams: Analysis of parametric uncertainties. *Composite Structures* 105: 14–28.

Poore, A.L., A. Barut and E. Madenci. 2008. Free vibration of laminated cylindrical shells with a circular cutout. *Journal of Sound and Vibration* 312(1-2): 55–73.

Press, W.H., S.A. Teukolsky, W.T. Vetterling and B.P. Flannery. 1992. *Numerical recipes in FORTRAN— The art of science computing.* Cambridge University Press, N.Y., p. 51.

Pronzato, L. and W.G. Müller. 2012. Design of computer experiments: space filling and beyond. *J. Statistics and Computing* 22(3): 681–701.

Prusty, B.G., C. Ray and S.K. Satsangi. 2001. First ply failure analysis of stiffened panels—a finite element approach. *Composite Structures* 51(1): 73–81.

Putter, H. and G.A. Young. 2001. On the effect of covariance function estimation on the accuracy of kriging predictors. *Bernoulli* 7: 421–438.

Qatu, M.S. and A.W. Leissa. 1991. Vibration studies for laminated composite twisted cantilever plates. *International Journal of Mechanical Sciences* 33(11): 927–940

Qatu, M.S. and A.W. Leissa. 1991. Natural frequencies for cantilevered doubly curved laminated composite shallow shells. *Composite Structures* 17: 227–255.

Qian, Z., C.C. Seepersad, V.R. Joseph, C.J.F. Wu and J.K. Allen. 2004. Building surrogate models based on detailed and approximate simulations. *In: ASME 2004 Design Engineering Technical Conferences and Computers and Information in Engineering Conference.* ASME, Salt Lake City, Utah, USA, September 28–October 2, DETC2004-57486.

Qiao, P., K. Lu, W. Lestari and J. Wang. 2007. Curvature mode shape-based damage detection in composite laminated plates. *Composite Structures* 80(3): 409–428.

Qin, X.C., C.Y. Dong, F. Wang and X.Y. Qu. 2017. Static and dynamic analyses of isogeometric curvilinearly stiffened plates. *Applied Mathematical Modelling* 45: 336–364.

Qiu, Z., X. Wang and M. Friswell. 2005. Eigenvalue bounds of structures with uncertain–but–bounded parameters. *J. Sound Vib.* 282(1-2): 297–312.

Qiu, Z. and J. Hu. 2008. Two non-probabilistic set-theoretical models to predict the transient vibrations of cross-ply plates with uncertainty. *Applied Mathematical Modelling* 32(12): 2872–2887.

Qu, H. and M.C. Fu. 2014. Gradient Extrapolated Stochastic Kriging. ACM Transactions on Modeling and Computer Simulation (TOMACS) 24(4): 23.1–23.25.

Rabitz, H. and Ö.F. Alis. 1999. General foundations of high dimensional model representations. *J. Math. Chem.* 25: 197–233.

Rabitz, H., Ö.F. Alis, J. Shorter and K. Shim. 1999. Efficient input output model representations. *Computer Phys. Comm.* 117: 11–20.

Radoslav, H. 2014. Multiplicative methods for computing D-optimal stratified designs of experiments. *Journal of Statistical Planning and Inference* 146: 82–94.

Rajamani, A. and R. Prabhakaran. 1977a. Dynamic response of composite plates with cut-outs, part I: simply supported plates. *Journal of Sound and Vibration* 54(4): 549–564.

Rajamani, A. and R. Prabhakaran. 1977b. Dynamic response of composite plates with cut-outs, part II: clamped clamped plates. *Journal of Sound and Vibration* 54(4): 565–576.

Rajmohan, T., K. Palanikumar and S. Prakash. 2013. Grey-fuzzy algorithm to optimise machining parameters in drilling of hybrid metal matrix composites. *Composites Part B: Engineering* 50: 297–308.

Rao, S.S. and K.K. Annamdas. 2008. Evidence-based fuzzy approach for the safety analysis of uncertain systems. *AIAA Journal* 46(9): 2383–2387.

Rasmussen, J. 1998. Nonlinear programming by cumulative approximation refinement. *Structural Optimization* 15: 1–7.

Ratko, P., K. Predrag, M. Snezana and P. Ivan. 2012. Influence of rotatory inertia on dynamic stability of the viscoelastic symmetric cross-ply laminated plates. *Mechanics Research Communications* 45: 28–33.

Ray, C., B.G. Prusty and S.K. Satsangi. 2004. Free vibration analysis of composite hat stiffened panels by finite element method. *Journal of Reinforced Plastic and Composites* 23(5): 533–547.

Rayleigh, J.W. 1877. Theory of sound. *Dover Publications*, New York, re-issue, 1945, second edition.

Reddy, J.N. 1985. A review of the literature on finite-element modeling of laminated composite plates. *The Shock and Vibration Digest* 17(4): 3–8.

Reddy, J.N. 1984. Exact solutions of moderately thick laminated shells. *Journal of Engineering Mechanics* 110: 794–809.

Reddy, J.N. 1984. A simple higher-order theory for laminated composite plates. *Journal of Applied Mechanics* 51(4): 745–752.

Reddy, J.N. 1982. Large amplitude flexural vibration of layered composite plates with cutouts. *Journal of Sound and Vibration* 1: 1–10.

Reddy, J.N. and A.K. Pandey. 1987. A first-ply failure analysis of composite laminates. *Computers & Structures* 25(3): 371–393.

Reddy, J.N. 2003. Mechanics of Laminated Composite Plates and Shells: Theory and Analysis, Second Edition, CRC Press.

Reissner, E. 1975. On transverse bending of plates, including the effect of transverse shear deformation. *International Journal of Solids and Structures* 11(5): 569–573.

Reissner, Eric. 1944. On the theory of bending of elastic plates. *Studies in Applied Mathematics* 23(1-4): 184–191.

Rezaeepazhand, J. and N. Jafari. 2005. Stress analysis of perforated composite plates. *Composite Structures* 71: 463–468.

Ribeiro, M.L., D. Vandepitte and V. Tita. 2015. Experimental analysis of transverse impact loading on composite cylinders. *Composite Structures* 133: 547–563.

Rikards, R., H. Abramovich, K. Kalnins and J. Auzins. 2006. Surrogate modeling in design optimization of stiffened composite shells. *Compos Struct.* 73: 244–51.

Rikards, R., A. Chate and O. Ozolinsh. 2001. Analysis for buckling and vibrations of composite stiffened shells and plates. *Composite Structures* 51(4): 361–370.

Rivest, M. and D. Marcotte .2012. Kriging groundwater solute concentrations using flow coordinates and nonstationary covariance functions. *J. Hydrol.* 472-473: 238–253.

Rodrigues, J.D., C.M.C. Roque, A.J.M. Ferreira, M. Cinefra and E. Carrera. 2012. Radial basis functions-differential quadrature collocation and a unified formulation for bending, vibration and buckling analysis of laminated plates, according to Murakami's zig-zag theory. *Computers & Structures* 90-91: 107–15.

Rodrigues, J.D., C.M.C. Roque, A.J.M. Ferreira, E. Carrera and M. Cinefra. 2011. Radial basis functions–finite differences collocation and a Unified Formulation for bending, vibration and buckling analysis of laminated plates, according to Murakami's zig-zag theory. *Composite Structures* 93: 1613–20.

Roque, C.M.C., A.J.M. Ferreira and R.M.N. Jorge. 2006. Free vibration analysis of composite and sandwich plates by a trigonometric layerwise deformation theory and radial basis functions. *J. Sandw. Struct. Mater.* 8: 497–515.

Rothman, A., T.-S. Ho and H. Rabitz. 2005. Observable-preserving control of quantum dynamics over a family of related systems. *Phys. Rev. A* 72: 023416.

Rozylo, P., H. Debski and T. Kubiak. 2017. A model of low-velocity impact damage of composite plates subjected to compression-after-impact (CAI) testing. *Composite Structures* 181: 158–170.

Ryu, J.-S., M.-S. Kim, K.-J. Cha, T.H. Lee and D.-H. Choi. 2002. Kriging interpolation methods in geostatistics and DACE model. *KSME International Journal* 16(5): 619–632.

Sacks, J., B.S.S. and W.J. Welch. 1989. Designs for computer experiments. *Technometrics* 31(1): 41–47.

Sacks, J., W.J. Welch, T.J. Mitchell and H.P. Wynn. 1989b. Design and analysis of computer experiments. *Statistical Science* 4(4): 409–423.

Sahoo, Rosalin and B.N. Singh. 2013. A new inverse hyperbolic zigzag theory for the static analysis of laminated composite and sandwich plates. *Composite Structures* 105: 385–397.

Sahoo, R. and B.N. Singh. 2014. A new trigonometric zigzag theory for static analysis of laminated composite and sandwich plates. *Aerospace Science and Technology* 35: 15–28.

Sahoo, S.S., S.K. Panda and T.R. Mahapatra. 2016. Static, free vibration and transient response of laminated composite curved shallow panel—an experimental approach. *European Journal of Mechanics-A/Solids* 59: 95–113.

Sai Ram, K.S. and P.K. Sinha. 1992. Hygrothermal effects on the free vibration of laminated composite plates. *Journal of Sound and Vibration* 158(1): 133–148.

Sai Ram, K.S. and P.K. Sinha. 1992. Hygrothermal effects on the buckling of laminated composite plates. *Compos Struct.* 21: 233–47.

Sai Ram, K.S. and T. Sreedhar Babu. 2002. Free vibration of composite spherical shell cap with and without a cutout. *Computers & Structures* 80(23): 1749–1756.

Sakata, S. and I. Torigoe. 2015. A successive perturbation-based multiscale stochastic analysis method for composite materials. *Finite Elements in Analysis and Design* 102-103: 74–84.

Sakata, S, F. Ashida and M. Zako. 2008. Kriging-based approximate stochastic homogenization analysis for composite materials. *Comput. Methods Appl. Mech. Eng.* 197(21): 1953–64.

Sakata, S., F. Ashida and M. Zako. 2004. An efficient algorithm for Kriging approximation and optimization with large-scale sampling data. *Comput. Methods Appl. Mech. Engg.* 193: 385–404.

Salim, S., N.G.R. Iyengar and. D. Yadav. 1998. Buckling of laminated plates with random material characteristics. *Applied Composite Materials* 5: 1–9.

Samaratunga, D., R. Jha and S. Gopalakrishnan. 2015. Wave propagation analysis in adhesively bonded composite joints using the wavelet spectral finite element method. *Composite Structures* 122: 271–283.

Santner, T.J., B. Williams and W. Notz. 2003. *The Design and Analysis of Computer Experiments* Springer, Heidelberg.

Sarangapani, G. and R. Ganguli. 2013. Effect of ply level material uncertainty on composite elastic couplings in laminated plates. *International Journal for Computational Methods in Engineering Science and Mechanics* 14(3): 244–261.

Sarangapani, G., R. Ganguli and C.R.L. Murthy. 2013. Spatial wavelet approach to local matrix crack detection in composite beams with ply level material uncertainty. *Applied Composite Materials* 20: 719–746.

Saraviaa, M., S.P. Machadoa and V.H Cortíneza. 2011. Free vibration and dynamic stability of rotating thin-walled composite beams. *European Journal of Mechanics—A/Solids* 30 (3): 432–441.

Sarrouy, E., O. Dessombz and J.J. Sinou. 2013. Stochastic study of a non-linear self-excited system with friction. *European Journal of Mechanics—A/Solids* 40: 1–10.

Sarvestan, V., H.R. Mirdamadi and M. Ghayour. 2017. Vibration analysis of cracked Timoshenko beam under moving load with constant velocity and acceleration by spectral finite element method. *International Journal of Mechanical Sciences* 122: 318–330.

Sarvestani, H.Y. and M. Hojjati. 2017. Failure analysis of thick composite curved tubes. *Composite Structures* 160: 1027–1041.

Sasena, M., M. Parkinson, P. Goovaerts, P. Papalambros and M. Reed. 2002. Adaptive experimental design applied to an ergonomics testing procedure. *In*: *ASME 2002 Design Engineering Technical Conferences and Computer and Information in Engineering Conference*. ASME, Montreal, Canada, September 29-October 2, DETC2002/DAC-34091.

Sasikumar, P., R. Suresh and S. Gupta. 2014. Stochastic finite element analysis of layered composite beams with spatially varying non-Gaussian inhomogeneities. *Acta Mech.* 225: 1503–1522.

Sayyad, A. and Ghugal, Y. 2015. On the free vibration analysis of laminated composite and sandwich plates: A review of recent literature with some numerical results. Composite Structures 129: 177–201.

Scarth, C., J.E. Cooper, P.M. Weaver and H.C. Silva Gustavo. 2014. Uncertainty quantification of aeroelastic stability of composite plate wings using lamination parameters. *Composite Structures* 116: 84–93.

Scarth, C. and S. Adhikari. 2017. Modelling spatially varying uncertainty in composite structures using lamination parameters. *AIAA Journal* 55(11): 3951–3965.

Seçgin, A. 2013. Modal and response bound predictions of uncertain rectangular composite plates based on an extreme value model. *Journal of Sound and Vibration* 332(5): 1306–1323.

Sepahvand, K., S. Marburg and H.J. Hardtke. 2011. Stochastic structural modal analysis involving uncertain parameters using generalized polynomial chaos expansion. *Int. J. Appl. Mechanics* 3(3): 587–606.

Sepahvand, K., S. Marburg and H.J. Hardtke. 2012. Stochastic free vibration of orthotropic plates using generalized polynomial chaos expansion. *Journal of Sound and Vibration* 331(1): 167–179.

Sepahvand, K., M. Scheffler and S. Marburg. 2015. Uncertainty quantification in natural frequencies and radiated acoustic power of composite plates: analytical and experimental investigation. *Applied Acoustics* 87: 23–29.

Sepahvand, K. 2016. Spectral stochastic finite element vibration analysis of fiber-reinforced composites with random fiber orientation. *Composite Structures* 145: 119–128.

Sepahvand, K. 2017. Stochastic finite element method for random harmonic analysis of composite plates with uncertain modal damping parameters. *Journal of Sound and Vibration* 400: 1–12.

Sepahvand, K. and S. Marburg. 2017. Spectral stochastic finite element method in vibroacoustic analysis of fiber-reinforced composites. *Procedia Engineering* 199: 1134–1139.

Sepe, R., L.A. De, G. Lamanna and F. Caputo. 2016. Numerical and experimental investigation of residual strength of a LVI damaged CFRP omega stiffened panel with a cut-out. *Composites Part B: Engineering* 102: 38–56.

Shaker, A., W.G. Abdelrahman, T. Mohammad and S. Edward. 2008. Stochastic finite element analysis of the free vibration of laminated composite plates. *Computational Mechanics* 41(4): 493–501.

Shankar, G., S.K. Kumar and P.K. Mahato. 2017. Vibration analysis and control of smart composite plates with delamination and under hygrothermal environment. *Thin-Walled Structures* 116: 53–68.

Shariyat, M. 2010. A generalized global–local high-order theory for bending and vibration analyses of sandwich plates subjected to thermo-mechanical loads. *International Journal of Mechanical Sciences* 52: 495–514.

Shariyat, M. 2007. Thermal buckling analysis of rectangular composite plates with temperature-dependent properties based on a layer wise theory. *Thin-Walled Struct.* 45(4): 439–52.

Shaw, A., S. Sriramula, P.D. Gosling and M.K. Chryssanthopoulos. 2010. A critical reliability evaluation of fibre reinforced composite materials based on probabilistic micro and macro-mechanical analysis. *Composites Part B: Engineering* 41(6): 446–453.

Shen, H.S. 2001. Thermal post buckling behaviour of imperfect shear deformable laminated plates with temperature-dependent properties. *Comput Methods Appl. Mech. Eng.* 190: 5377–90.

Shimoyama, K., S. Kawai and J.J. Alonso. 2013. Dynamic adaptive sampling based on kriging surrogate models for efficient uncertainty quantification. 54th AIAA/ASME/ASCE/AHS/ASC structures, structural dynamics, and materials conference, Boston, Massachusetts, USA.

Shin, Y.S. and R.V. Grandhi. 2001. A global structural optimization technique using an interval method. *Structural and Multidisciplinary Optimization* 22: 351–363.

Shinozuka, M. and G. Deodatis. 1988. Response variability of stochastic finite element systems. *Journal of Engineering Mechanics* 114(3): 499–519.

Shorter, J.A., P.C. Ip and H. Rabitz. 1999. An efficient chemical kinetics solver using high dimensional model representation. *J. Phys. Chem. A* 103: 7192–7198.

Shu, C., W.X. Wu, H. Ding and C.M. Wang. 2007. Free vibration analysis of plates using least-square-based finite difference method. *Computer Methods in Applied Mechanics and Engineering* 196(7): 1330–1343.

Shu, D. and C.N. Della. 2004. Free vibration analysis of composite beams with two non-overlapping delaminations. *International Journal of Mechanical Sciences* 46(4): 509–526.

Simpson, T.W., T.M. Mauery, J.J. Korte and F. Mistree. 1998. Comparison of response surface and kriging models for multidisciplinary design optimization. Proc. 7th AIAA/USAF/NASA/ISSMO Symp. on Multidisciplinary Analysis & Optimization (held in St. Louis, MO), Vol. 1. pp. 381–391. AIAA.

Simpson, T.W., T.M. Mauery, J.J. Korte and F. Mistree. 2001. Kriging metamodels for global approximation in simulation-based multidisciplinary design optimization. *AIAA Journal* 39(12): 2233–2241.

Simpson, T.W., J. Peplinski, P.N. Koch and J.K. Allen. 1997. On the use of statistics in design and the implications for deterministic computer experiments. Design Theory and Methodology – DTM'97 (held in Sacramento, CA), Paper No. DETC97/DTM-3881, ASME.

Singh, A., S. Panda and D. Chakraborty. 2015. A design of laminated composite plates using graded orthotropic fiber-reinforced composite plies. *Composites Part B: Engineering* 79: 476–493.

Singh, B.N., D. Yadav and N.G.R. Iyengar. 2001. Natural frequencies of composite plates with random material properties using higher-order shear deformation theory. *International Journal of Mechanical Sciences* 43(10): 2193–2214.

Singh, B.N, D. Yadav and N.G.R. Iyengar. 2002. Free vibrations of composite cylindrical panels with random material properties. *Composite Structures* 58: 435–442.

Singh, B.N., A.K.S. Bisht, M.K. Pandit and K.K. Shukla. 2009. Nonlinear free vibration analysis of composite plates with material uncertainties: a Monte Carlo simulation approach. *Journal of Sound and Vibration* 324(1): 126–138.

Singh, S.K. and A. Chakrabarti. 2013. Static vibration and buckling analysis of skew composite and sandwich plates under thermo mechanical loading. *Int. J. Appl. Mech. Eng.* 18(3): 887–98.

Singha, M.K. and R. Daripa. 2009. Non-linear vibration and dynamic stability analysis of composite plates. *Journal of Sound and Vibration* 328: 541–554.

Sivakumar, K., N.G.R. Iyengar and K. Deb. 1999. Free vibration of laminated composite plates with cutout. *Journal of Sound and Vibration* 221(3): 443–470.

Slawomir, K. and X.-S. Yang. 2011. Computational Optimization, Methods and Algorithms, ISBN: 978-3-642-20858-4 (Print) 978-3-642-20859-1.

Smith, R. 2014. Uncertainty Quantification: Theory, Implementation, and Applications, Society for Industrial and Applied Mathematics.

Smith, M. 1993. *Neural Networks for Statistical Modeling*. NY: Von Nostrand Reinhold.

Sobieszczanski-Sobieski, J. and R.T. Haftka. 1997. Multidisciplinary aerospace design optimization: survey of recent developments. *Struct. Optim.* 14: 1–23.

Soize, C. 2013. Stochastic modeling of uncertainties in computational structural dynamics-recent theoretical advances. *Journal of Sound and Vibration* 332(10): 2379–2395.

Song, Y., S. Kim, I. Park and U. Lee. 2015. Dynamics of two-layer smart composite Timoshenko beams: frequency domain spectral element analysis. *Thin-Walled Structures* 89: 84–92.

Splichal, J., A. Pistek and J. Hlinka. 2015. Dynamic tests of composite panels of an aircraft wing. *Progress in Aerospace Sciences* 78: 50–61.

Sreenivasamurthy, S. and V. Ramamurti. 1981. Coriolis effect on the vibration of flat rotating low aspect ratio cantilever plates. *Journal of Strain Analysis* 16(2): 97–106.

Srikanth, G. and A. Kumar. 2003. Post buckling response and failure of symmetric laminates under uniform temperature rise. *Compos. Struct.* 59: 109–18.

Sriramula, S. and K.M. Chryssanthopoulos. 2009. Quantification of uncertainty modelling in stochastic analysis of FRP composites. *Composites Part A: Applied Science and Manufacturing* 40: 1673–1684.

Stefanou, G. and M. Papadrakakis. 2004. Stochastic finite element analysis of shells with combined random material and geometric properties. *Computer Methods in Applied Mechanics and Engineering* 193(1-2): 139–160.

Stein, A. and C.A. Corsten. 1991. Universal Kriging and cokriging as a regression procedure on JSTOR. *Biometrics* 47: 575–587.

Steuben, J., J. Michopoulos, A. Iliopoulos and C. Turner. 2015. Inverse characterization of composite materials via surrogate modeling. *Composite Structures* 132: 694–708.

Strife, J.R. and K.M. Prewo. 1979. The thermal expansion behavior of unidirection and bidirectional kevlar/epoxy composites. *J. Compos. Mater*. 13: 265–277.

Sudjianto, A., L. Juneja, A. Agrawal and M. Vora. 1998. Computer aided reliability and robustness assessment. *Int. J. Reliability, Quality, and Safety* 5: 181–193.

Symp. on Multidisciplinary Analysis & Optimization (held in St. Louis, MO), 1: 392–404. AIAA-98-4758, 1998.

Szebényi, G., B. Magyar and T. Iványicki. 2017. Comparison of static and fatigue interlaminar testing methods for continuous fiber reinforced polymer composites. *Polymer Testing* 63: 307–313.

Taguchi, G., Y. Yokoyama and Y. Wu.1993. Taguchi methods: design of experiments. American Supplier Institute, Allen Park, Michigan.

Taibi, F. Z., S. Benyoucef, A. Tounsi, R.B. Bouiadjra, E.A.A. Bedia and S.R. Mahmoud .2015. A simple shear deformation theory for thermo-mechanical behaviour of functionally graded sandwich plates on elastic foundations. *Journal of Sandwich Structures and Materials* 17(2): 99–129.

Talha, Md. and B.N. Singh, 2014. Stochastic perturbation-based finite element for buckling statistics of FGM plates with uncertain material properties in thermal environments. *Composite Structures* 108: 823–833.

Tan, P. and G.J. Nie. 2016. Free and forced vibration of variable stiffness composite annular thin plates with elastically restrained edges. *Composite Structures* 149: 398–407.

Tang, B. 1993. Orthogonal array-based latin hypercubes. *Journal of American Statistical Association* 88(424): 1392–1397.

Thai, C.H., V.N.V. Do and H. Nguyen-Xu. 2016. An improved moving Kriging-based meshfree method for static, dynamic and buckling analyses of functionally graded isotropic and sandwich plates. *Engineering Analysis with Boundary Elements* 64: 122–136.

Thinh, T.I. and M.C. Nguyen. 2016. Dynamic Stiffness Method for free vibration of composite cylindrical shells containing fluid. *Applied Mathematical Modelling* 40(21): 9286–9301.

Thornburgh, R.P. and M.W. Hilburger. 2006. A numerical and experimental study of compression-loaded composite panels with cutouts, Technical Papers—*AIAA/ASME/ASCE/AHS/ASC Structures, Structural Dynamics and Materials Conference* 7 · May 2006, DOI: 10.2514/6.2006–2004.

Thorsson, S.I., S.P. Sringeri, A.M. Waas, B.P. Justusson and M. Rassaian. 2018. Experimental investigation of composite laminates subject to low-velocity edge-on impact and compression after impact. *Composite Structures* 186: 335–346.

Tonkin, M.J. and S.P. Larson. 2002. Kriging water levels with a regional-linear and point-logarithmic drift. *Ground Water* 40: 185–193.

Tonkin, M.J., J. Kennel, W. Huberand J.M. Lambie. 2016. Multi-event universal Kriging (MEUK). *Advances in Water Resources* 87: 92–105.

Torabi, K., M. Shariati-Nia and M. Heidari-Rarani. 2016. Experimental and theoretical investigation on transverse vibration of delaminated cross-ply composite beams. *International Journal of Mechanical Sciences* 115: 1–11.

Tornabene, F. 2011. Free vibrations of laminated composite doubly-curved shells and panels of revolution via the GDQ Method. *Computer Methods in Applied Mechanics and Engineering* 200(9): 931–952.

Tornabene, F., N. Fantuzzi, M. Bacciocchi, A.M.A. Neves and A.J.M. Ferreira. 2016. MLSDQ based on RBFs for the free vibrations of laminated composite doubly-curved shells. *Composites Part B: Engineering* 99: 30–47.

Tornabene, F., S. Brischetto, N. Fantuzzia and E. Viola. 2015a. Numerical and exact models for free vibration analysis of cylindrical and spherical shell panels. *Composites Part B: Engineering* 81: 231–250.

Tornabene, F., N. Fantuzzi, M. Bacciocchi and E. Viola. 2015b. A new approach for treating concentrated loads in doubly-curved composite deep shells with variable radii of curvature. *Composite Structures* 131(1): 433–452.

Tornabene, F., N. Fantuzzi and M. Bacciocchi. 2014a. Free vibrations of free-form doubly curved shells made of functionally graded materials using higher-order equivalent single layer theories. *Composites Part B: Engineering* 67: 490–509.

Tornabene, F., N. Fantuzzi, E. Viola and J.N. Reddy. 2014b. Winkler-Pasternak foundation effect on the static and dynamic analyses of laminated doubly-curved and degenerate shells and panels. *Composites Part B: Engineering* 57: 269–296.

Tornabene, F., N. Fantuzzi E. Viola and A.J.M. Ferreira. 2013. Radial basis function method applied to doubly-curved laminated composite shells and panels with a general higher-order equivalent single layer formulation. *Composites Part B: Engineering* 55: 642–659.

Tornabene, F., A. Liverani and G. Caligiana. 2011. FGM and laminated doubly-curved shells and panels of revolution with a free-form meridian: a 2-D GDQ solution for free vibrations. *International Journal of Mechanical Sciences* 53(6): 446–470.

Tripathi, V., B.N. Singh and K.K. Shukla. 2007. Free vibration of laminated composite conical shells with random material properties. *Composite Structures* 81(1): 96–104.

Tsai, S.W. and H.T. Hahn. 1980. Introduction to Composite Materials. Westport, Connecticut: Technomic.

Turner, C.J. and R.H. Crawford. 2005. Selecting an appropriate metamodel: The case for NURBS metamodels, *In: ASME 2005 Design Engineering Technical Conferences and Computers and Information in Engineering Conference*. ASME, Long Beach, California, September 24–28, DETC2005–85043.

Umesh, K. and R. Ganguli. 2013. Material uncertainty effect on vibration control of smart composite plate using polynomial chaos expansion. *Mechanics of Advanced Materials and Structures* 20(7): 580–591.

Unal, R., D.O. Stanley and R.A. Lepsch. 1996. Parametric modeling using saturated experimental designs. *Journal of Parametrics* XVI(1): 3–18.

Van Beers, W. and J.P.C. Kleijnen. 2004. Kriging interpolation in simulation: a survey. pp. 113–121. *In*: Ingalls, R.G., M.D. Rossetti, J.S. Smith and B.A. Peters (eds.). Proceedings of the 2004 Winter Simulation Conference, Washington D.C., USA, December 5–8.

Vandervelde, T. and A.S. Milani. 2009. Layout optimization of a multi-zoned, multi-layered composite wing under free vibration. *In: Proceedings of SPIE, the International Society for Optical Engineering.* San Diego, CA, USA.

Vapnik, V.N. 1998. *Statistical Learning Theory*, Wiley.

Varadarajan, S., W. Chen and C.J. Pelka. 2000. Robust concept exploration of propulsion systems with enhanced model approximation capabilities. *Engineering Optimization* 32(3): 309–334.

Venini, P. and C. Mariani. 1997. Free vibrations of uncertain composite plates via stochastic Rayleigh-Ritz approach. *Computers & Structures* 64(1): 407–423.

Venkatachari, A., S. Natarajan, M. Haboussi and M. Ganapathi. 2016. Environmental effects on the free vibration of curvilinear fibre composite laminates with cutouts. *Composites Part B: Engineering* 88: 131–138.

Venkatram, A. 1988. On the use of Kriging in the spatial analysis of acid precipitation data, *Atmospheric Environment* (1967), 22(9): 1963–1975.

Viana, F.A.C., C. Gogu and R.T. Haftka 2010. Making the most out of surrogate models: tricks of the trade. pp. 587–98. *In: ASME Conference Proceedings*.

Viola, E, F. Tornabene and N. Fantuzzi. 2013. General higher-order shear deformation theories for the free vibration analysis of completely doubly-curved laminated shells and panels. *Composite Structures* 95(1): 639–666.

Vo, T.P., H.-T. Thai and M. Aydogdu. 2017. Free vibration of axially loaded composite beams using a four-unknown shear and normal deformation theory. *Composite Structures* 178: 406–414.

Wang, B., J. Bai and H.C. Gea. 2013. Stochastic kriging for random simulation metamodeling with finite sampling. *In*: Vol. 3B 39th Des. Autom. Conf. ASME, p V03BT03A056.

Wang, C.H. and A.J. Gunnion. 2008. On the design methodology of scarf repairs to composite laminates. Composites Science and Technology 68(1): 35-46.

Wang, C.M., K.K. Ang and L. Yang. 2000. Free vibration of skew sandwich plates with laminated facings. *J. Sound Vib.* 235(2): 317–340.

Wang, D., F.A. DiazDelaO, W. Wang, X. Lin, E.A. Patterson and J.E. Mottershead. 2016. Uncertainty quantification in DIC with Kriging regression. *Optics and Lasers in Engineering* 78: 182–195.

Wang, G.G. 2003. Adaptive response surface method using inherited latin hypercube design points, transactions of ASME. *Journal of Mechanical Design* 125: 210–220.

Wang, G.G. and T.W. Simpson. 2004. Fuzzy Clustering based hierarchical metamodeling for space reduction and design optimization. *Journal of Engineering Optimization* 36(3): 313–335.

Wang, J., K.M. Liew, M.J. Tan and S. Rajendran. 2002. Analysis of rectangular laminated composite plates via FSDT meshless method. *International Journal of Mechanical Sciences* 44(7): 1275–1293.

Wang, K., X. Chen, F. Yang, D.W. Porter and N. Wu. 2014. A new stochastic Kriging method for modeling multi-source exposure-response data in toxicology studies. *ACS Sustain Chem. Eng.* 2: 1581–1591.

Wang, L.P., R.V. Grandhi and R.A. Canfield. 1996. Multivariate hermite approximation for design optimization. *Int. J. of Numerical Methods in Engineering* 39: 787–803.

Wang, S.W., H. Levy II, G. Li and H. Rabitz. 1999. Fully equivalent operational models for atmospheric chemical kinetics within global chemistry-transport models. *J. Geophys. Res.* 104(D23): 30417–30426.

Wang, X., Y. Liu and E.K. Antonsson. 1999. Fitting functions to data in high dimensional design spaces. Advances in Design Automation (held in Las Vegas, NV), Paper No. DETC99/DAC-8622. ASME.

Wang, H.R., S.C. Long, X.Q. Zhang and X.H. Yao. 2018. Study on the delamination behavior of thick composite laminates under low-energy impact. *Composite Structures* 184: 461–473.

Wang, X.F. and J.H. Zhao. 2001. Monte-Carlo simulation to the tensile mechanical behaviors of unidirectional composites at low temperature. *Cryogenics* 41(9): 683–691.

Warnes, J.J. 1986. A sensitivity analysis for universal kriging. *Math Geol.* 18: 653–676.

Watkins, R.J. and O. Barton. 2009. Characterizing the vibration of an elastically point supported rectangular plate using eigensensitivity analysis. *Thin Walled Struct.* 48: 327–33.

Wattanasakulpong, N. and A. Chaikittiratana. 2015. Exact solutions for static and dynamic analyses of carbon nanotube-reinforced composite plates with Pasternak elastic foundation. *Applied Mathematical Modelling* 39(18): 5459–5472.

Wei, S.U.N., G.U.A.N. Zhidong, L.I. Zengshan, M. Zhang and Y. Huang. 2017. Compressive failure analysis of unidirectional carbon/epoxy composite based on micro-mechanical models. *Chinese Journal of Aeronautics* 30(6): 1907–1918.

Whiteside, M.B., S.T. Pinho and P. Robinson. 2012. Stochastic failure modelling of unidirectional composite ply failure. *Reliability Engineering & System Safety* 108: 1–9.

Whitney, J.M. and J.E. Ashton. 1971. Effect of environment on the elastic response of layered composite plates. *AIAA Journal* 9: 1708–1713.

Wiener, N. 1938. The homogeneous chaos. *American Journal of Mathematics* 60(4): 897–936.

Witteveen, J.A.S. and H. Bijl. 2006. Modeling arbitrary uncertainties using Gram-Schmidt polynomial chaos, AIAA paper, AIAA 2006–896. *44th AIAA Aerospace Sciences Meeting and Exhibit*, 9–12 January 2006, Reno, Nevada, USA.

Wu, L., H. Li and D. Wang. 2005. Vibration analysis of generally laminated composite plates by the moving least squares differential quadrature method. *Composite Structures* 68: 319–330.

Wu, J., X. Liu, H. Zhou, L. Li and Z. Liu. 2018. Experimental and numerical study on soft-hard-soft (SHS) cement based composite system under multiple impact loads. *Materials & Design* 139: 234–257.

Xiang, Y. and G.W. Wei. 2004. Exact solutions for buckling and vibration of stepped rectangular Mindlin plates. *Int. J. Solids Struct.* 41: 279–94.

Xiang, S., Y.-x. Jin, Z.-y. Bi, S.-x. Jiang and M.-s. Yang. 2011. A n-order shear deformation theory for free vibration of functionally graded and composite sandwich plates. *Composite Structures* 93(11): 2826–2832.

Xu, M., Z. Qiu and X. Wang. 2014. Uncertainty propagation in SEA for structural–acoustic coupled systems with non-deterministic parameters. *Journal of Sound and Vibration* 333(17): 3949–3965.

Yadav, D. and N. Verma. 1992. Free vibration of composite circular cylindrical shell with random material properties. *Computer and Structures* 43: 331–338.

Yang, C., G. Jin, Z. Liu, X. Wang and X. Miao. 2015a. Vibration and damping analysis of thick sandwich cylindrical shells with a viscoelastic core under arbitrary boundary conditions. *International Journal of Mechanical Sciences* 92: 162–177.

Yang, P.C., C.H. Norris and Y. Stavsky. 1966. Elastic wave propagation in heterogeneous plates. *International Journal of Solids and Structures* 2: 665–684.

Yang, R.J., L. Gu, L. Liaw, C. Gearhart, C.H. Tho, X. Liu and B.P. Wang. 2000. Approximations for safety optimization of large systems. ASME Design Automation Conf. (held in Baltimore, MD), Paper No. DETC-00/DAC-14245.

Yang, X., Y. Liu, Y. Zhang and Z. Yue. 2015b. Probability and convex set hybrid reliability analysis based on active learning Kriging model. *Applied Mathematical Modelling* 39(14): 3954–3971.

Yang, Y., X. Fei, X. Gao, G. Liu and M. Zhang. 2018. Two failure modes of C/SiC composite under different impact loads. *Composites Part B: Engineering* 136: 158–167.

Yao, J.C. 1965. Nonlinear elastic buckling and parametric excitation of a cylinder under axial loads. *Journal of Applied Mechanics* 32: 109–115.

Ye, K.Q., W. Li and A. Sudjianto.2000. Algorithmic construction of optimal symmetric latin hypercube designs. *Journal of Statistical Planning and Inferences* 90: 145–159.

Yeh, M.K. and Y.T. Kuo. 2004. Dynamic instability of composite beams under parametric excitation. *Composites Science and Technology* 64: 1885–1893.

Yin, S., T. Yu, Q. Bui Tinh, S. Xia and S. Hirose. 2015. A cutout isogeometric analysis for thin laminated composite plates using level sets. *Composite Structures* 127: 152–164.

Ying, Z.G., Y.Q. Ni and S.Q. Ye. 2013. Stochastic micro-vibration suppression of a sandwich plate using a magneto-rheological visco-elastomer core. *Smart Materials and Structures* 23: 025019.

Youn, B.D. and K.K. Choi. 2004. A new response surface methodology for reliability-based design optimization. *Computers and Structures* 82: 241–256.

Yu, T., S. Yin, T.Q. Bui, S. Xia, S. Tanaka and S. Hirose. 2016. NURBS-based isogeometric analysis of buckling and free vibration problems for laminated composites plates with complicated cutouts using a new simple FSDT theory and level set method. *Thin-Walled Structures* 101: 141–156.

Yuan, X., Z. Lu, C. Zhou and Z. Yue. 2013. A novel adaptive importance sampling algorithm based on Markov chain and low-discrepancy sequence. *Aerosp. Sci. Technol.* 19: 253–261.

Yue, R.-X., X. Liu and K. Chatterjee. 2014. D-optimal designs for multi-response linear models with a qualitative factor, *J. Multivariate Analysis* 124: 57–69.

Zadeh, L.A. 1975. Concept of a linguistic variable and its application to approximate reasoning-I. *Information Sciences* 8(3): 199–249.

Zadeh, L.A. 1965. Fuzzy sets. *Information and Control* 8(3): 338–353.

Zaman, K., M. McDonald and S. Mahadevan. 2011. A probabilistic approach for representation of interval uncertainty. *Reliability Engineering and System Safety* 96(1): 117–130.

Zappino, E., A. Viglietti and E. Carrera. 2017. Analysis of tapered composite structures using a refined beam theory. *Composite Structures*.

Zhang, F., Z.Z. Lu, L.J. Cui and S.S. Song. 2010. Reliability sensitivity algorithm based on stratified importance sampling method for multiple failure modes systems. *Chin. J. Aeronaut.* 23: 660–669.

Zhang, J.C., T.Y. Ng and K.M. Liew. 2003. Three-dimensional theory of elasticity for free vibration analysis of composite laminates via layerwise differential quadrature modelling. *International Journal for Numerical Methods in Engineering* 57: 1819–1844.

Zhang, Q.J. and M.G. Sainsbury. 2000. The Galerkin element method applied to the vibration of rectangular damped sandwich plates. *Computers & Structures* 74(6): 717–30.

Zhang, C., E.A. Duodu and J. Gu. 2017. Finite element modeling of damage development in cross-ply composite laminates subjected to low velocity impact. *Composite Structures* 173: 219–227.

Zhang, Z., P. Klein and K. Friedrich. 2002. Dynamic mechanical properties of PTFE based short carbon fibre reinforced composites: experiment and artificial neural network prediction. *Composites Science and Technology* 62(7): 1001–1009.

Zhao, W., J.K. Liu and Y.Y. Chen. 2015. Material behavior modeling with multi-output support vector regression. *Applied Mathematical Modelling* 39(17): 5216–5229.

Zhao, G., H. Hu, S. Li, L. Liu and K. Li. 2017. Localization of impact on composite plates based on integrated wavelet transform and hybrid minimization algorithm. *Composite Structures* 176: 234–243.

Zhao, X., K.M. Liew and T.Y. Ng. 2003. Vibration analysis of laminated composite cylindrical panels via a meshfree approach. *International Journal of Solids and Structures* 40(1): 161–180.

Zhen, W., C. Wanji and R. Xiaohui. 2010. An accurate higher-order theory and C0 finite element for free vibration analysis of laminated composite and sandwich plates. *Composite Structures* 92: 1299–307.

Zhigang, S., C. Wang, X. Niu and Y. Song .2016. A response surface approach for reliability analysis of 2.5D C/SiC composites turbine blade. *Composites Part B: Engineering* 85: 277–285.

Zhou, X.Y., P.D. Gosling, C.J. Pearce, L. Kaczmarczyk and Z. Ullah. 2016. Perturbation-based stochastic multi-scale computational homogenization method for the determination of the effective properties of composite materials with random properties. *Computer Methods in Applied Mechanics and Engineering* 300: 84–105.

Zhu, P.H. and L.H. Chen. 2014. A novel method of dynamic characteristics analysis of machine tool based on unit structure. *Sci. China Tech. Sci.* 57: 1052–1062, doi: 10.1007/s11431-014-5524-2.

Zienkiewicz, O.C. and R.L. Taylor. 1991. *The Finite Element Method. McGraw-Hill*, UK.

Index